International Scientific and Research Conference on Knowledge-based Technologies in Development and Utilization of Mineral Resources (KTDMUR 2017)

IOP Conference Series: Earth and Environmental Science
Volume 84

Novokuznetsk, Russia
6 - 9 June 2017

Editor:

Viktor Fryanov

ISBN: 978-1-5108-4863-4
ISSN: 1755-1307

Printed from e-media with permission by:

Curran Associates, Inc.
57 Morehouse Lane
Red Hook, NY 12571

Some format issues inherent in the e-media version may also appear in this print version.

Printed by Curran Associates, Inc. (2017)

For permission requests, please contact the Institute of Physics
at the address below.

Institute of Physics
Dirac House, Temple Back
Bristol BS1 6BE UK

Phone: 44 1 17 929 7481
Fax: 44 1 17 920 0979

techtracking@iop.org

Additional copies of this publication are available from:

Curran Associates, Inc.
57 Morehouse Lane
Red Hook, NY 12571 USA
Phone: 845-758-0400
Fax: 845-758-2633
Email: curran@proceedings.com
Web: www.proceedings.com

Table of contents

Volume 84

International Scientific and Research Conference on Knowledge-based Technologies in Development and Utilization of Mineral Resources (KTDMUR2017)

6–9 June 2017, Novokuznetsk, Russian Federation

Accepted papers received: 6 September 2017
Published online: 21 September 2017

Preface

Papers

Preface

Since 1995 the Siberian State Industrial University has been organizing an international scientific conference *Knowledge-based technologies in development and utilization of mineral resources* which is devoted to the issues of critical technologies in exploration, development of mineral deposits and their extraction, search for original knowledge-based directions of mining production diversification providing a comprehensive extraction of mineral resources, their deep processing and use. The conference is held annually within the framework of the International Specialized Exhibition of Mining Technologies *Russian Coal and Mining*.

In 2017 the conference was held from 6 to 9 June and was dedicated to the 400th anniversary of the city of Novokuznetsk. The conference was organized by Siberian State Industrial University (Novokuznetsk, Russia), Federal Research Center for Coal and Coal Chemistry of the Siberian Branch of the Russian Academy of Sciences (Kemerovo, Russia), National Mine Rescue Center (Novokuznetsk, Russia), the University of Science and technology Liaoning (Anshan, China), Expo center “Kuzbass Fair” (Novokuznetsk, Russia). The event was actively supported by the Administration of Kemerovo Region and Novokuznetsk, the Russian Foundation for Basic Research.

The conference is a recognized scientific and practical forum for professional discussion of problems related to the state and prospects for the development of new technologies, machinery and equipment that ensure the rational use of mineral resources, industrial and environmental safety of mining enterprises. The event was attended by scientists and specialists from different institutions, companies and coal mining enterprises from Russia, Germany, India, Italy, the Donetsk People’s Republic. Active participation of PhD students and young scientists is particularly noteworthy.

More than 100 papers were presented at the conference by authors and research teams. The present issue of *IOP Conference Series: Earth and Environmental Science* contains 45 specially selected research papers covering the following topics: knowledge-based technologies for the development of mineral deposits, geomechanical and gas-dynamic processes in the integrated development of underground resources, innovative solutions for improvement of the technical level of mining equipment production, automation of control systems in mining, new directions for energy supply and energy saving at mining enterprises, provision of industrial and environmental safety in mining, deep processing of coal and coal products.

The Organizing Committee would like to thank the Russian Foundation for Basic Research for its financial support (project No. 17-05-20150). We also appreciate contribution of all conference participants into development of mining science and practice.

Vice-chair of the Organizing Committee
and Editor-in-chief
Prof. Dr. Viktor Fryanov

IOP Conf. Series: Earth and Environmental Science **84** (2017) 011002 doi:10.1088/1755-1315/84/1/011002

Peer review statement

All papers published in this volume of *IOP Conference Series: Earth and Environmental Science* have been peer reviewed through processes administered by the proceedings Editors. Reviews were conducted by expert referees to the professional and scientific standards expected of a proceedings journal published by IOP Publishing.

IOP Conf. Series: Earth and Environmental Science **84** (2017) 012001 doi:10.1088/1755-1315/84/1/012001

Theoretical approaches to creation of robotic coal mines based on the synthesis of simulation technologies

V N Fryanov, L D Pavlova and M V Temlyantsev

Siberian State Industrial University, 42 Kirova Street, Novokuznetsk, 654007, Russia

E-mail: ld_pavlova@mail.ru

Abstract. Methodological approaches to theoretical substantiation of the structure and parameters of robotic coal mines are outlined. The results of mathematical and numerical modeling revealed the features of manifestation of geomechanical and gas dynamic processes in the conditions of robotic mines. Technological solutions for the design and manufacture of technical means for robotic mine are adopted using the method of economic and mathematical modeling and in accordance with the current regulatory documents. For a comparative performance evaluation of technological schemes of traditional and robotic mines, methods of cognitive modeling and matrix search for subsystem elements in the synthesis of a complex geotechnological system are applied. It is substantiated that the process of technical re-equipment of a traditional mine with a phased transition to a robotic mine will reduce unit costs by almost 1.5 times with a significant social effect due to a reduction in the number of personnel engaged in hazardous work.

1. Introduction

At present, the competition of oil, natural gas, coal as natural energy sources leads to a reduction in coal production in many countries. In the developed countries of Europe (France, Belgium, Spain, Great Britain, etc.) coal mining is practically stopped. The tendency for reduction in consumption of coal as an environmentally hazardous energy source will lead to a drop in production at coal mining enterprises. First of all, the volume of production in coal mines, which are dangerous production facilities with the first class of danger, will decrease. However, the use of large coal reserves (for example, in Kuzbass balance reserves to the depth of 600 m is 67.1 billion tonnes [1]), can make the basis of the economy of coal-producing regions now and in the future.

The possibility of mining these coal reserves by traditional technologies is limited by complex mining and geological conditions: depth of development up to 1800 m, high methane content and propensity of coal seams to spontaneous combustion, dynamic forms of rock pressure manifestation. One of the options for extending the period of operation of coal mines is the creation of robotic coal mine, that includes methods and means of underground geotechnology for manless mining and rational use of mineral resources. In accordance with the urgency, the program of research into the underground geotechnology of manless coal mining with underground utilization of waste products in the gob and remote control of technological processes was developed.

Taking into consideration the complexity of the issue and labour inputs, the research program includes several stages. The aim of the research at the first stage is to create scientific foundations of a manless technology for underground mining of coal deposits in complex mining and geological conditions.

The goal of the research can be achieved by solving the following problems (figure 1):

1) Conducting a systematic analysis of fundamental science and production experience achievements in the development of coal deposits for scientific justification of technological scheme of a robotic mine with a manless mining of coal seams.

2) Identification of regularities in manifestation of geomechanical and gas dynamic processes in the remote mining of coal seams by the results of numerical modeling.

3) Justification and optimization of robotic mine parameters by the economic criterion with the maximum coefficient of coal extraction.

4) Justification of technological requirements for the design and manufacture of technical means of a robotic mine.

5) Preparation of a robotic mine draft design for development of coal seams in difficult mining and geological conditions.

2. Methods of research

The solution of the scientific tasks is carried out on the basis of the methodological approaches indicated in figure 1.

To evaluate the state and directions of development of traditional and advanced coal mining technologies, it is proposed to use system analysis by decomposition of the known complex of geotechnologies, characterized by the method of selecting adaptive elements for a wide range of mining and geological conditions, the synthesis of which provides the optimal variants of technological schemes of robotic mines based on the results of cognitive modeling.

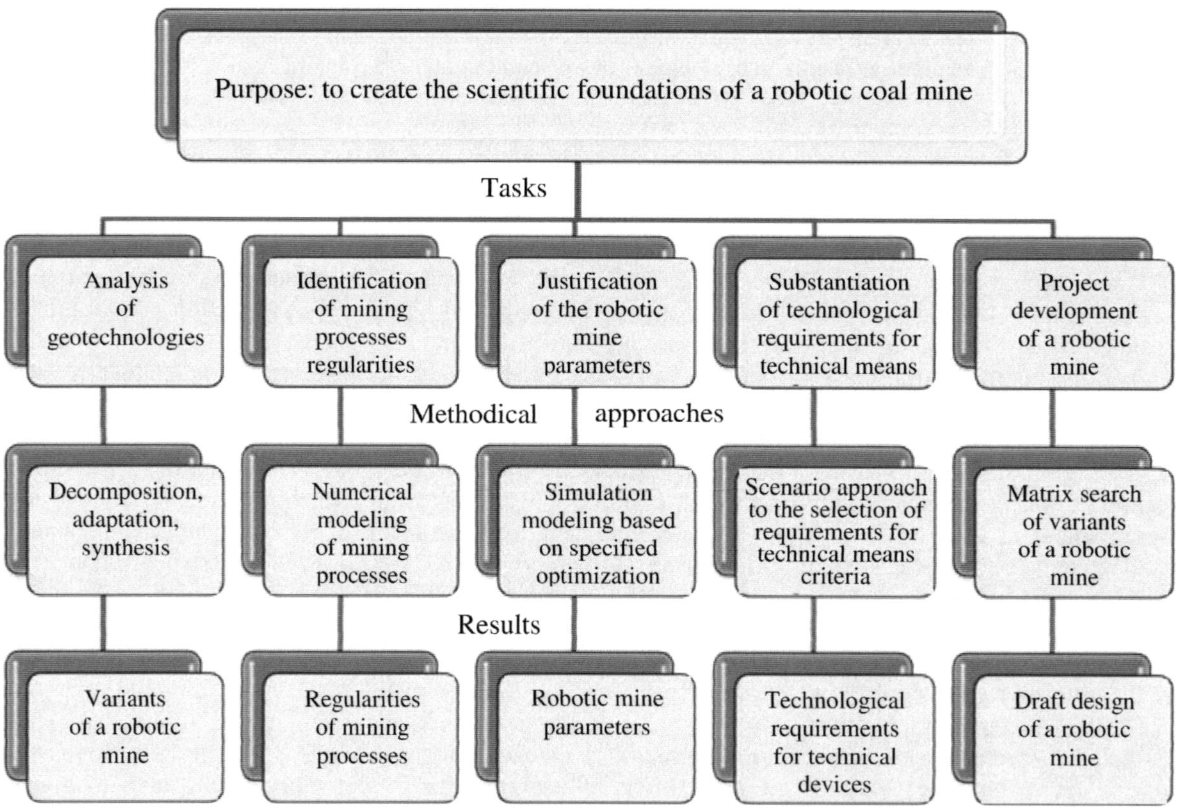

Figure 1. System structure of development of theoretical foundations for design of robotic coal mines.

Identification of regularities in the manifestation of geomechanical and gas-dynamic processes and justification of robotic mine parameters are planned to be carried out by mathematical and numerical simulation methods using the developed set of problem-oriented programs that differ from analogs by taking account of rocks transition from elastic to elastoplastic, limiting and transverse states in their nonlinear deformation.

When justifying the robotic mine parameters, it is planned to use new approaches that are distinguished by the application of methods for geomechanical processes simulation to establish the dependencies of main technical and economic indicators of a mine on geological and technological parameters of the mining system.

3. Results and discussion

Creation of scientific foundations of a robotic mine, where the development of coal seams is carried out in complex mining and geological conditions, is planned to be performed on the basis of a scenario approach characterized by a variety of variants of robotic mine technological schemes adapting to complex geological and mining conditions, by choose of an optimal option according to integral criterion including minimum production costs and a risk of a dangerous production situation not higher than acceptable.

Justification of technological requirements for the design and manufacture of technical means of a robotic mine is supposed to be carried out using the method of economic and mathematical modeling, which is distinguished by the development of bounding algorithm of parameters of robotic mine technological system in accordance with the requirements of existing regulatory documents.

New approaches in the development of a robotic mine draft design consist in the use of dynamic programming, matrix searching of scenarios and the method of constructing options for coal mining with remote control over machines and units, taking into account the variability of mining and geological conditions.

Implementation of these methodological approaches has been carried out by the authors for the last 7 years [2 - 6]. The novelty of the results of scientific research obtained during development of the main principles of robotic mine is protected by patents for invention and inventor's certificates. The efficiency of results is confirmed by the cost-effective operation of mines in the south of Kuzbass, where separate technological solutions have been implemented.

More than 10 patents for inventions have been received that include methods of extracting coal from chambers with methane production (patent No. 2415266) and hydraulic mining of steep seams by boreholes with methane production (patent No. 2473806). The originality of technological solutions in these patents consists in combining the processes of coal mining in a face and extraction of methane from the gas reservoir in the gob. The efficiency of coal extraction is achieved by applying a hydraulic method of coal destruction and hydrotransport without presence of people in the face, while simultaneous methane extraction is carried out by evacuation through pipes to the soil surface. The use of a hydraulic mining method reduces the possibilty of endogenous fire in the gob due to the phlegmatization of methane and coal dust in a humid atmosphere.

The original solution of the method coal mining from interseam strata during development of thick seams is proposed in patent No. 2416721, which is distinguished by the use of a slicing method with coal extraction by a drilling complex and supporting the roof rocks by a hydroficated support with a re-entry of the drilling complex onto the lower seam under the cantilevered part of the support.

Promising for the robotic coal mining technology is the method proposed in patent No. 2381364 consisting in the developing steep seams by a sublevel hydraulic winning with the release of coal from the upper-level chambers. Originality of technological solutions lies in preliminary drilling of wells and coal fracturing with its subsequent muck draw into the face of the lower layer and fencing of this zone by the coal pillars.

The theoretical foundations of robotic coal mines design were developed based on the results of R&D projects: "Development of measures aimed at preventing soil heave in the mine workings of the branch *Mine Yerunakovskaya-VIII* of *Yuzhkuzbassugol JSC*", "Recommendations for strengthening the support of the transport slope of seam 45 and adjacent workings taking into account the forecast of the state of the host rocks in the vicinity of these excavations for accident-free conduct and maintenance

IOP Conf. Series: Earth and Environmental Science **84** (2017) 012001 doi:10.1088/1755-1315/84/1/012001

of capital mine workings along seam 45 in the conditions of the branch *Mine Yerunakovskaya-VIII* of *Yuzhkuzbassugol JSC*". The originality of technological solutions developed and implemented in the conditions of this mine consists in justifying the parameters of multi-entry preparation and conducting mining of superimposed seams 45 and 48 by fully-mechanized longwalls. A positive effect was achieved due to group preparation of seams with the thickness of interseam rocks up to 10 m.

Recommendations developed on the basis of full-scale research results of R&D project "Justification of recommendations for the safe completion of reserves mining on site 6-1-11 in the conditions of *Mine Alardinskaya LLC*" are distinguished by original solutions concerning the reduction of number of unloading wells in the process of prevention of gas-dynamic phenomena in seam 6.

Two- and three-dimensional mathematical models of the stress-strain state of a gas-bearing geomass were developed for numerical modeling of the processes of non-linear rock deformation, modification of the finite element method was performed, allowing the transition of rocks from elastic to elastoplastic, limit and beyond the limit states to be taken into account.

Complexes of authorial programs, characterized by the possibility of 3D modeling of geomechanical processes in the massif during underground mining of coal seams (Certificate No. 17997), modeling of dynamic block rock caving with successive accumulation of damage (Certificate No. 326723), determination of the parameters of the bulk geomechanical state of a layered rock massif when mining a series of shallow or inclined coal seams (Certificate No. 6605) ensure the prediction of geomechanical parameters of workings in the conditions of a robotic mine.

The developed software packages are registered in the Joint Fund of Electronic Resources "Science and Education", they received certificates confirming the novelty and priority of developments.

The originality of authors' developments designs was repeatedly awarded at international specialized exhibitions, including "Software and hardware complex for monitoring and prediction of parameters of interacting gas-dynamic and geomechanical processes during underground mining of coal deposits" which was awarded in 2015 Diplomas and Gold medals at the International Specialized Exhibition "Coal of Russia and Mining", and the International Fair "Expo-Siberia".

For the generalization and development of theoretical foundations of robotic mine design principles, the results of the research in geotechnologies conducted by other scientific schools have been used

In article [14], in addition to traditional technologies, a group of high geotechnologies is distinguished, among which the underground well leaching, underground well drilling, heap underground and ground leaching, etc., can be identified. This group can also include underground technologies using robotic technical means for performing basic technological processes.

Synthesis of a coal mine technological scheme in works [7 - 9] is proposed to be implemented in accordance with the theory of adaptive control. Mining technology is considered as a three-component complex system, including subsystems S_1 personnel; S_2 – unit as a set of artificial objects; S_3 – environment, as a combination of mining, geological, technological and climatic conditions. Each subsystem is characterized by a set of elements and connections between the elements of one subsystem and neighboring subsystems. Elements and connections of subsystems can be quantitative or qualitative. It is customary to evaluate the complexity of geotechnology as a single system by the number of connections on a quasi-ordered graph. For the mathematical description of such systems with weakly structured objects and situations in recent years a cognitive approach is widely used that is associated with a certain family of formal models and is based on the use of cognitive maps, including the mathematical apparatus of signed and weighted graphs.

When designing the robotic mine technological scheme in comparison with the technological scheme of a traditional mine, the number of such connections is smaller, since personnel in working shifts do not contact with the unit elements and the environment. For a comparative evaluation of the complexity of technological schemes of a traditional and and robotic mine, methods of cognitive modeling and matrix search for variants of subsystem elements synthesis into a complex geotechnological system are applied.

IOP Conf. Series: Earth and Environmental Science **84** (2017) 012001 doi:10.1088/1755-1315/84/1/012001

To form scenarios for the creation of a robotic coal mine, taking into account the novelty of robotic elements of an enterprise and the risks of their inefficiency, the principle of phased integration of individual or groups of robotic elements into the technological scheme of the mine (TSM) was adopted. The choice of progressive elements of technological schemes of robotic mines was made using domestic and foreign experience of development of coal deposits in difficult conditions [10 - 14].

To select the optimal scenario for the phased conversion of a traditional mine into a robotic mine, a simulation using matrix search for variants of elements of this system was carried out according to a given quantitative optimality criterion [15].

In the matrix (table 1) for each element of the mine technological scheme, two levels of costs are allocated: capital costs for the creation or acquisition of technical devices and construction of facilities and operational ones. Alternative options of costs for each element of the mine technological scheme are formed by experts.

Table 1. Initial data for the evaluation of option effectiveness of the mine technological scheme.

Number and name of the TSM element	Expenses, rub/t (capital / operational)		
	traditional mine (basic version)	automated mine with robotic elements	robotic mine
1. Technological complex on the surface	360/60	360/50	330/50
2. Method and scheme of opening the mine field	150/135	180/110	200/90
3. Method and scheme for preparation of the mine field	75/60	80/50	100/40
4. Method of mine ventilation	60/75	60/60	70/40
5. Mining system	120/600	140/500	200/300
6. Degassing	150/90	150/80	100/60
7. Industrial Safety	240/150	240/100	300/50
8. Environmental Safety	90/150	90/100	120/50

Specialists from mines, design and expert organizations, as well as university professors in the field of *Mining* were involved as experts. Experts evaluated the costs for three options of mine technological scheme: a traditional, automated mine with robotic elements and a robotic mine (Table 1). One of the options is adopted as a basic one, as a rule, it is an operating mine with traditional coal mining technology.

For the formation of alternative options of mine technological scheme and the choice of an optimal variant, a computer program has been developed which, through the matrix search, determines and ranks all options according to the given criterion and chooses the one optimal from them. The results of the program implementation with the accumulation of costs according to the levels of the mine technological scheme are given in table 2.

Table 2. Results of modeling with accumulation of costs for the base and optimal options of mine technological scheme.

Number and name of the TSM element	TSM options	Costs distribution by TSM elements, rub/t		
		capital	maintenance	total
1. Technological complex on the surface	Basic	360.00	60.00	420.00
	Optimal	331.20	49.80	381.00
2. Method and scheme of opening the mine field	Basic	510.00	195.00	705.00
	Optimal	511.20	160.50	671.70
3. Method and scheme for preparation of the mine field	Basic	585.00	255.00	840.00
	Optimal	591.45	160.50	751.95

4. Method of mine ventilation	Basic	645.00	330.00	975.00
	Optimal	651.45	220.50	871.95
5. Mining system	Basic	765.00	930.00	1695.00
	Optimal	791.85	220.50	1012.35
6. Degassing	Basic	915.00	1020.00	1935.00
	Optimal	941.85	300.60	1242.45
7. Industrial Safety	Basic	1155.00	1170.00	2325.00
	Optimal	1181.85	401.10	1582.95
8. Environmental Safety	Basic	1245.00	1320.00	2565.00
	Optimal	1298.86	450.60	1749.45

The analysis of the obtained results make it possible to state that the main economic effect can be achieved in the structure of the mining system due to the use of manless coal mining with remote control over the excavating unit figure 2).

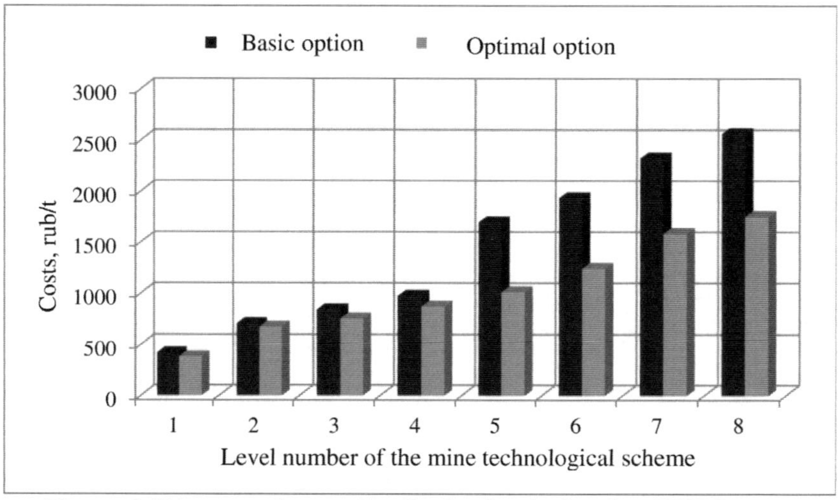

Figure 2. Diagram of costs distribution by the levels of TSM.

In the process of modeling 109 options of combinations of mine elements with traditional technology and automated mine with the elements of robotization were considered. The structure analysis of optimal scenario showed that of the 8 elements only 2 elements of robotic mine are included into the optimal option, and the remaining elements are taken from an automated mine with robotic elements. The reasons for this structure formation of the optimal option are the following: experts do not have information about the parameters of robotic mine and the overestimated capital costs for the acquisition of robotic technical devices that have not been produced serially yet.

4. Conclusions

1. The main theoretical approaches to the creation of robotic coal mines are:

- system analysis of the achievements of fundamental science and production experience in the development of coal deposits for the scientific justification of robotic mine technological scheme of a with a manless mining of coal seams;
- identification of regularities in the manifestation of geomechanical and gasdynamic processes based on the results of numerical modeling in the case of remote mining of coal seams;
- parameters optimization of robotic mine by the economic criterion with the maximal coal mining coefficient;

- justification of technological requirements for design and manufacture of technical means of a robotic mine;
- development of a robotic mine draft design for mining coal seams in difficult mining and geological conditions.

2. The simulation results confirm that the creation of a robotic mine should be carried out in stages according to the scheme: traditional mine → automated mine with robotic elements → robotic mine. Such a process of technical re-equipment of a traditional mine with a phased transition to a robotic mine will reduce unit costs by almost 1.5 times with a significant social effect due to reduction in the number of personnel engaged in hazardous work.

References

[1] Shaklein S V and Pisarenko M V 2013 *Mountain Information and Analytical Bulletin* **OB6** 186–93
[2] Pavlova L D, Fryanov V N and Kornev E S 2011 *Proc. of All-Russian Conf. on Geodynamics and Stress State of the Earth Interior* (Novosibirsk: Inst. of Mining, SB RAS) vol 2 pp 163–9
[3] Kornev E S, Pavlova L D and Fryanov V N 2013 *Bulletin of KuzSTU* **2** 65–9
[4] Kornev E S, Pavlova L D and Fryanov V N 2014 *Numerical Modeling of Geomechanical Processes in Shortwall Mining of Coal Seams* (Prague: Vědecko vydavatelské centrum "Sociosféra-CZ") p 206
[5] Fryanov V N and Pavlova L D 2016 *IOP Conference Series: Earth and Environmental Science* **45** 012002
[6] Fryanov V N and Pavlova L D 2017 *IOP Conference Series: Earth and Environmental Science* **53** 012028
[7] Rogov E I and Rogov A E 2015 *Mining Journal of Kazakhstan* **6** 4–6
[8] Vylegzhanin V N, Vitkovskiy E I and Potapov V P 1987 *Adaptive Management of Underground Coal Mining Technology* (Novosibirsk: Nauka) p 232
[9] Rogov E I, Gritsko G I and Vylegzhanin V N 1979 *Mathematical Models of Adaptation of Processes and Subsystems of a Coal Mine* (Alma-Ata: Science of KazSSR) p 240
[10] Tokalic R, Vidanovic N and Savic L 2013 *Underground Mining Engineering* **23** 11–20
[11] Vidanovic N, Vujanovic M, Ognjanovic S and Nikolic D 2013 *Underground Mining Engineering* 2013 **23** 59-68
[12] Denic M, Stojadinovic S, Vusovic N, Kokeric S and Denic N 2014 *Underground Mining Engineering* **24** 1-10
[13] Wexler Yu A, Outer M R and Kuch Sh 2002 *Coal* **5** 32–34
[14] Klishin V I, Zvorygin L V, Lebedev AV and Savchenko A V 2011 *Problems of Safety and New Technologies for Underground Mining of Coal Deposits* (Novosibirsk: IGD SB RAS) p 247–50
[15] Pegat A 2011 *Fuzzy Modeling and Control* (Moscow: BINOM. Laboratoriya znaniy) p 798

IOP Conf. Series: Earth and Environmental Science **84** (2017) 012002 doi:10.1088/1755-1315/84/1/012002

Robotic complex for the development of thick steeply-inclined coal seams and ore deposits

M S Nikitenko[1], Yu V Malakhov[1], Biswarup Neogi[2], Pritam Chakraborty[2] and Dipesu Banerjee[2]

[1]Federal Research Center of Coal and Coal Chemistry, Siberian Branch of the Russian Academy of Sciences, 18 Sovetskiy Ave., Kemerovo, 650000, Russia
[2]CoE & CII Centre for Innovation , JIS College of Engineering (JIS Group India), Block A, Phase III, Dist. Nadia, Kalyani, West Bengal, 741235, India

E-mail: schum24@rambler.ru

Abstract. Proposal for the formulation of robotic complexes for steeply inclined coal seams as a basis of the supportive-enclosing walking module and power support with a controlled outlet for mining industry has been represented in this literature. In mining industry, the available resource base reserves and mineral deposits are concentrated deep down the earth crust leading towards a complicated geological condition i.e. abrupt ore bedding and steeply inclined strata with the high gas content and fire hazard of thick coal stratum, heading against an unfavorable and sometimes human labor life risk during subversive mining. Prevailing towards the development of effective robotic complexes based on the means of "unmanned technologies" for extraction of minerals from hard-to-reach deposits and make sure the safety of underground staff during sublevel mining technology.

1. Introduction

Advancement of mining industry and the improvement of mining safety structure depend critically not only on the use of new different types of equipment and technologies, dispatching and automation of mining complexes' control, but also on the accelerated conception and widespread introduction of unique automated mining control systems based on fully or partially robotic technology provides a platform for implementation of "unmanned" technology [1, 2].

On contemporary, existing equipment used in mines are ineffective for the extraction of mineral deposits from abruptly inclined stratums, and it does not work without supervision of constant human labor at extraction and preparatory faces, thus there is an urge need of modern technology keeping in view the stimulating geological circumstances, along with subversive mine gas and dust generating a need for the development of "unmanned" mineral extraction technology [3]. In connection with the foregoing, there is also a need to develop new technical solutions for the extraction of minerals from steeply-inclined coal strata and ore materials [4]. The newly developed technological schemes [5] and technical facilities [6 - 8] will allow creating robotic complexes for mining and mineral deposits (coal, diamond ore, etc.) with fewer hominid labor and with high efficacy with a control system that works considering the rock conditions and its dynamic manifestations.

Presently in coal industry lacks high-performance coal extraction mechanized complexes for effective, safe underground method development of thick steep and steeply-inclined coal seams. Therefore, due to the lack of appropriate mining equipment subversive mining in Russia is still old-fashioned leading to unutilized coal stratums, although they contain more than 50 % of industrial coal reserves, including valuable coking grades.

As per Order of the Ministry of Industry and Trade of the Russian Federation No. 654 of March 31, 2015, approved plan of measures for import substitution in the heavy engineering industry of the Russian Federation, in which the creation of domestic automated extraction complexes was recognized as the priority among mining equipment. The share of imports in their consumption in 2016 was 100 %. In addition, the development of complexes and conveyor transport control systems was recognized as a priority, with the share of imports in consumption of which in 2016 was 80 %. As part of the approved long-term program for the development of the Russian coal industry until 2030, approved by the RF Government Decree of 24.01.2012 No. 14-p, the main direction of technological development with respect to the underground mining method is the implementation of measures to develop and implement systems of "unmanned" minerals extraction on the basis of complex mechanization and automation [9].

In 2016, forecast of the scientific and technological development of the fuel and energy complex of Russia for the period of 2035 developed by the Ministry of Energy of Russia, and in the list of 24 priority critical technologies were included - coal mining technologies without the constant presence of people in the working space (robotic complexes for development of thin and steep seams with high selectivity).

2. Methods of research

Proposed robotic complexes based on supportive-enclosing walking module and KPV power support with controlled outlet as an effective technological solution capable of ensuring the steep and steeply inclined coal seams underground method development. The weakening of the let out coal massif is supposed to be carried out by safe methods - by means of its stepwise directed hydraulic fracturing or by vibro-seismic method (seismic vibrators). In this method of working out coal seams, explosives are not used. The proposed technological solutions create the transition basis to "unmanned" mining technologies.

The actuality of the proposed technical solutions is caused by the following:

- the inconsistency of the used traditional shielding steeply inclined seam coal extraction technology to modern requirements of the Federal Safety Standards and Regulations for mines with high seam gas content [10];
- the orientation of the coal industry and domestic coal machinery on the technology of "unmanned" productions. In comparison with the low-efficient traditional shield method for working out thick, steeply-inclined coal seams and alluvial ore bodies (materials), the robotic complex based on the supportive-enclosing walking module and the KPV power support with controlled outlet has the following main advantages:

1. Creates the basis for the introduction of a robotic method of mineral mining by underground method (unmanned technologies).
2. Provides a high level of safety (there is no manual labor, no explosives are used to weaken the massif of minerals).
3. High degree (up to 80 %) of minerals extraction from seams (otherwise it can be said that it reduces the mineral resources extraction losses by half).
4. Rapidly readjustable, complex design reusable layout. Convenient transportation.

An additional direction of using a supportive-enclosing walking module can be the creation on their basis of a robotic mine rescue complex for the MChS Militant Mine Rescue units.

"Unmanned" technologies – this is a conventional name for ways to organize technological processes without the direct presence of a person in the place of their implementation.

The main advantages of using "unmanned" technologies:

- exclusion of people from places with dangerous and harmful working conditions;

- the opportunity to work with such parameters of mine workings that are prohibited under safety conditions when people are located there, which makes it possible to ensure a higher efficiency of mining;
- possibility to control machinery being at a very large distance from it, which also allows solving the concomitant problem of delivering a large number of people to remote areas;
- equipment productivity increase due to more accurate and optimal control of equipment.

The use of "unmanned" technologies for the mineral extraction is provided by the use of robotic complexes' automated control systems.

Concerning the depth of automation and functionality of underground mining robotic complexes control system it is proposed to implement at several levels:

- "dispatching system" – a set of software and hardware means that allows remote monitoring and control of engineering systems of one or more robotic complexes (remote control);
- "system based on adaptive technologies"
- automatic control (maintenance in specified ranges) process parameters functions transfer to the software and hardware complex, with the operator's ability to remotely perform their direct control from the control station;
- "Brain computer interface (BCI) control system" – a perspective system of adaptive automated control, with the possibility of remote control based on the mind control technology "Mind control" or "Concentration control". Human thought, the interaction between brain neurons, creates various electrical waves, generating a unique electrical signal. These electrical signals are determined by the brain wave sensors, the data is converted into packets and transmitted via Bluetooth to the level analyzer unit (LAU).

The analyzer block receives raw brain data, extracts and processes the signal using the Matlab platform. The control commands are then transferred to the robotic module for processing, thus controlling the movement and transferring functions. The application of these technologies will allow providing direct communication and control between the human brain and physical devices by transferring the brain activity structures and commands in real time and implementing on their basis the control systems for mechanized complexes and control of auxiliary systems (initiation of outlet, loading, manipulators, etc.) depending on the tasks assigned.

Several researchers familiarize the future aim of the BCI technology for the different kinds of application aspects. Joseph N. Mak et. al [11], Mikhail, A. Lebedev et. al [12], Petrov et.al [13] subsidized future prospects of BCI considering industrial and ergonomic advancement. One phase in advance, the stage will be commenced for sophisticated Internet of Things (IoT) based technology and Brain Computer Interface (BCI) amalgamation research. The present research illustrated with two parts introducing high end mining device association with brain control interface (BCI). The platform of these two annexes is presented as novelty research theme. This extent will carry its importance as interlinked stage of IoT and BCI implemented in mining industry [14 - 16]. Limited researches are even now presented, allied mining technology and BCI.

Application of BCI technology has found a wide span from medical prosthetics to artificial robotics in industrial automation. Presently researchers are widely interested to extend the support of BCI technology in the mining industry where the geological surroundings are incompatible for extraction of minerals deep down the core [17].

In JIS Collage of Engineering (West Bengal, India), Neogi et. al contributed some of ideas involving BCI technology which can be incorporated together towards the Mining Industry Automation [18]. The specialists have already mastered the basic modules of the motion control system [18]. The intellectual device Concentration control Switch, Cognitively velocity controlled car and Mind control artificial arm are mainly ergonomic devices developed. In the framework of joint work of JIS College of Engineering and Federal Research Center for Coal and Coal Chemistry of the Russian Academy of Sciences Siberian Branch started to integrate them into the control system of the robotic complexes based on the support-enclosing walking module and KPV power support with controlled outlet.

3. Results and discussion

3.1. Concentration Control Switching System

In view of contemporary sovereign technological procreation this innovative discovery relates to a concentration control switch system based brain computer interface (BCI). Furthermore, this innovation relates to the switching of various electrical appliances and electronic industrial devices through concentration wherein brain computer interface can circumvent predictable channels of communication to arrange for direct communication and control between the human brain and physical devices by decoding diverse configurations of brain activity into directives in tangible interval.

The prime embodiment of the innovation is to provide a concentration controlled switching system based brain computer interface (BCI) in procuring volatile, cultivated and technically manufactured contemporary customary appliances that could end up to the switching of various electrical appliances and electronic industrial devices through concentration. The further focal object of the innovative discovery is to provide the concentration controlled switching system wherein brain computer interface can bypass conventional channels of communication (i.e., muscles and thoughts) to provide direct communication and control between the human brain and physical devices by translating different patterns of brain activity into directives in real time.

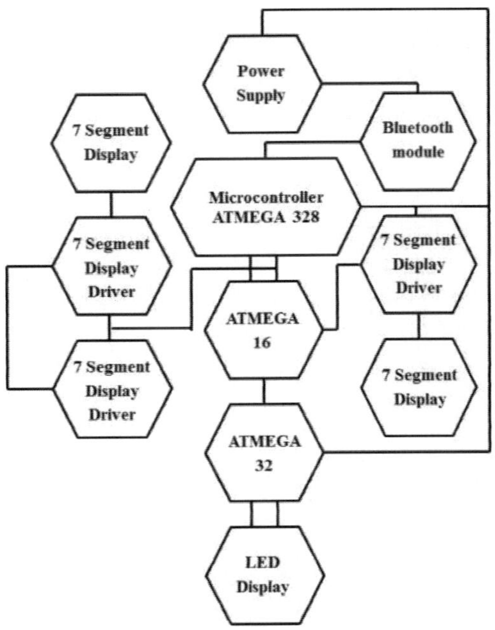

Figure 1. Circuitry block representation of concentration control switching system.

The proposed technology can be utilized in underground mining lighting of robotic complex for proper navigation during mineral extraction.

3.2. Cognitively Velocity Controlled Vehicle

This innovation relays a cognitively velocity controller vehicle which is a mind controlled cognitive vehicle based brain computer interface (BCI). Additionally, this also relays cognitively velocity controller vehicle wherein brain computer interface can sidestep conservative networks of communication (i.e., muscles and thoughts) to provide direct communication and control between the human brain and physical devices by translating different patterns of brain activity into commands in real time.

The foremost purpose of the idea is to afford a cognitively velocity controller vehicle. The additional incarnation of the innovation is to offer a cognitively velocity controller vehicle which is a mind controlled cognitive vehicle based brain computer interface (BCI).The origination is to provide the cognitively velocity controller vehicle wherein brain compute interface can bypass conventional channels of communication (i.e., muscles and thoughts) to provide direct communication and control between the human brain and physical devices by translating different patterns of brain activity into commands in real time. The supplementary embodiment of the invention is to provide the cognitively velocity controller vehicle which can control up to four types of orientations of movements-right, left, forward, backward but if there is a suitable arrangement it can control clock wise and anti-clock wise rotation also.

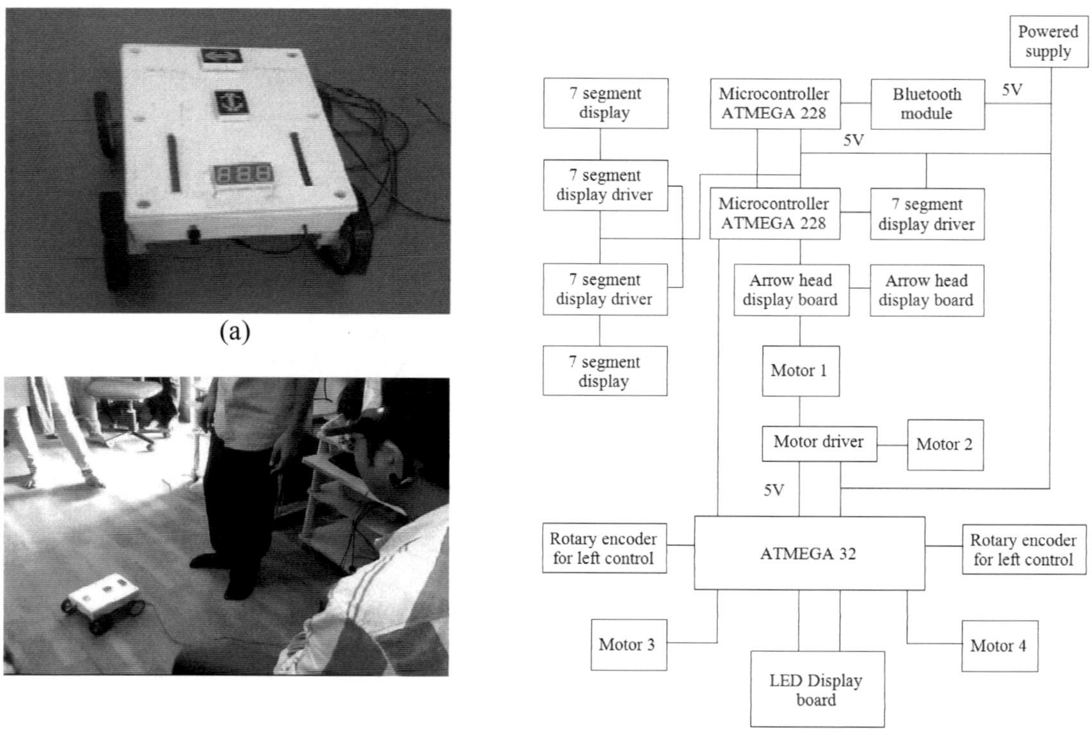

(a) (b) (c)

Figure 2. Prototype of a cognitively velocity controlled vehicle (a), working overview at JIS University Lab (b), schematic flow-chart representation of the cognitively velocity controlled vehicle (c).

Incorporation of the proposed methodology in relation to mining industry would enable to reach the steeply inclined coal seams for extraction and diagnosis of the stratum geological conditions for extraction of minerals during sublevel mining.

3.3. Thought Concentration Controlled Dexterous Mechanic Arm

This origination shares to a thought controlled dexterous mechanic arm. This shares to a thought controlled dexterous mechanic arm which can be operated by a brain-computer interface (BCI) that enables communication between brain activity and an external device (arm).

IOP Conf. Series: Earth and Environmental Science **84** (2017) 012002 doi:10.1088/1755-1315/84/1/012002

This creation narrates to thought controlled dexterous mechanic arm which can be activated by a brain-computer interface (BCI) that empowers communication between brain activity and an external device. Entity of the invention is to provide a thought controlled dexterous prosthetic arm which consists of a) the gripping with concentration, b) A lever appliance is used for opening and closing the gripper; c) a single DC motor for the gripper control.

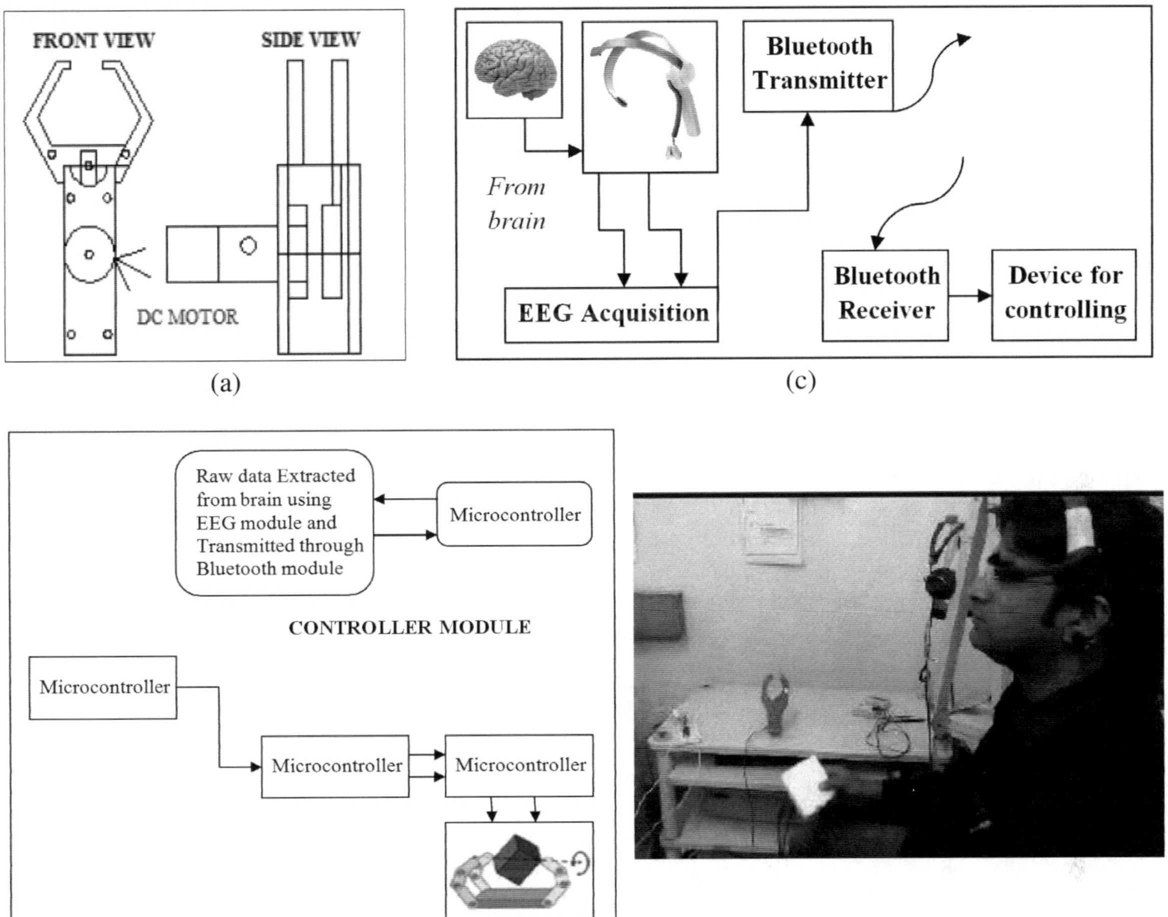

(b) (d)

Figure 3. Schematic diagram of thought controlled mechanic arm (a), BCI interface schematic module (b-c), working overview at JIS University Lab (d).

The epitome of the invention is to provide a thought controlled dexterous mechanic arm which has the beneficial effects of having saving manpower cost, reducing labor intensity, and having safety and reliability which can be further applied in mining industry during underground mining mineral extraction with high efficacy. For example development of a manipulator to intensify the top coals caving process and other tasks.

4. Conclusions

The support-enclosing walking module and the KPV power support with controlled outlet are the basis for creating robotic complexes for the safe development of stratified mineral deposits with the underground method, which provide working out of seams with complicated mining and geological bedding conditions (steep and steeply inclined thick seams). The control system of the robotic complex provides adaptive-program systems of the complex functional elements without people constant presence at the extraction face.

The technological theme of the Indian project attempt is to introduce in coal mining part / others mining parts in Russia with technology transfer vertical. Safety is one of the major consideration points for mining. The safety matter will be instigated with effort of BCI technological upbringing. All the coal mining technologies are under review stage en route for the intention of suitable implication of IoT based BCI skills. Robotic based unmanned application of walking module and the KPV power support with controlled outlet are already initiated as joint collaboration. Control strategy with performance analysis introducing controllability & observability testing are additional objective of the future for this research. In addition, the nonlinearity analysis will be the best challenge of this research in upcoming future after developing error less system based MIMO approach. Moreover, the research as interlinked with different interdisciplinary engineering concepts will produced better future path to extent.

Acknowledgements
The research was carried out due to the grant of Russian Science Foundation (project No. 17-17-01143).

References
[1] Nikitenko S M 2006 *Innovation* **4(91)**
[2] Nikitenko S M, Goosen Ye V and Klishin V I 2013 *Innovation* **9(179)**
[3] Yermekov T Ye, Nesipbaev Zh S and Sarsembaev T U 2001 *Proceedings of the University* **1** pp 46–7
[4] Malakhov Yu V 2017 Proc. *Int. Sci. Conf. on Innovations in the Fuel and Energy Complex and in Engineering* pp 62–8
[5] Klishin V I 2013 *GIAB* **6** 36–47
[6] Klishin V I and Kokoulin D I 2015 *Support for Development of Thick Steep Coal Seams with Sub-level* Extraction Patent for utility model RU 1604472 U1
[7] Stepanova L N, Kabanov S I, Bekher S A and Nikitenko M S 2013 *Automation and Remote Control* **74(5)** 891 –7
[8] Nikitenko M S 2016 *IOP Conference Series: Earth and Environmental Science.* 012007
[9] The Russian Federation Government Order of January 24, 2012 N 14-r "On Approval of the Russian Coal Industry Long-Term Development Program for the Period till 2030"
[10] Order of the Federal Service for Ecological, Technological and Nuclear Supervision of November 19, 2013 N 550 "On Approval of Federal Norms and Rules in the Field of Industrial Safety "Safety Rules in Coal Mines".
[11] Joseph N Mak and Jonathan R Wolpaw 2009 *IEEE Reviews in Biomedical Engineering* **2** 187–99
[12] Lebedev M A and Nicoleli Miguel A L 2006 *Trends in Neuroscience*
[13] Petrov B B et al 2016 *Int. Sci. Conf. Electronics - ET2016* (Sozopol, Bulgaria)
[14] *Hwaiyu Geng and McKeeth J 2016 Internet of Things and Data Analytics Handbook* (Wiley Publications)
[15] *Eirini Mathe and Evaggelos Spyrou 2016 PETRA Conference* (Corfu Island, Greece)
[16] Song Xiaoyana and Xie Zhongpengb 2014 *International Symposium on Safety Science and Technology* pp 87–92
[17] Sharma R, Pavlovic V I and Huang T S *Proc. IEEE* **86(5)** 853–69
[18] *Neogi B et al 2016 Cognitively Velocity Controlled Vehicle Patent IND File no. 201631017186 Indian Patent Journal* no. 26/2016 http://www.ipindia.nic.in/writereaddata /Portal/IPO Journal/ 1_350_1/part1.pdf

IOP Conf. Series: Earth and Environmental Science **84** (2017) 012003 doi:10.1088/1755-1315/84/1/012003

3D Discrete element approach to the problem on abutment pressure in a gently dipping coal seam

S V Klishin and A F Revuzhenko

Chinakal Institute of Mining, Siberian Branch of the Russian Academy of Sciences, 54 Krasnyi pr., Novosibirsk, 630091, Russia

E-mail: sv.klishin@gmail.com

Abstract. Using the discrete element method, the authors have carried out 3D implementation of the problem on strength loss in surrounding rock mass in the vicinity of a production heading and on abutment pressure in a gently dripping coal seam. The calculation of forces at the contacts between particles accounts for friction, rolling resistance and viscosity. Between discrete particles modeling coal seam, surrounding rock mass and broken rocks, an elastic connecting element is introduced to allow simulating coherent materials. The paper presents the kinematic patterns of rock mass deformation, stresses in particles and the graph of the abutment pressure behavior in the coal seam.

1. Introduction

Abutment pressure in a gently dipping coal seam is represented by normal compressive stresses affecting the seam along its length and generated under re-distribution of stress state by a discontinuity in rock mass. Abutment pressure is created by the weight of the overlying strata in combination with the sum of bending moments of rocks above a mined-out void or another cavity [1 - 5].

Abutment pressure induces diverse processes in rocks that impose serious threat to mining. Unexpected and uncontrollable dynamic rock falls are highly harmful for miners, machines and excavations. Furthermore, roof overhangs create concentrated ground pressure on coal in the zone of production headings and junctures with roadways, which provokes rockbursting. Under such conditions, pillars and supported development headings also experience high abutment pressure due to roof overhang over large areas.

Despite appreciable advance in the description of the abutment pressure events, selection of the most adequate characterization of the post-limit rock mass behavior in the boundary value problems on rock stability and failure in underground mines yet presents a difficulty [6 - 8].

Figure 1 depicts the problem to be discussed in this paper. Let rock mass (*1*) contain a gently dipping coal seam (*2*) with a thickness h_c and caved roof rocks (*3*) at a depth of h_m from the free surface (*5*). The contour of a heading (*4*) is free from normal and shear stresses. The main approach to solving this problem is based on the theory of plasticity and elasticity and uses the finite element method. The boundary conditions in the considered domain are governed by the initial stress state of rock mass in the field of gravity. Many researchers describe these conditions using Dinnik's hypothesis. Usually, conditions of continuous stresses and displacements are assumed at the interfaces of enclosing rocks, caved rocks and coal seam. The elastic and elastoplastic models use, as a rule, geometrically linear formulations of problems. For this reason, displacements are assumed small. This constraint is essential and not always conformable with a real situation.

The two-dimensional model of rock mass constructed in [8] based on the concept of rock as a medium with the internal sources and sinks of energy took into account block structure, anisotropic properties, inter-block sliding and strength loss in rock mass. The model was used to solve the problem on deformation of a rock pillar. The authors described successive development of zones of local strength loss and residual strength. The effect of unstable deformation was demonstrated in the form of development of long narrow bands of shearing that can result in disastrous failure of a pillar.

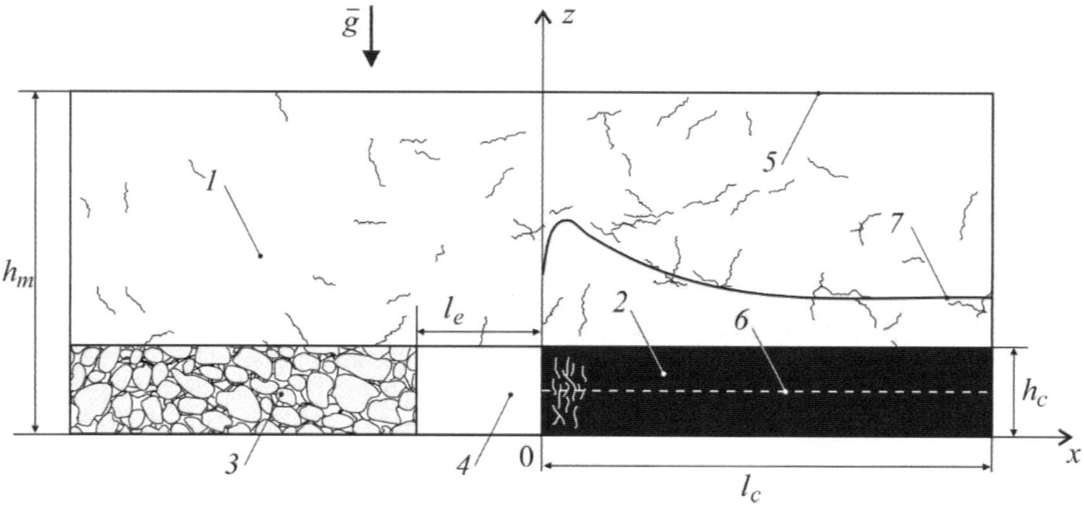

Figure 1. Analytical model of mining: 1 – enclosing rocks; 2 – coal seam; 3 – caved rocks; 4 – production heading; 5 – free surface; 6 – central cross section of the seam to measure abutment pressure; 7 – epure of abutment pressure distribution.

To study large deformations, it seems sufficiently adequate to use an approach based on the discrete element method (DEM). This method enables determining evolution of stress state in rocks both at the pre-limit stage of loading and at the stage of transition to localization of shearing and failure. Introduced in [9], DEM currently is widely popular and used to solve many geomechanical problems [10–12]. In this method, a real medium is replaced by a package of discrete particles with the certain interaction conditions assumed at the particle contacts. Most researches use spherical discrete elements with the pre-set distribution of radii. This method involves no difficulties when solving problems on large (finite) deformations and rotations. Thus, DEM is the main alternative to the classical methods based upon conventional provisions of continuum mechanics.

2. Discrete element method

In accordance with DEM, a medium to be studied is represented as a package of N spherical particles (discrete elements). Each i-th element possesses a radius r_i and the known set of physical and contact properties of the particle material: density, elastic and viscous moduli, friction, cohesion etc. ($i=1,\ldots,N$). Motion of a discrete element is composed of translation and rotation and is given by the equations below:

$$m_i \ddot{\boldsymbol{x}}_i = \sum_j \boldsymbol{f}_{ij} + m_i \boldsymbol{g}, \tag{1}$$

$$I_i \ddot{\boldsymbol{\theta}}_i = \sum_j (\boldsymbol{r}_{ic} \times \boldsymbol{f}_{ij} + \boldsymbol{M}_{r,ij}) \cdot \tag{2}$$

Here, points imply differentiation with respect to time t; $\boldsymbol{x}_i$ is the radius-vector of the center of gravity of an i-th particle; $\boldsymbol{\theta}_i$ is the turn of the i-th particle relative to the coordinate axes; m_i and I_i are the mass and inertia moment of the i-th particle, respectively; $\boldsymbol{g}$ is the acceleration of gravity; $\boldsymbol{r}_{ic}$ is the vector from the center of the i-th particle to the contact point; $\boldsymbol{f}_{ij}$ is the contact force applied to the i-th particle by a j-th particle (or boundary), depends on the overlapping of the particles, as well as on the elastic and viscous moduli; $\boldsymbol{M}_{r,ij}$ is the rolling resistance moment of the i-th and j-th particles (or boundary). Summation in (1) and (2) is carried out with respect to all elements and boundaries that contact the i-th particle. It is assumed that discrete elements preserve their shape for the period of contact; for this reason, the degree of their deformation is only described by the value of overlap between the particles in contact.

The contact force between particles is calculated from the Kelvin–Voigt viscoelastic model that accounts for normal and shear components. This model is schematically depicted in figure 2a. The normal component of the contact force is found using the Hertz law [13], and the shear component is calculated from the scheme by Mindlin and Deresiewicz [14 - 15].

Since rocks are a cohesive material capable to accumulate and release elastic energy, the described deformation model should be added with a connecting element between the particles in contact. This connecting element is introduced in the model in the form of extra forces (elastic springs) of pre-set stiffness, which prevent separation of the contact elements (figure 2b).

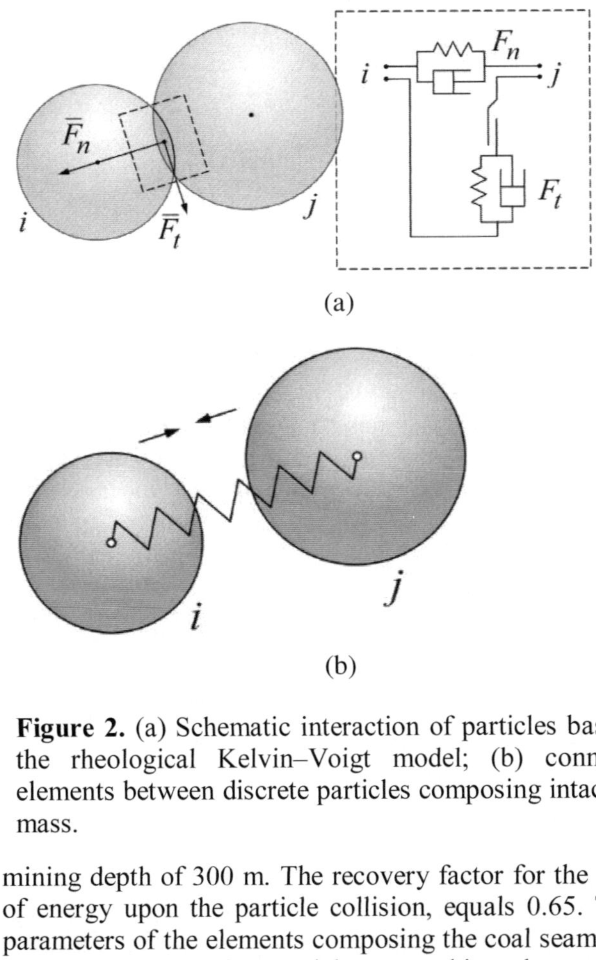

Figure 2. (a) Schematic interaction of particles based on the rheological Kelvin–Voigt model; (b) connecting elements between discrete particles composing intact rock mass.

3. Numerical experiment

Having the data on the law of motion and interaction of particles, i.e. Eqs. (1) and (2), as well as initial and boundary conditions, it is possible to solve the equations and determine evolution of stress state in a medium. The Chinakal Institute of Mining has developed DEM-based software for the numerical analysis of motion modes in granular materials composed of individual spherical particles [16, 17].

Figure 3 presents the scheme of the 3D numerical experiment. The objective domain is presented by a package of $N = 150\,000$ discrete elements having radii selected from a uniform distribution between 0.025 and 0.05 m. The particles fill a domain in the form of a parallelepiped with a length of 34 m, height of 18 m and width of 1 m along the axes Ox, Oz and Oy, respectively. The sides AB and CD, and the faces oriented in parallel to the coordinate plane Oxz are perfectly smooth. At the upper boundary AD the pressure F_z is set in conformity with the mining depth of 300 m. The recovery factor for the velocity between particles, describing dissipation of energy upon the particle collision, equals 0.65. The table below (table 1) describes the physical parameters of the elements composing the coal seam (*1*), rock mass (*2*) and caved area (*3*). The initial package of untouching particles was subjected to settlement under the force of gravity with regard to contact interaction. Upon reaching the equilibrium, elastic nondestructive connections (spring) with the stiffness of 10 MPa were created between the particles being in contact. The connections between the discrete elements representing coal and enclosing rocks are uniformly spread in the volume, while the connections between the particles in the caved area are created within a mesh of cubic cells 0.5×0.5×0.5 m, with formation of separate blocks.

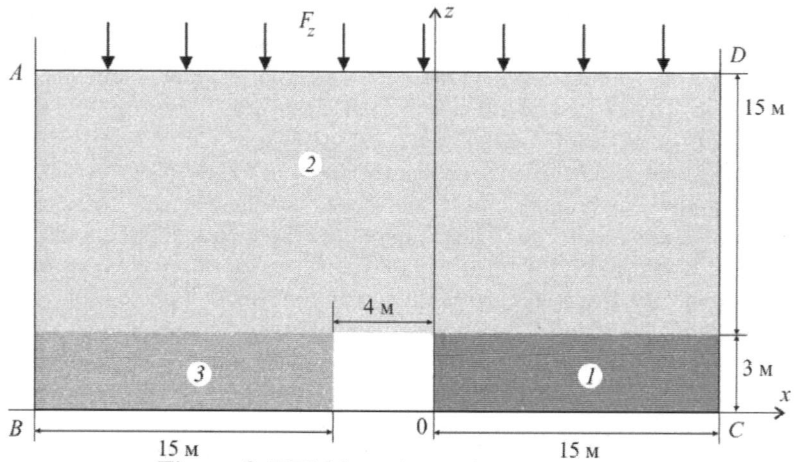

Figure 3. DEM-based experiment scheme.

Table 1. Physical parameters of particles.

	Coal	Rock mass	Caved rocks
Elasticity modulus, MPa	$5 \cdot 10^3$	$1.5 \cdot 10^4$	$1.5 \cdot 10^4$
Poisson's ratio	0.16	0.25	0.25
Density, kg/m	1300	2400	2400

The external dry friction angle φ_{ij} varies from 26 to 33° for the test materials. The present studies use φ_{ij}= 30°. Selection of the rolling resistance angle ψ_{ij}, having large influence on the contact interaction and relative rotations of discrete elements, needs additional numerical experimentation. The specified parameter is determined as follows. In Eq. (2) the moment upon the contact of two particles is represented as two summands. The first summand is a moment that appears under relative sliding of particles over one the other given dry friction is present. The second summand is a moment that impedes relative rotation of discrete elements:

$$\boldsymbol{M}_{r,ij} = -\frac{\boldsymbol{\omega}_i - \boldsymbol{\omega}_j}{\left|\boldsymbol{\omega}_i - \boldsymbol{\omega}_j\right|} \mu_{ij} r_{ij} \left|\boldsymbol{f}_{ij} \cdot \boldsymbol{n}_{ij}\right|, \tag{3}$$

where $\boldsymbol{\omega}_i$ and $\boldsymbol{\omega}_j$ are the angular velocities of the contact particles; $\boldsymbol{n}_{ij}$ is the unit vector lying in the straight line between the centers of the contact particles; r_{ij} is the reduced radius:

$$\frac{1}{r_{ij}} = \frac{1}{r_i} + \frac{1}{r_j}.$$

The dimensionless parameter μ_{ij} is the rolling resistance coefficient. By analogy with the dry friction angle between two bodies, it is readily shown that $\mu_{ij} = \mathrm{tg}\,\psi_{ij}$ [18]. The angle of rolling resistance is the maximum slope of a plane, at which a spherical particle occurs in this plane in the equilibrium condition under the action of gravity.

The authors have carried out a test experiment on numerical determination of slope of granular material in the three-dimensional formulation (figure 4). Particles were placed in a parallelepiped box with the perfectly smooth walls (figure 4a). When the particles reached equilibrium, the side walls of the box were moved upwards at a constant low velocity. The particles showered from the box and were withdrawn from the calculations. By the end of the experiment, the granular medium acquired

IOP Conf. Series: Earth and Environmental Science **84** (2017) 012003 doi:10.1088/1755-1315/84/1/012003

the triangular form as in figure 4b. The tests were performed for the constant dry friction angle φ_{ij}= 30° and varied angles of rolling resistance. It is seen that the latter has an essential influence on the formation of slopes of the granular material. Based on the calculated data, the rolling resistance angle was selected for all media as ψ_{ij}= 15°.

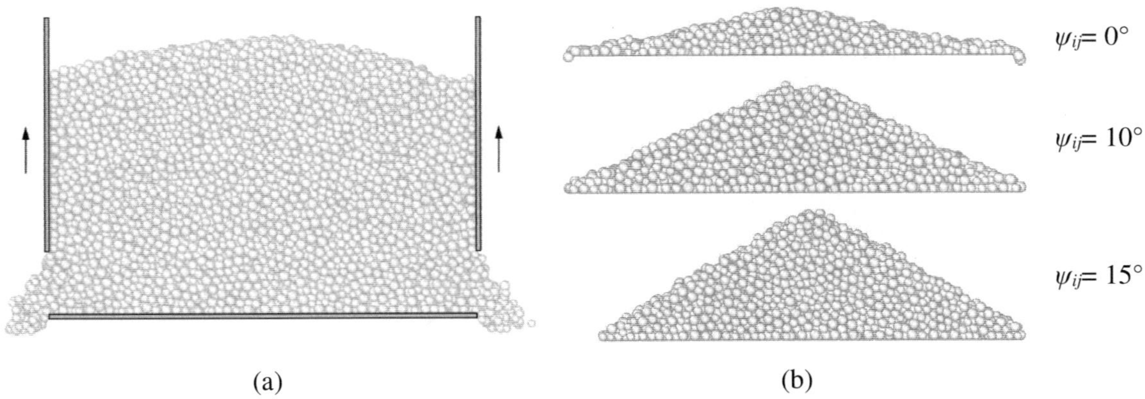

Figure 4. Numerical experiment to find natural slope of granular material at different rolling resistance angles ψij and constant dry friction angle φij= 30°: (a) schematic of the experiment; (b) final configuration of material.

Turn back to the problem. At the upper boundary *AD* of the initially equilibrium medium (figure 3), we assign the vertical pressure F_z that gradually increases in value from zero until pressure conformable with the mining depth of 300 m. Figure 5 depicts the state of the medium under deformation at the final stage of loading. It is seen that as a consequence of the zero boundary conditions at the contour of the heading, squeezing of edge coal towards the heading takes place, and high subsidence of the heading roof is visible. The blocks composing the caved material deform under the weight of the overlying rocks and fall in the mined-out area.

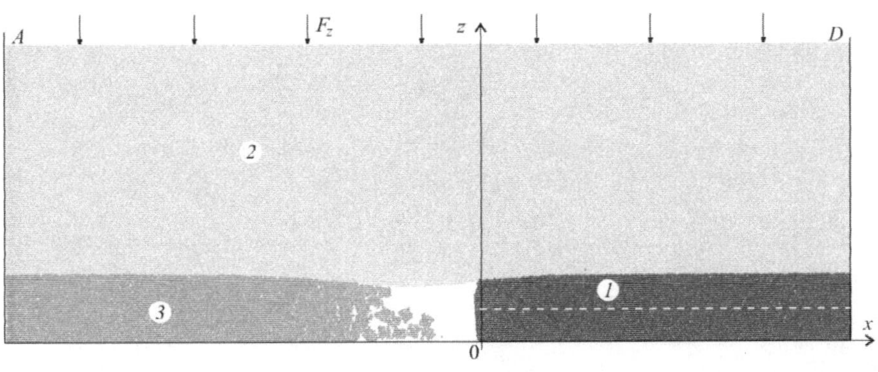

Figure 5. State of the medium at the last stage of deformation.

The distribution of horizontal and vertical stresses in discrete elements is illustrated in figure 6. Here, particles having higher compression modulus are shown by the darker color, and the particles with the lower compression modulus — by the lighter color.

In the epure of abutment pressure in a coal seam in figure 7, the vertical compressive stresses on the particles in the central cross section (refer to figure 5) are shown. It is seen that the peak pressure is at the distance of 0.5 to 2 m from the face breast.

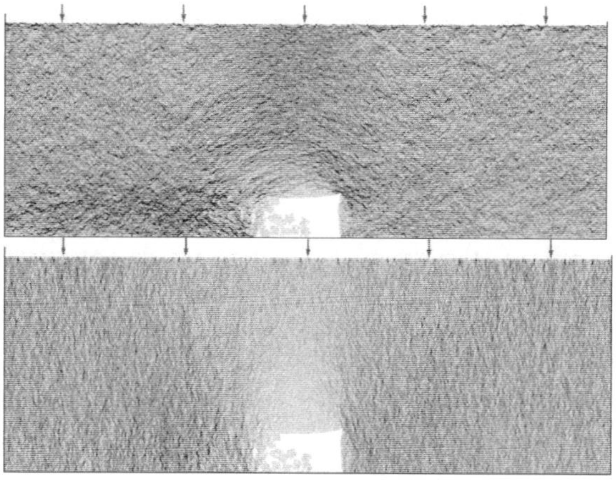

Figure 6. Distribution of (top) horizontal and (bottom) vertical stresses in the particles at the end of loading.

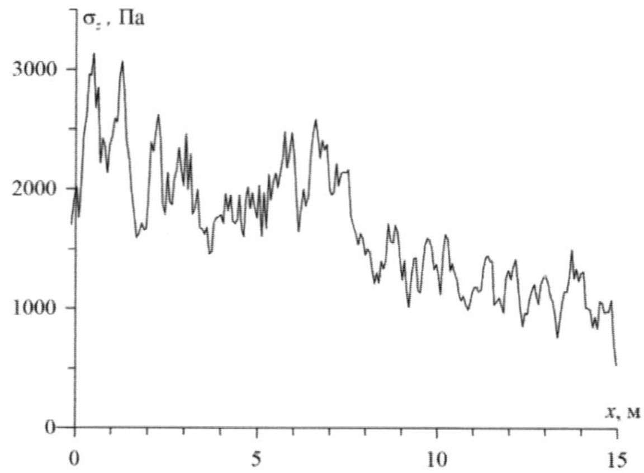

Figure 7. Vertical stresses on the particles in the central cross section of coal seam.

4. Conclusion

It seems the most appropriate to study problems on abutment pressure events in gently dipping coal seams using the discrete element method for its allows accounting for large deformations and rotations. The introduction of additional connections in the form of elastic potentials between particles enables describing the field of finite deformation during interaction of gently dipping coal seams and enclosing rock mass as well as the fall of broken rocks in roadways.

Acknowledgements

The study has been supported by the Russian Science Foundation, Project No. 16-17-10121.

References

[1] Gritsko G I, Vlasenko B V and Posokhov G E et al 1980 *Prediction and Calculation of Rock Pressure Events* (Novosibirsk: Nauka)
[2] Borisov A A, Matantsev V I, Ovcharenko B P and Voskoboev F N 1983 *Ground Control* (Moscow: Nedra)
[3] Ardashev K A, Krylov V F and Kuksov N I et al 1967 *Improvement of Ground Control in Mining Inclined and Steeply Dipping Seams* (Moscow: Nedra)
[4] Bieniawski Z T 1987 *Strata Control in in Mineral Engineering* (Wiley)
[5] Litvinsky G G 2000 *Proc. Int. Conf. on Propsects for Coal Industry Development at the Edge of XXI Century* (Alchevsk: DGMI) pp 72–8
[6] Pavlova L D 2006 *GIAB* **4** 57–60
[7] Kuznetsova A V and Smolin I Yu 2010 *Vestn. Tomsk Gos. Univer. Matem. Mekh.* **2(10)** 79–87
[8] Lavrikov S V and Revuzhenko A F 2014 *Calculation of Deformation and Stability of Pillars in Rocks AIP Conference Proceedings* **1623** 335
[9] Cundall P A and Strack O D L 1979 *Géotechnique* **29(1)** 47–65
[10] Klishin S V, Klishin V I and Opruk G Yu 2013 *J. Min. Sci.* **49(6)** 932–40
[11] Kramadzhyan A A et al 2015 *J. Min. Sci.* **51(6)** 1077–84
[12] Klishin S V and Revuzhenko A F 2015 *J. Fundament. Appl. Min. Sci.* **2(1)** 235–9

[13] Johnson K 1985 *Contact Mechanics* (Cambridge University Press)
[14] Mindlin R D 1949 *J. Appl. Mech.* **16** 259–68
[15] Mindlin R D and Deresiewicz H 1953 *J. Appl. Mech.* **20** 327–44
[16] Klishin S V and Revuzhenko A F 2016 *Proc. Int. Sci. Conf on Knowlwdge-based Technologies in Development of Use of Mineral Resources* (Novokuznetsk: SibSIU) **2** pp 39–46
[17] Lavrikov S V and Revuzhenko A F 2016 *J. Min. Sci.* **52(4)** 632–7
[18] Klishin S V and Revuzhenko A F 2015 *J. Min. Sci.* **51(5)** 1070–6

IOP Conf. Series: Earth and Environmental Science **84** (2017) 012004 doi:10.1088/1755-1315/84/1/012004

Modelling limit stress of a seam roof ahead of a working face

N V Cherdantsev

Institute of Coal of the Federal Research Centre of Coal and Coal Chemistry, Siberian Branch of the Russian Academy of Sciences, 10 Leningradsky Prospect, Kemerovo, 650065, Russia

E-mail: cherdancevn@bk.ru

Abstract. Solving of boundary problem of geomechanic state of a roof working developed along the seam is introduced in the article. The rock mass works under the condition of a plane-strain deformation with marginal seam zones are in a limit stress state. The problem is solved by boundary element method using Coulomb – Mohr and Mohr – Kuznetsov strength indices.

1. Introductions

Quantitative evaluation of a stress state of a rock mass with hard rock enclosing a coal seam and its workings with limit stress zones on its sides is an important and challenging research and production issue as different geodynamic phenomena take place in such zones [1].

The solutions of several problems on limit stress state of a coal seam sides with hard walls together with the presence of a soft layer in workings are introduced in [2].

Depending on strength characteristics of a coal seam and enclosing wall rocks the transition of this rock into a limit state during the seam development can take place in different ways. In one case a limit state of a seam roof with its subsequent failure can take place under relatively small size of a goaf. In other cases, before the roof failure, the size of a goaf can be of rather significant volume and could result in rock heaving in a seam soil that in case of a roof failure can cause accidents while ventilating mine openings or bring about a failure of the operating mining equipment [3]. Such cases lead to intentional roof breaking to prevent sizable overhanging of a roof [4].

The researching results of a coal mass state for several characteristics of a working seam structure and size where the seam roof goes into limit state ahead of the face are presented in the article.

2. Problem setting and solutions

It is known that in the seam sides, during deep mining, limit stress zones with inelastic deformations are formed. And one can define only approximate dimension of such zones [5].

This research adopts a model of a massif where the presence of limit stress marginal zones in a coal seam with the size of LOT is taken into an account.

Solving the problem, we assume:

1) the path of the workings and its dimensions along x-axis are vastly superior then the dimensions in its cross-sectional plane 0yz , hence we may assume that rock mass and slots in the vicinity of the workings are under the state of plane strain.

2) a seam strength is significantly below the strength of its enclosing rock;

3) compressive normal stresses are positive.

Further, the following task can be set. In a rock mass modelled by weightless plane there is a rectangular cross-section working with the dimensions b× h which was made in a depth H along the seam with the thickness of m = h. Support reaction p is in effect in the roof and sill of the working. The massif is loaded with gravitational pressure γH (γ – average volume density of overlaying rock), and λγH is pressure on sides (λ – coefficient of lateral thrust) (figure 1).

While solving the problem we define stress field in the vicinity of the working, providing that limit stress zones of a seam are formed prior.

A defined problem demands solving of several other problems. Firstly, we set a problem on the state of the massif in a limit stress zone where the sliding lines and three section appear (up and down from the y-axis) (see figure 1). Unique features of a seam state reveal themselves on each of them. Section VCN (V1C1N1) – is a section of a limit uniaxial compression with the length Lk. Sections V1BCNGDC1 and VCDG1N1C1B with the dimension Lpr are Prandtl zones, and section GDG1P1P with the dimension Lp – is a section of seam inelastic deformation.

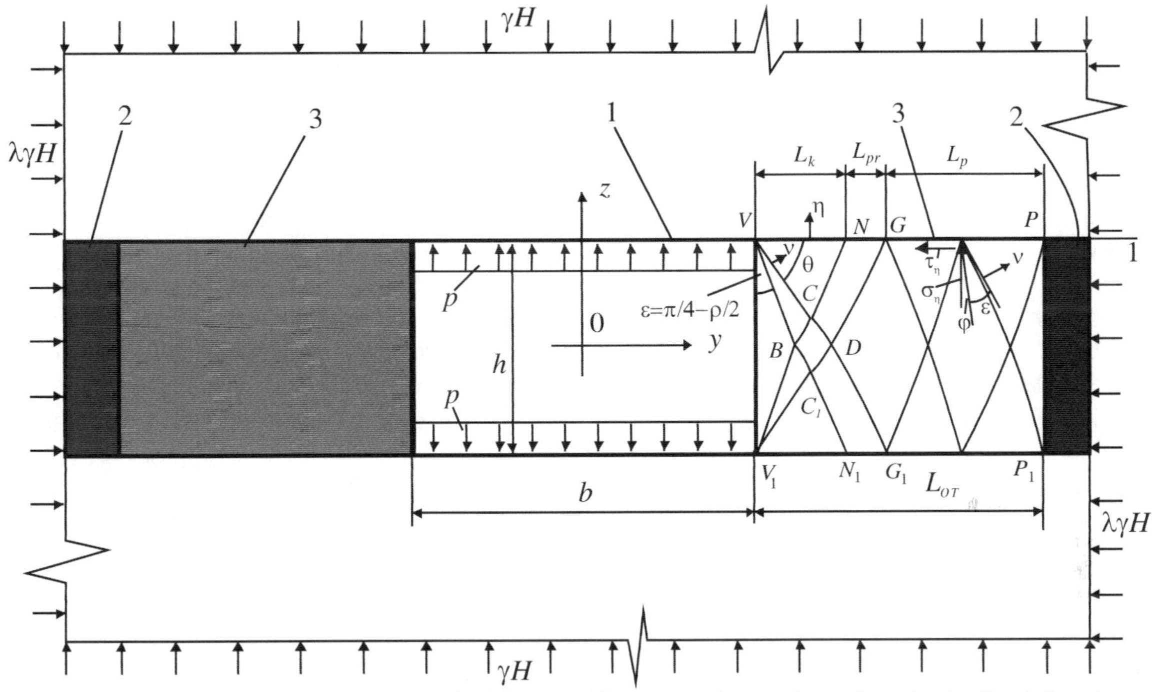

Figure 1. Calculation model of massif with a working seam 1 – working, 2 – elastically deforming seam parts, 3 – limit stress seam zones.

The seam experiences uniaxial compression in the area of protrusion in a shape of prism VBV_1. Principal stress σ_1 has an effect along the edge of seam outcorp forming an angle ε with system of isogonal sliding lines direction. It is equal to limit seam strength on uniaxial compression σ_0.

Considering that along the seam contacts with neighboring rocks the rock discontinuity is possible and it is manifested in the form of a seam sliding motion, so both on the seam and on its contacts with side rocks there will exist two limit state simultaneously – normal (for a seam) and unique (for the seam contacts) [5-7]. It can be demonstrated by Mohr's stress circle (figure 2). It is seen from the figure that the first area element of sliding surface with normal ν according to the Coulomb-Mohr criteria is in the contacting point of a stress circle and the rectilinear envelope 1 of the seam limit state with the angle of gradient ρ to X-axis.

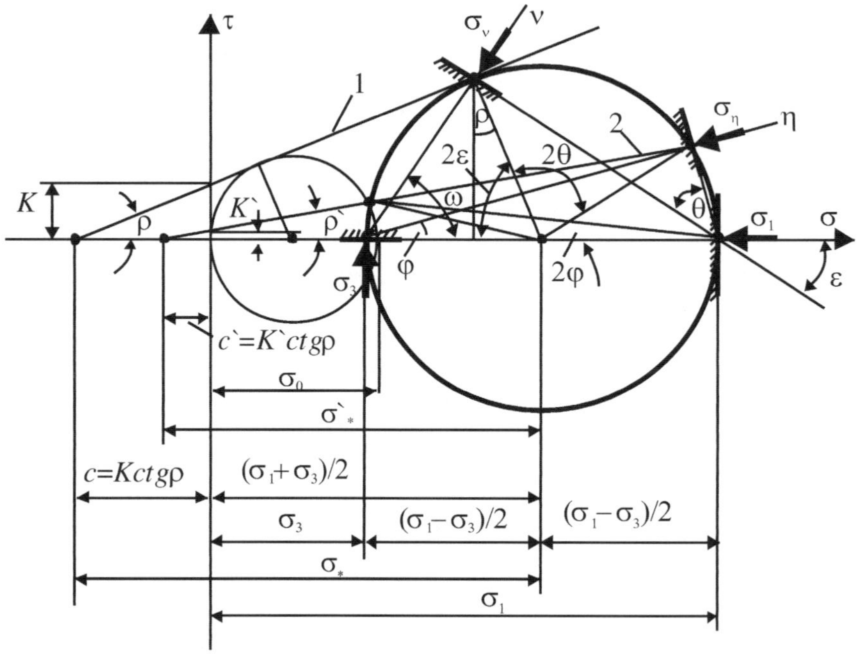

Figure 2. Scheme of defining angles ε, ω, φ, θ in limit stress zones.

The position of the second area with normal η is calculated according to Mohr-Kuznetsov criteria as a result of crossing with the stress circle of rectilinear envelope 2 of the limit state along the surface of weakness making an angle of gradient ρ' with X-axis. The area with normal η coincides with seam contacting with surrounding rock. Angles ρ and ρ' are angles of internal friction, K and K' are the cohesion coefficients, respectively for the seam and its contact with side rocks, σ_0 – is a strength limit on the seam uniaxial compression.

Angle θ between the areas with normal between ν and η we calculate in figure 2

$$\theta = \frac{\pi}{4} + \frac{1}{2}(\rho - \rho') - \frac{1}{2}\arcsin\left[\frac{\sin\rho'}{\sin\rho}\left(1 - \frac{c}{\sigma_*}\right)\right],$$

where σ_* и σ'_* values are defined according to the formulas

$$\sigma_* = c + \frac{\sigma_1 + \sigma_3}{2}, \quad \sigma'_* = c' + \frac{\sigma_1 + \sigma_3}{2},$$

Resulted from figure 2 and they are presented as reduced stress in a seam and on the surface of weakness, respectively [2, 5]. Values c and c' are expressed through K, ρ, K', ρ' using the formulas (figure 2).

The angles ε and ω are formed by the sliding surface with normal ν and primary stress σ_1, σ_3. As it is seen from figure 2 they are calculated by the formulas

$$\varepsilon = \frac{\pi}{4} - \frac{\rho}{2}, \quad \omega = \frac{\pi}{4} + \frac{\rho}{2}.$$

IOP Conf. Series: Earth and Environmental Science **84** (2017) 012004 doi:10.1088/1755-1315/84/1/012004

Two systems of sliding lines are formed in Prandtl zones (figure 1). The first system it is the family of logarithmic spirals and the second system is a bunch of radial lines.

Angle $\Delta\varphi$ of the sectors and radial lines length are defined according to [5]

$$\Delta\varphi = \omega - \theta, \quad r = r_0 \cdot e^{-\Delta\varphi \cdot \mathrm{tg}\rho},$$

where r_0 – is an interval VB (V_1B) adjacent to the area of protrusion VBV_1, r – border (marginal) interval of Prandtl sector equal to the interval VC (or V_1C_1).

Stresses in Prandtl zones along the respective radial line are constant and they change exponentially along the logarithmic spirals [5]. For example, the stress in points C and D is defined as

$$\sigma_C = \sigma_0 \cdot \mathrm{e}^{2\Delta\varphi \cdot \mathrm{tg}\rho}, \quad \sigma_D = \sigma_C \cdot \mathrm{e}^{2\Delta\varphi \cdot \mathrm{tg}\rho}. \tag{1}$$

In the areas of limit stress zone positioned to the right of the sliding line DG (figure 1) to define angles θ it is necessary to know reduced stress in junction points on a slope contacts with enclosing rocks. To define them it is recommended to use the function of reduced stresses σ_* acting along Y-axis that change exponentially [5]

$$\sigma_* = \sigma_{0*} \cdot e^{k \cdot \left(y - \frac{b}{2}\right)},$$

where σ_{0*} – reduced stress on the edge of the seam and k paramete are defined according to the formulas [5].

$$\sigma_{0*} = \frac{\sigma_0}{2 \cdot \sin\rho}, \quad ,k = \frac{2}{m} \cdot \mathrm{tg}^2\omega \cdot \mathrm{tg}\rho'$$

Normal and shear stresses in limit stress zones on the contact of the seam with a massif are expressed by evident relations arise from figure 2

$$\sigma_\eta = \frac{\sigma_1 + \sigma_3}{2} + \frac{\sigma_1 - \sigma_3}{2} \cdot \cos 2\varphi, \quad \tau_\eta = \frac{\sigma_1 - \sigma_3}{2} \cdot \sin 2\varphi,$$

which results in obtaining formulas for calculating primary stresses.

$$\sigma_1 = \sigma_*(1 + \sin\rho) - c, \quad \sigma_3 = \sigma_*(1 - \sin\rho) - c,$$

$$\sigma_\eta = \sigma_*(1 - \sin\rho \cdot \cos 2\varphi) - c, \quad \tau_\eta = \sigma_* \cdot \sin\rho \cdot \sin 2\varphi,$$

where φ – is an angle between σ_1 and a normal towards the surface of weakness.

As it follows from figure 2 there is a link between angles φ, θ, ε, and two more important relations

$$\varphi = \frac{\pi}{2} - (\varepsilon + \theta), \left(K \cdot \mathrm{ctg}\rho + \frac{\sigma_0}{2}\right)\sin\rho = \frac{\sigma_0}{2}, \left(\mathrm{K} \cdot \mathrm{ctg}\rho + \frac{\sigma_1 + \sigma_3}{2}\right)\sin\rho = \frac{\sigma_1 - \sigma_3}{2}. \tag{2}$$

The transformation of the relations (2) allows us to get the connection between K and σ_0, and the limit state according to Coulomb-Mohr strength theory

$$\sigma_0 = \frac{2K \cdot \cos\rho}{1-\sin\rho},\ \sigma_1 - \beta_n\sigma_3 = \sigma_0, \tag{3}$$

where β_n – the parameter of bulk stress is calculated according to the following formula.

$$\beta_n = \frac{1+\sin\rho}{1-\sin\rho}. \tag{4}$$

As it follows from the formula (3) the left side which is called equivalent stresses σ_i according to Coulomb-Mohr theory is constant on all the sections of limit stress zone. Hence the diagram of conventional stresses σ_i – ε_i (ε_i –equivalent deformations) corresponds to Prandtl diagram (line 0-1-2 in figure 3). Section 0-1 of the diagram is an area of a seam elastic deformation, and its slope ratio to X-axis is equal to a seam elasticity module E.

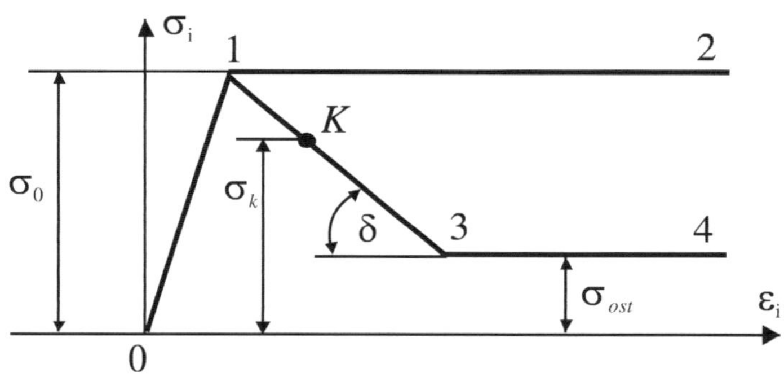

Figure 3. Prandtl and Linkov diagrams on deformed seam conventional stresses.

Under real conditions, a seam can partially deform on the sections of limit stress zones and in its turn lessen the limit strength to the value of residual strength σ_{ost} [1]. So, to evaluate a massif stress state Linkov idealized diagram of a seam deformation is used (line 0-1-3-4 in figure 3)[1]. There is a section 1-3 of a seam weakening (drop down) and section 3-4 – its residual strength. Tangent of an angle δ numerically equals to weakening (dropdown) module M.

As it is assumed from the problem of a massif with a seam working where the seam is deforming according to Linkov diagram, the first section of limit stress zone VN (figure 1) corresponds to a horizontal section 3-4 of Linkov diagram. On the second section, NG stresses are also changed according to the formula (1) but substituting σ_0 for σ_{ost}. On the third section of this GP zone the parameters σ_{0*} and c in each point change proportionally to equivalent stresses on weakening section 1-3 of the diagram σ_i - ε_i. If n – is number of points on sections 1-3 and GP and k is a number of points from 3 to K then

$$\sigma_{0*} = \frac{\sigma_K}{2\sin\rho} = \frac{1}{2\sin\rho}\left[\sigma_{ost} + (\sigma_0 - \sigma_{ost})\cdot\frac{k}{n}\right], \qquad c = \frac{1}{2}\left(\frac{1}{\sin\rho} - 1\right)\left[\sigma_{ost} + (\sigma_0 - \sigma_{ost})\cdot\frac{k}{n}\right].$$

All abovementioned functions describe fully the seam state in limit stress zones allowing us to formulate boundary states on the seam contact with surrounding rocks.

The condition of excessive maximum primary stresses σ_1 acting in the seam over σ_1 values of its roof can be taken as a transition of a seam roof into a limit state [5]. The comdition forms the formula of seam rock strength index f_k that corresponds to the initial transition into a limit state [5]

$$f_k \leq \frac{k_\sigma \cdot \gamma H - \beta_k^{\ 2} \cdot p}{10 \cdot (1+\beta_k)}, \qquad (5)$$

where k_σ - is a coefficient of accumulation of stresses equals to the relation between maximum primary stress in a seam to γH value, β_k is a parameter of a bulk strength of a seam roof calculated from the formula (4) with the substitution of ρ for ρ_k, where ρ_k is an angle of a roof rock internal friction.

k_σ is defined out of the solution of elastic boundary value problem.

In this boundary problem boundary states are stated about a closed path that includes the roof, the soil of a working and the contact of a seam with side rocks on the area of limit zone with a length L_{OT}. Thus, on horizontal sections of this maximum stress projection contour p_z, p_y are equal respectively to stresses σ_z and τ_{yz}. So then, in boundary integral equation of a boundary problem [1, 8, 9] integration is made about a closed contour that covers the roof, the working soil and the limit stress zone contour. There apart from the vector of distributed imaginary load the length L_{OT} is also unknown variable and to define it one can use method of successive approximations. According to this method, the L_{OT} value is being given and after that an integral equation of distributed imaginary load vectors attached to each boundary element is solved. After that summing up, the initial field stresses with the stresses resulted from the influence of defined, imaginary load made on the basis of Kelvin solution [8 - 10], we can find combined stress in the points on the line 1 that crosses the seam roof (figure 1). Here upon, on the margin of elastic and limit stress zones we compare the respective stresses. In case they differ, the calculations are repeated but with other L_{OT} values to the point the stresses coincide.

The introduced approach to solving the problem is a model of geomechanic massif state enclosing a coal seam together with the mine working driven along it. It describes the parameters of a bearing pressure (maximum stress and the width of a limit stress zone) in the walls of a working in exact approximation and is introduced in [11].

3. Performed calculations and results analysis

In the framework of the described model, together with the developed program a computational experiment on defining the bearing pressure indices: accumulation of stress index, the width of a limit stress zone was performed. The following massif and working parameters were taken as initial data: γ=25 kN/m^3, λ=1, H=800 m, h=3 m, seam stress index fp=1, σost=0.5σ0, M=0.25E, an angle of a seam and roof rock internal friction ρp=ρk=20°; K′=0, ρ′=10°, p=0.5MPa. Other parameters were being changed during the experiment.

Figure 4 demonstrates graphic results of the method of successive approximations. Thus, in figure 4 a the diagrams built with parameters b=6 m, p=0.5 MPa introduce stress envelopes σz and τyz in the seam along line 1 (figure 1), limit stress zone (diagrams 1 and 3) and in elastic field of a seam (diagrams 2 and 4). Points V, N, G, P in the diagrams are the bordering sections of limit stress zone of a seam (figure 1). As it can be deduced from the figure, the stress values in elastic field and in a limit stress zone coincide in point P. A measured horizontal interval between V and P points corresponds to the volume of a limit stress zone LOT. The diagrams in figure 4 b corresponds to the primary stress σ1 in a limit stress zone (figure 1) and to the elastic field of a seam (figure 2). Vertical axis referred to γH in P point equals kσ in a seam roof. Certain values LOT=4.81 m and kσ=2,125 are deduced from diagrams 2. It can be assumed from the equation (5) that the seam roof strength indices to correspond to their transition into a limit state ahead (behind) of the working should be not higher 1.33.

The parameters of a bearing pressure in the absence of a support (p=0) are defined. They slightly differ from the previous values and equal to LOT=4.81 m, kσ=2.14. The volume of a limit stress zone did not change and the stress accumulation indices differ from each other less than 0.5 percent.

It is important to note that the support reaction value p = 0.5 MPa in the introduced sample corresponds to the force in hydraulic cylinders of a powered support unit in 2.4 MN (240 tons) and its width of 0.8 m. Increasing the working distance the influence of a support will be less meaningful and in further calculations that influence was not taken into account. Moreover, in significantly long distances the powered support unit is not able to provide the roof and soil backing of the working along its whole distance.

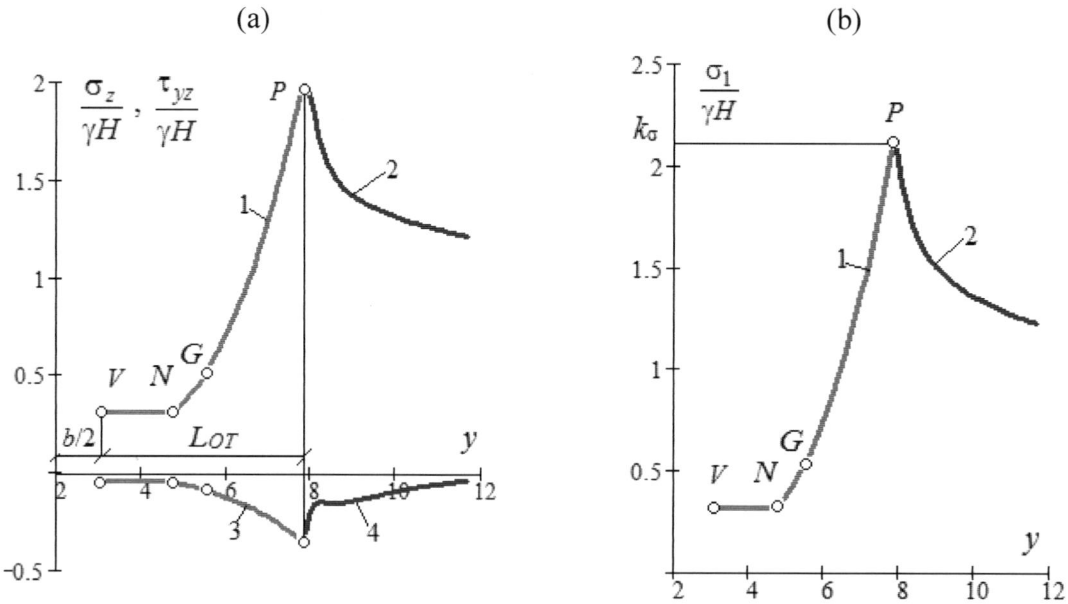

Figure 4. Stress envelopes σ_z, τ_{yz}, σ_1 along the seam roof.

In figure 5, graphic results of a performed computational experiment are presented. According to the state (5) the dependency diagrams of the roof rock strength index f_k, corresponding to its stress state and the width of a face working for several values of a seam residual strength are built. In figure 5*b* the dependency diagrams of the stress accumulation index k_σ in a limit seam zone and the width of a face working for several residual seam strength values are built.

Diagrams 1, 2 corresponds to Linkov diagram, diagram 1 is built when σ_{ost}=0,5σ_0, and diagram 2 – when σ_{ost}=0,05σ_0. Diagram 3 which corresponds to Prandtl diagram where σ_{ost}=σ_0 is given as an example.

Both series of diagrams present smooth low-arched curves. They prove that limit state of a seam roof ahead of a face working happens only in rocks with relatively small strength index even in significantly long distances of a face working. Reducing the strength limit of a seam in marginal zone change the research parameters in 20 times but not more then on 17 percent. The limit value of roof strength does not prevail over 3 units and the index for the accumulation of stresses in a bearing pressure zone is not more than 4.5 units.

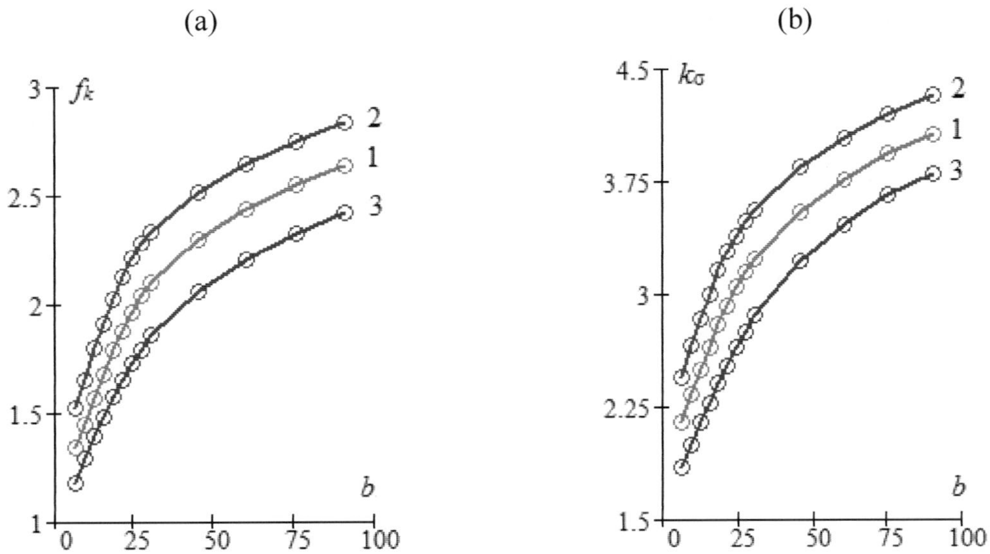

Figure 5. Dependency diagrams of a roof strength index f_k (*a*) and accumulation of stresses index k_σ in a seam limit zone (b) from the face working.

4. Conclusions

1. The problem of stress field distribution in a rock massif with seam working is solved in the framework of the model of geomechanic state of anisotropic massif. The parameters of a bearing pressure in pre-contour zone of a seam are defined while solving an elastic boundary value problem where through considering limit stress zones under the boundary states an elastic-plastic problem was brought up.

2. Reducing residual strength of a seam can increase parameters of a bearing pressure and the hardness of the roof rock when they go into a limit state ahead of the face working. The influence of the support reaction on a bearing pressure in a face working is negligible.

3. In isotropic rock mass that exists in equally component stress field and enclosing a seam developing by a stope, limit state of its structurally intact monolithic roof ahead of the face manifests itself under the relatively small strength index. However, if the side pressure coefficient is less than 1 and a massif is of bedded structure then a limit state of the roof can appear due to significant horizontal stresses and stratification of the mass with its further failure directly above a goaf (mined-out space).

Acknowledgements
The study was performed by the grant of Russian Scientific Foundation (project No. 17-17-01143).

References

[1] Petukhov I M and Linkov A M 1983 *Mechanics of Rock Bumps and Oubursts* (Moscow: Nedra) p 280
[2] Cherdantsev N V and Cherdantsev S V 2014 *Labour Safety in Industry* **11** 41–5
[3] Krivolapov V G and Paleev 2011 *Bulletin of VOSTEC* **1** 77–80
[4] Klishin V I 2002 *Adaptation of Powered Roof Supports to the Dynamic Loading States* (Novosibirsk: Nauka) p 200
[5] Fisenko G L 1976 *Limit State of Rocks Around an Excavation* (Moscow: Nedra) p 272
[6] Kuznetsov G N 1961 *Proc. VNIMI L* **43** 98–112
[7] Sokolovski V V 1990 *Statics of Granular Media* (Moscow: Nauka) p 272
[8] Lurie A I 1970 *Theory of Elasticity* (Moscow: Nauka) p 940

[9] Brebbia K, Telles F and Wrobel L 1987 *Boundary Element Methods* (Moscow: Nauka) p 525
[10] Rabotnov Yu N 1988 *Mechanics of Deformable Solids* (Moscow: Nauka) p 712
[11] Cherdantsev N V and Cherdantsev S V 2004 *Appl. Mech. and Tech. Physics* **3** 141–8

IOP Conf. Series: Earth and Environmental Science **84** (2017) 012005 doi:10.1088/1755-1315/84/1/012005

Numerical simulation of geomechanical state of coal massif in the vicinity of underground workings in the superimposed seams

A B Tsvetkov, L D Pavlova and V N Fryanov

Siberian State Industrial University, 42 Kirova Street, Novokuznetsk, 654007, Russia

E-mail: ld_pavlova@mail.ru

Abstract. The results of numerical modeling of stresses distribution in the formation of coal seams, mined successively in a descending order are presented. The variant is considered, in which the working of the upper seam is carried out in advance and the selvages of the mine workings in the tapped and overworked seam are located in zones of increased rock pressure and unloading, which creates dangerous conditions for mining operations. A mathematical model of stress-strain state of the geomassif in the form of a boundary-value problem, which was solved by the finite element method. Computational experiments were carried out to assess the mutual influence of excavations in the superimposed seams. The zones of increased rock pressure are determined within which the most dangerous geomechanical situation arises. The evaluation of the conformity of the numerical simulation results with the requirements of normative documents is performed. The proposed approach is recommended for the development of project documentation.

1. Introduction

The principles of controlling geomechanical processes in rock massifs are based on the use of the revealed regularities in the distribution of geomassif stress-strain state parameters in the process of mining coal seams lying in different mining and geological conditions.

Coal mining technologies and technical capabilities of mine equipment allow the face advancing speed to be increased up to 20 m per day. However, intensive technogenic impact on the geomassif leads to the formation of geodynamic phenomena in the of coal seam selvages and zones of increased rock pressure such as rock bumps, sudden coal, rock and gas burstings, cavings and dynamic sediments of roof rocks with the release of hazardous gases in the form shock air waves [1 - 3]. These phenomena limit the productivity of mechanized faces equipped with expensive equipment.

To ensure the safety of coal mining and the prevention of these dangerous phenomena, it is necessary to prevent pre-emergency situations by system analysis of mathematical and numerical modeling results of geomassif stress-strain state, the stress field of which is formed under the influence of natural and technogenic factors [4 - 6].

When mining the formation of seams, the presence of workings in the superimposed seams leads to stresses redistribution in the zone of influence of both workings as a result of their mutual underworking or overworking [7]. Such anthropogenic impact can lead to the formation of zones of increased rock pressure and unloading in the geomassif. Therefore, in order to justify the order and sequence of mining of seams in the formation, it is necessary to establish the parameters of these zones.

IOP Conf. Series: Earth and Environmental Science **84** (2017) 012005 doi:10.1088/1755-1315/84/1/012005

2. Methods of research

When carrying out computational experiments the mathematical model of the geomassif stress-strain state was used [8] that takes into account the effects of gravitational forces $\vec{F}$, formulated in the form of a boundary value problem in a general formulation: find the displacement vector $U=(u,v)$, $u=u(x,y)$, $v=v(x,y)$, which satisfies the system of differential equations within the computational domain

$$\mu\Delta\vec{u} + (\lambda + \mu) grad\, div \vec{u} = \vec{F} \tag{1}$$

and the boundary conditions shown in figure 1, where u and v are horizontal and vertical displacements; σ_y – vertical stresses; λ, μ are lame parameters.

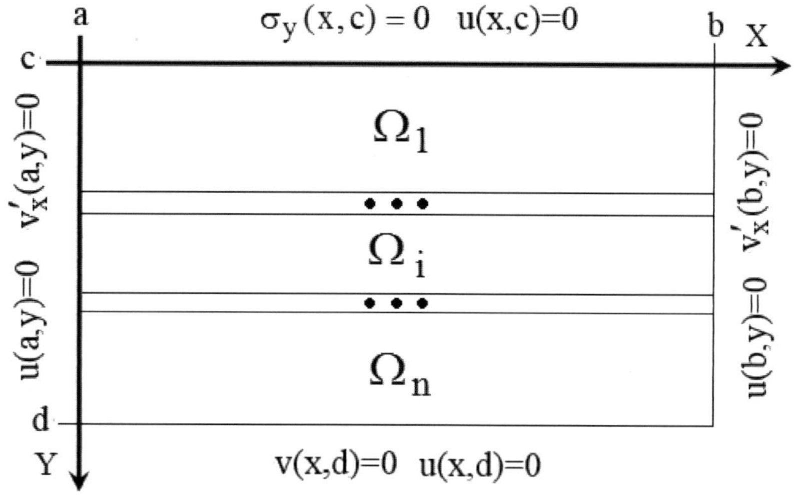

Figure 1. Boundary conditions accepted for mathematical model of the investigated geomassif section.

The conditions on the vertical and horizontal boundaries of mathematical model were determined as follows:

- horizontal displacements $u(a,y)=0$, $u(b,y)=0$, $u(x,c)=0$, $u(x,d)=0$ are given on the vertical and horizontal boundaries of computational domain at $x=a$, $x=b$, $y=c$ и $y=d$;
- on the vertical boundaries of computational domain at $x=a$, $x=b$ the following derivatives are given: $v'_x(a,y)=0$, $v'_x(b,y)=0$;
- vertical stresses are set on the upper base $\sigma_y(x,c)=0$;
- vertical displacements $v(x,d)=0$ are set on the bottom base.

The boundary problem was solved under the condition that the mass forces are directed along the vertical axis and were created by the intrinsic weight of the rocks.

The calculated domain of the investigated section of geomassif Ω (figure 2) is represented by argillites (Ω1, Ω3), siltstones (Ω5) and contains formation of coal seams (Ω2, Ω4). In the underworked seam (Ω2) the working boundaries in figure 2 are denoted by points B1 and B2, and in the mined seam (Ω4) – by points B3 and B4. Under the basic conditions of computational experiment mining is carried out on the upper seam of the formation.

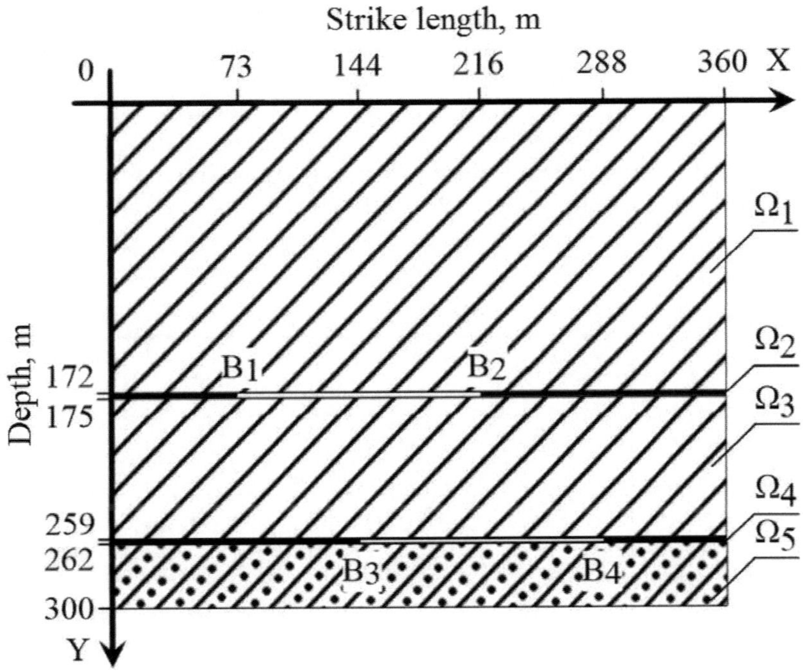

Figure 2. Estimated area of the geomassif section which includes workings in the superimposed coal seams.

Physical and mechanical properties of coal and rock are specified in table 1. The mathematical model was studied by the finite element method [10 - 12]. For carrying out computational experiments the authorial complex of problem-oriented programs NDSolverV1.0, developed in the Mathematica system in Wolfram Language [13], was used.

Table 1. Physical and mechanical properties of rocks [9].

Rock	Density ρ, kg/m^3	Elastic modulus E, MPa	Poisson coefficient, ν	Legend
Argillite	2600	26000	0.28	
Aleurolite	2700	28000	0.27	
Coal	1380	3000	0.34	

3. Results and discussion

The estimation of mutual influence of workings in the superimposed seams of formation was made on the basis of study results of stresses distribution obtained for two mining variants of the investigated section in the mining massif:

- mining of the coal seam in the formation by a single face without taking into account the influence of workings in the superimposed seams;
- simultaneous mining of two layers by mechanized faces taking into account their mutual influence in the form of underworking and overworking of seam selvages (figure 2).

Figure 3 shows the isolines of vertical stresses distribution (the minus sign corresponds to compressive stresses). The dotted isoline -4.47 MPa, corresponding to the position of vertical stresses equal to the stresses in the unmined massif for the upper seam, is distinguished. The presence of the working led to the unloading of the roof rocks and seam soil, which is confirmed by the change in the values of the vertical stresses in the zone of influence of gob relative to the gravitational stress field. In the roof of the seam tensile stresses are observed, which can lead to the rock destruction, the formation of caving and cracking zones. The height of arches of these zones can be approximately half the length of the working.

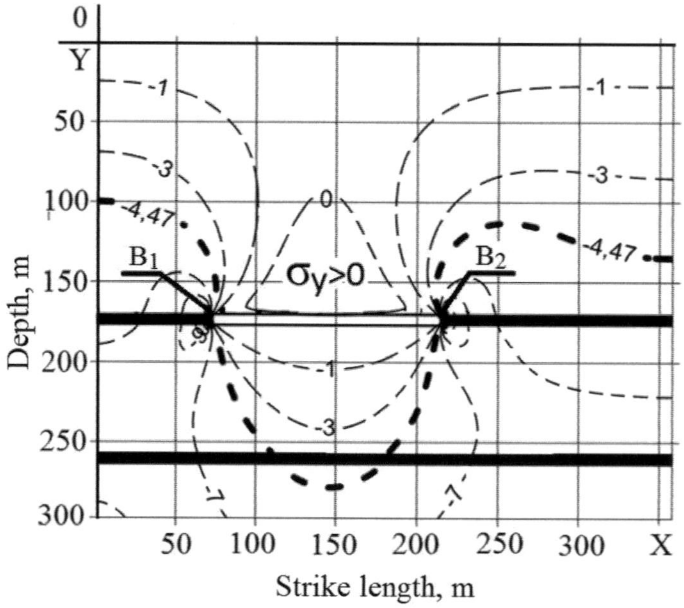

Figure 3. Isolines of vertical stresses distribution during mining in the upper seam of formation, MPa.

In the selvages of the coal seam a bearing pressure zone is formed, the presence of which was proved by the results of a mine experiment, physical and numerical modeling [14 - 17]. According to figure 3, the maximal compressive stresses are 9 MPa, i.e. the coefficient of vertical stresses concentration is equal to 2. Since the maximal compressive stresses do not exceed the coal strength limit of 11 MPa, coal spall in the selvage of the seam does not occur.

Figure 4 shows the stresses distribution in the presence of mutual influence of two faces in the superimposed seams 1 and 2. Vertical stresses in the unmined massif for the upper seam 1 at a depth of H=172 m make -4.47 MPa, and for the lower seam 2 at H=259 the stresses reach the value of -6.73 MPa.

In order to compare the results of numerical modeling of vertical stresses under mutual influence of two faces with the base variant, in figure 3 the isoline from figure 4 is dotted, corresponding to the values of vertical stresses equal to -4.47 MPa. Due to the extra underworking of the upper seam 1 by works in seam 2 in the selvage of the mine in the vicinity of point B_2, the unloading took place that confirms the isoline displacement towards the unmined massif corresponding to the vertical compressive stresses of -4.47 MPa in figure 4. In the vicinity of point B_1 there is an increase in vertical stresses by 1.4 times in comparison with the corresponding values of stresses in figure 3.

As a result of overworking of lower seam 2 in its selvage (in the vicinity of point B_4) the zone of increased rock pressure is formed, within which the coefficient of stress concentration reaches 1.93.

The selvage of the lower seam 2 (in the vicinity of point B_3) underwent partial unloading under the influence of gob in the upper seam; the coefficient of vertical stress concentration is 1.34.

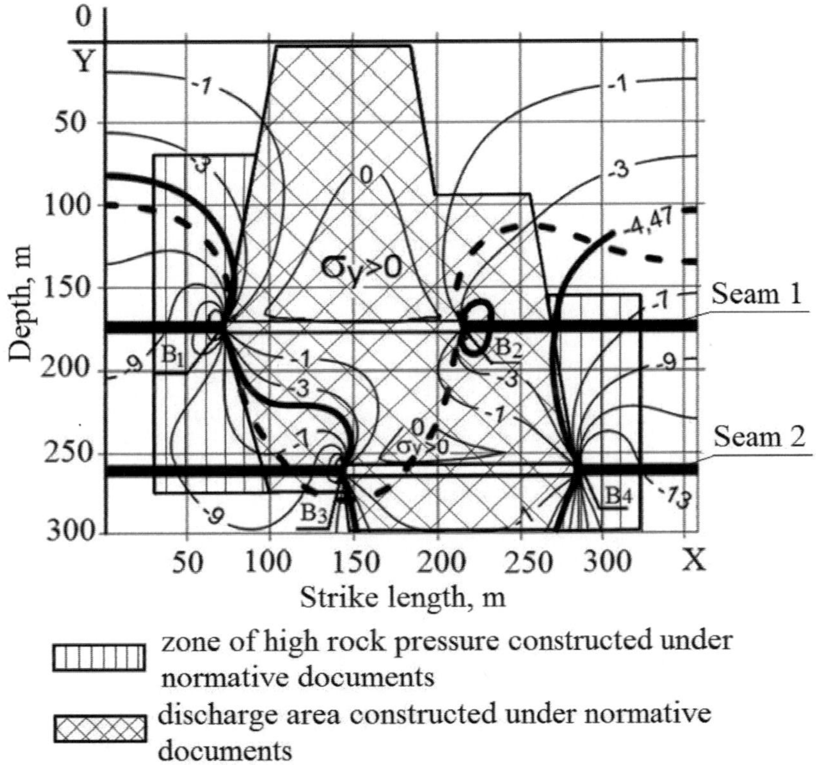

Figure 4. Isolines of vertical stresses distribution under mutual influence of two faces in the superimposed seams, MPa.

The maximal coefficient of stress concentration is revealed in the vicinity of point B_4, in which the vertical stresses of two bearing pressure zones in the selvages of the overworking face of the upper seam 1 and the inferior lower seam 2 are superimposed. The zones of caving and gradual descent are clearly delineated by the boundaries of zones of partial unloading, where the values of vertical stresses are essentially less than the stresses in the unmined massif.

To determine the correspondence between the distribution of calculated stresses and the recommendations of normative documents [18], zones of increased rock pressure and unloading are constructed in figure 4, the boundaries of which are qualitatively coincide with the areas of vertical stress distributions obtained from the results of computational experiments.

4. Conclusions

It is substantiated that under mutual influence of gobs in the superimposed seams, the stress fields are superimposed with the formation of unloading zones and increased rock pressure. Based on the results of numerical modeling of geomechanical state of massif, which includes mutual impacts of workings in the superimposed seams, the shapes and sizes of zones of increased rock pressure and unloading were determined.

It was established that the most dangerous situation in terms of geomechanical conditions occurs when the zones of reference pressure of the upper and lower seams are superimposed. The suggested approach is recommended for the prediction of quantitative parameters of the stressed-deformed state of geomassif at the stage of development of project documentation and justification of spatial position of faces and development workings in the formation of superimposed seams.

References

[1] Bolshinskiy M I 2003 *Gas Dynamic Phenomena in Mines* (Sevastopol: Weber) p 285
[2] Petukhov I M and Linkov A M 1983 *Mechanics of Rock Bursts and Emissions* (Moscow: Nedra) p 280
[3] Ayruni A T 1987 *Prediction and Prevention of Gas-dynamic Phenomena in Coal Mines* (Moscow: Nauka) p 310
[4] Seryakov V M 1995 *Modeling of the Deformation Process of the mined massif* (Novosibirsk) pp 184–9
[5] Kurlenya M V and Mirenkov V E 1994 *Methods of Mathematical Modeling of Underground Facilities* (Novosibirsk: Nauka) p 188
[6] Ershov L V, Iofis I M and Neyman I B 1983 *Mathematical Models of the Rock massif* (Moscow: MGI) p 138
[7] Fryanov V N and Pavlova L D 2006 *Information and Analytical Bulletin on Mining. Physics of Rocks* pp 245–51
[8] Tsvetkov A B, Pavlova L D and Fryanov V N 2015 *Information and Analytical Bulletin on Mining* **1** 365–70
[9] Shtumpf G G, Ryzhkov Yu A, Shalamanov V A and Petrov A I 1994 *Physical and Technical Properties of Rocks and coals of the Kuznetsk Basin: Handbook* (Moscow: Nedra) p 447
[10] Fadeev A B 1987 *Method of Finite Elements in Geomechanics* (Moscow: Nedra) p 221
[11] Amusin B Z and Fadeev A B 1975 *Method of Finite Elements in Solving Problems of Mining Geomechanics* (Moscow: Nedra) p 144
[12] Erzhanov Zh S and Karimbayev T D 1975 *Finite Element Method in Problems of Rock Mechanics* (Alma-Ata: Science) p 241
[13] Tsvetkov A B 2014 *Software Package NDSolver V1.0 for Mathematical Modeling of Stress-strain State of Gas-bearing Geomassif.* Reg. No. 19867 from 01.10.2014
[14] Borisov A A 1980 *Mechanics of Rocks and Massifs* (Moscow: Nedra) p 360
[15] Pavlova L D and Fryanov V N 2005 *Bulletin of TPU* **308(3)** 43–6
[16] Pavlova L D 2006 *Information and Analytical Bulletin on Mining* **4** 57–60
[17] Pavlova L D and Fryanov V N 2008 *Proc. Int. Conf. on Geodynamics and Stress State of the Earth Interior* (Novosibirsk: IGD SB RAS) pp 457–465
[18] *Prevention of Gas Dynamic Phenomena in Coal Mines: Collection of Documents 2000* (Moscow: State Enterprise of NTTs on Safety in Industry of Gosgortechnadzor of Russia) p 320

IOP Conf. Series: Earth and Environmental Science **84** (2017) 012006 doi:10.1088/1755-1315/84/1/012006

Algorithm of probabilistic assessment of fully-mechanized longwall downtime

A N Domrachev[1], S V Rib[1], Yu M Govorukhin[2] and V G Krivopalov[2]

[1]Siberian State Industrial University, 42 Kirova Street, Novokuznetsk, 654007, Russia
[2]National Mine Rescue Center, 5 Gornospasatelnaya Street, Novokuznetsk, 654028, Russia

E-mail: Govorukhin_YM@mail.ru

Abstract. The problem of increasing the load on a long fully-mechanized longwall has several aspects, one of which is the improvement of efficiency in using available stoping equipment due to the increase in coefficient of the machine operating time of a shearer and other mining machines that form an integral part of the longwall set of equipment. The task of predicting the reliability indicators of stoping equipment is solved by the statistical evaluation of parameters of downtime exponential distribution and failure recovery. It is more difficult to solve the problems of downtime accounting in case of accidents in the face workings and, despite the statistical data on accidents in mine workings, no solution has been found to date.
The authors have proposed a variant of probability assessment of workings caving using Poisson distribution and the duration of their restoration using normal distribution. The above results confirm the possibility of implementing the approach proposed by the authors.

1. Introduction

At present, most of the mines in Russia carry out the mining of coal seams according to the "mine-longwall" scheme, when all coal mining is concentrated in one high productive longwall. At the existing concentration of mining operations, any longwall stops due to failures in one of the face technological element lead to significant financial losses. In Kuzbass, the highest indices of load on a fully-mechanized longwall were achieved by mining the medium thickness and thick seams developed by parallel workings leaving protective pillars between them. With an increase in the depth of mining operations at the mines the problems associated with the negative effect of rock pressure on the state of workings, the mining and geological conditions of which are characterized as complex, is becoming more urgent.

The attempts made to group the deposits (and their sites) in terms of complexity of mining and geological conditions for the implementation of efficient technologies for carrying out, walling and protecting the development workings do not completely exclude emergency situations. [1, 2]. Considering the fact that objective geological information on the coal seam and host rocks characteristics is formed on the basis of a limited database of exploratory wells and operational exploration in the process of mining, there is a difficulty in making reliable technological decisions about accident-free maintenance of workings due to the variability of geological conditions [3, 4].

To ensure the control over the state of these excavations at the mines of OJSC SUEK Kuzbass, the technological service of workings walling is functioning, with the help of which in-situ studies are conducted and recommendations are being developed to improve the quality of the construction of the anchor support with increasing working depth [5], however, the decisions often fall behind the change in mining and geological conditions. In connection with that, the task of simulating the load on the face as a random variable that depends on these complicating factors, including those related to the change in the state of workings [6 - 8], is very relevant.

2. Methods of research

For the solution of the set problem it is proposed to use the methods of probability theory to estimate the number of roof cavings in the working, the main principles of which are discussed in [7 - 9]. The roof caving is seen as a rare random event that obeys Poisson distribution, which is modeled using a random number generator based on the inverse function method [10]. The use of the analytical dependences given in [7 - 9] allows the probability of roof caving in the development workings to be determined and the parameter λ of Poisson distribution to be estimated for further use in determining the number of emergency situations during the simulation period. Such an estimate is made by means of analytical or numerical solution of the equation, which is obtained by substituting the value of caving probability in the dependence describing Poisson law.

The task of downtime estimation (restoration of the working) is solved using a random number generator with a normal distribution [10], the parameters of which are determined on the basis of statistical data on the mines of the south of Kuzbass.

3. Results and discussion

As a result of the performed studies the integrated structural scheme of the estimation algorithm of downtime due to roof cavings in the model of technological scheme of a fully-mechanized longwall was developed. It is given figure 1.

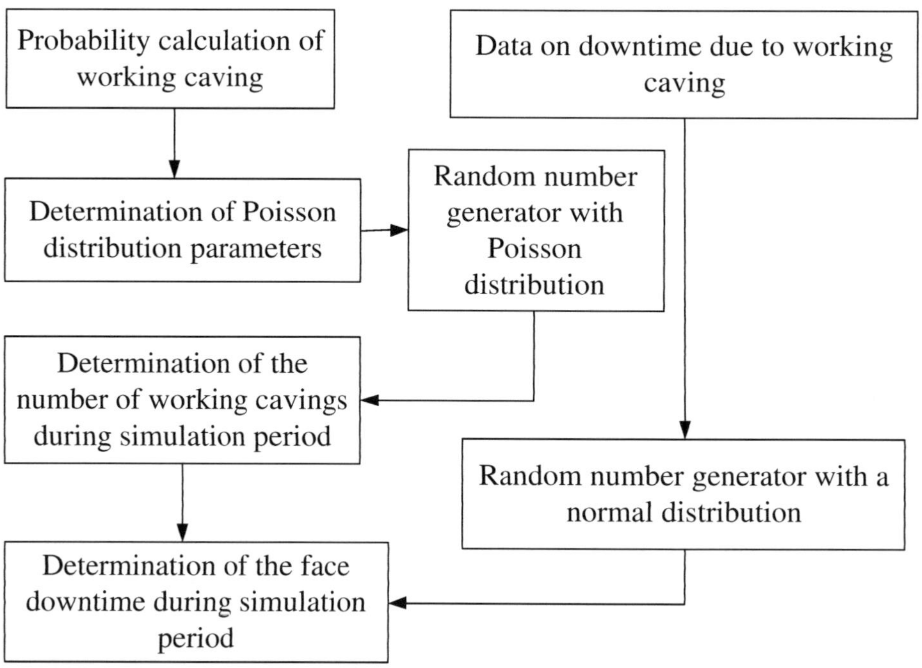

Figure 1. Enlarged structural scheme of estimation of downtimes caused by roof cavings in the workings.

On the basis of the proposed algorithm, an additional module was implemented to the model of a fully-mechanized longwall [11, 12] and the assessment of the state of mine workings was made with a calculated probability of roof caving p = 0.04. The duration of observation period was 7 months, the results of modeling the actual frequency of working caving are given in table 1. The average frequency of roof caving in the workings for all simulation periods was 0.0398, which is close to the declared probability of p = 0.04.

Table 1. Probabilistic assessment of the state of mine workings.

Observation period, months	1	2	3	4	5	6	7
Frequency of roof cavings in the workings	0.0356	0.0356	0.0356	0.0600	0.0438	0.0300	0.0384

4. Conclusions

1. In connection with the increase in the concentration of mining operations and the complication in the mining, geological and technological conditions of the work in the faces, the problem of modeling the load on the face as a random variable that depends on the random complicating factors associated with the change in the state of workings is very relevant.

2. Using the probabilistic estimate of the number of working cavings and the duration of their restoration, it is possible to realize the imitation model of the face operation, taking into account the influence of the state of working site on the coefficient of machine time of a shearer.

3. The obtained results confirm the possibility of using the approach proposed by the authors for estimating the actual operational time of a working site.

References

[1] Kazanin O I, Kozulin V V, Barabash M V and Yutyaev E P 2010 *Coal* **6** 24–8
[2] Sidorenko A A, Sidorenko S A and Kozharsky P V 2013 *Notes of the Mining Institute* **201** 236–40
[3] Kopylov A B, Savin I I, Sokolov E M and Kharlamov A E 2013 *Bulletin of TulSU* **12-2** 264–72
[4] Smirnov A V And Remezov A V 2015 *Bulletin of KuzSTU* **4.1** 36–40
[5] Kazanin O I, Yutyayev E P and Ermakov A Yu 2013 *Information and Analytical Bulletin on Mining* **4** 253–6
[6] Rib S V, Fryanov V N, Voloshin V A and Cherepov A A 2015 *Mining Informational and Analytical Bulletin* **7** 23–9
[7] Domrachev A N, Nikitina A M and Rib S V 2016 *Bulletin of TulSU. Earth Sciences* **4** 81–9
[8] Ghasemi E, Ataei M, Shahriar K, Sereshki F, Esmaeil S J and Ramazanzadeh A 2012 *Int. J. Rock Mechanics and Mining Sc.* **54** 2–9
[9] Mark C, Papas D M and Chase F E 2003 *Soc. Mining Metallurgy and Exploration* **314** 153–60
[10] Press W, Teukolsky S et al 2002 *Numerical Recipes in C++* (New York: Cambridge University Press) p 976
[11] Domrachev A N and Kuttsar T M 2009 *Int. Conf. on High Technology of Development and Use of Mineral Resources* (Novokuznetsk: SibSIU) pp 103–7
[12] Domrachev A N and Rib S V 2016 *Bulletin of the SibSIU* **3(17)** 8–10

IOP Conf. Series: Earth and Environmental Science **84** (2017) 012007 doi:10.1088/1755-1315/84/1/012007

The peculiarities of structurizing enclosing rock massif while developing a coal seam

E N Kozyreva and M V Shinkevich

Institute of Coal of the Federal Research Centre of Coal and Coal Chemistry, Siberian Branch of the Russian Academy of Sciences, 10 Leningradsky Prospect, Kemerovo, 650065, Russia,

E-mail: gas_coal@icc.kemsc.ru

Abstract. Different concepts of the development of geo-mechanical processes during longwall mining of a seam which are fundamentally different from the conventional ones are introduced in the article. Fundamental principles of the model for structurizing enclosing rock mass while longwall mining along the strike are described. The model was developed on the bases of non-linear geomechanical laws. According to the model, rock mass in the area of mining operation is organized as rock geomechanical layers with shifting arches. And the formation period of shifting arches in disintegrated rock mass is divisible by the length of the stope. Undulate characteristic of a massif as a peculiarity of man-made structurization of a massif is defined. It is shown that structuring the broken massif causes the formation of block-structured system and it can be detected while monitoring the ground pressure in powered support props. The results of the research allow decreasing the negative influence of a ground pressure and can be applied to specify parameters for controlling the roof, defining geometrical dimensions of a mining section and positioning of holing chute (face entry).

1. Introduction

Mining extraction leads to the change of a rock stress state and even effect on forming shift troughs on the ground surface. When the stope is moving different geomechanical processes take place in coal-gas massif in significant volumes. Resulting from the disintegration of a massif gas, which coal seams contain, evolves actively. Its intensive flows disperse in different directions: into a goaf (mine openings) and on the surface. This can explain gas outburst danger in mines and in surface buildings. Zones of its dangerous influence, volumes, character and the length of the disintegration, rock- and ground surface displacement depend on:

- thickness, angle of gradient, depth and gas-bearing capacity of working seams and their adjacent seams;
- size of stopes, stable pillars size and layout;
- ways to ground pressure control;
- face advance (movement) rate;
- presence of earlier worked out areas in the vicinity of a stope;
- physical and mechanical properties of a rock;
- rock mass structural features (thickness of the faces, geological dislocations etc.).

German research scientists when applying physical modelling method found out [1] that above the goaves (mined-out face) of long stopes not one but several arches in underworking rock mass are

IOP Conf. Series: Earth and Environmental Science **84** (2017) 012007 doi:10.1088/1755-1315/84/1/012007

formed during the face movement. One of the reasons of arch formation is horizontal thrust in rock layers. The phenomena of formation and disappearance of the thrusts in the roof caving is in their periodical alternation. The consequences of the process manifest themselves in periodical roof-caving. Later, German research scientists [2, 3] obtained the results on measuring pressure in powered support props of a long face. It was proved that the distribution of ground pressure along the long face is periodic and undulate. However, the parameters of the results have not been identified yet.

The research of abovementioned features started in Institute of Coal of Siberian Branch of the Russian Academy of Science (now it is Institute of Coal of The Federal Research Center of Coal and Coal Chemistry of Siberian Branch of the Russian Academy of Sciences) in 2000. Nowadays this research is in its full progress [4 - 11]. The researches of man-made geomechanical processes in coal and gas-bearing rocks are being done and the ways of decreasing negative influence on mine safety are being developed [13 - 24]. The researches of gas-geomechanical process allowed us to develop a model of structuring the enclosing rock while developing the working area by longwall mining along the strike. The description of the model is published in papers [9 - 11].

2. Fundamental principles of an enclosing rock structuring model

To specify the size of the zones of a mass geostatic stress relief and dynamics of its development current knowledge about the massif disintegration while stope (working face) moving were applied. The laws of non-linear geomechanical were used for that [12]. The method of mine aero-gas-dynamics was taken as a basic method for the development of the model. The ground for it was formed on the basis of a widely-known rule that firedamp (methane) evolution in coal seams is possible only while softening the stresses, in other words outer contour of gas depletion zone for coal and gas-bearing rocks is a front of their man-made relief from natural stress.

The model is based on the rules of zonal mass disintegration ahead of a stope (radial stress relief) and on a hierarchy of non-linear block structures in a goaf (vertical stress relief). These processes develop undulatory. Geometric boundaries of this development are defined by the length of stopes (working faces), stratification depth of a working out seam and mechanical features of the massif. The rock mass is introduced as a combination of rock geomechanical layers in the area of mining activity. Their thickness is equal to arch-paraboloid heights that are forming in corresponding layers. General picture of the displacement in the rock is introduced by the system of nested undulations with the parameters divisible by a square root of two raised to the n-th power (n- level of structural hierarchy). In its turn, this undulation is a combination of imbed periodical functions which describes adequately the contour of upper half of arches displacements. That is why to make the algorithm presentation of geomechanical features more convenient sinusoidal functions are used.

3. Model application case study

Vivid confirmation of a rock pressure undulation while the stope moving is in experimental observations of the hydraulic pressure in powered support props during mining of a working area in "Aldardinskaya" coalmine in Kuzbass region (figure 1). Site features: the length of a stope is 220 m, stratification depth of a working seam – 520-650 m, extracting seam thickness– 4.5 m. with abandoning coal patches at the seam floor, moving velocity about 3 m. per sec, hard seam roof. Natural observations are made by two of the authors of the article together with the mine engineering staff. The quantitative evaluation of the results revealed the periodical change of the pressure divisible by the length of a stope (Figure 1).

Proving the adequacy of the concept about the development of geo-mechanical processes while underground mining was fulfilled in a complex environment – directly in a massif bottom-hole area, i.e. on the border transition from radial stress relief ahead of the face into vertical stress relief above the goaf (Figure 2). Assuming that the layer's weight corresponds to the support response and average density of a roof rock is 2.5 ton per m^3, the values of the layers' height that press on the bars of the powered support are taken as a pressure arch thickness.

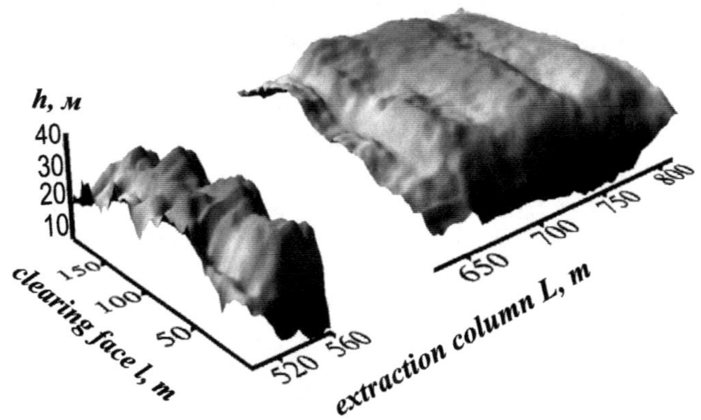

Figure 1. *h* height of a pressure arch on the sections of a powered support while mining a seam with a hard roof in "Aldarinskaya" coalmine.

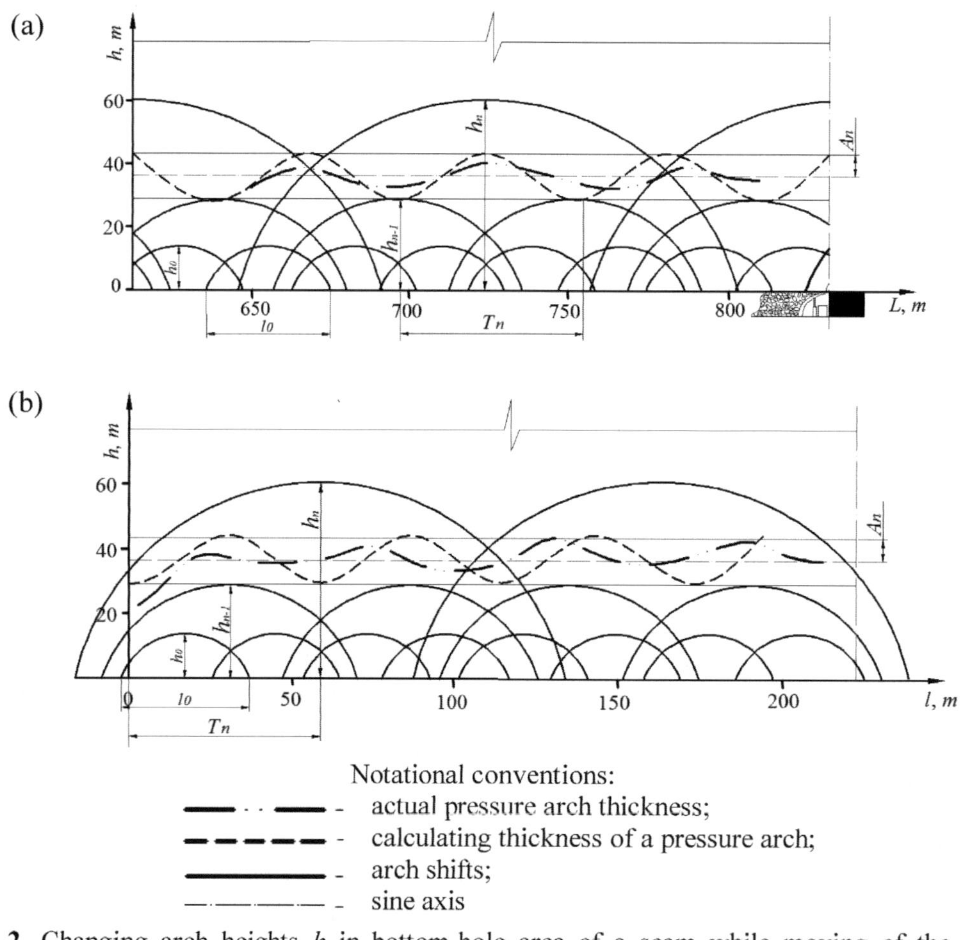

Figure 2. Changing arch heights *h* in bottom-hole area of a seam while moving of the stope in "Aldarinskaya" coalmine: a) along the length of the extraction panel *L* at the interval 610–830 m; б) along the length of a stope (working face) *l* at the distance of 548 m from installation chamber

Calculating thickness of a pressure arch is introduced in figure 2 as a sine wave indicated by a dashed line with T_n period corresponds to

$$T_n = l_0 \cdot 2^n, \text{ m,} \tag{1}$$

where l_0 – base diameter of a minimal geo-environment element, m; n –structural hierarchy level.

Pressure arch amplitude A_n corresponds to

$$A_n = \frac{h_n - h_{n-1}}{2}, \text{ m,} \tag{2}$$

where h_n and h_{n-1} are the thickness of n and n-1 layers of geomechanical structurization of the underworking massif, respectively, m;

Arch-paraboloid parameters l_0 and h_0 in the closest to a working seam geomechanical layer are taken as the size of minimal geo-environment element:

l_0 – base diameter of a minimal arch is taken as the closest to the step of a secondary caving or calculated according to the formula

$$l_0 = \frac{l}{2^n} \approx \frac{r_1}{4} \approx r_2, \text{ m,} \tag{3}$$

where l – is the length of a stope (working face), m; n – the level of structural hierarchy, is calculated according to the assumption that the relation r_1 to r_2 is about 4; r_1 and r_2 – are the steps of primary and secondary roof caving, respectively (they are calculated according to known formulas or elaborated according to geotechnological data), m; h_0 – minimal arch height (or the height of minimal geomechanical layer) is calculated according to a formula

$$h_0 = l_0/2, \text{ m.} \tag{4}$$

Actual thickness of a pressure arch is indicated in figure 2 by dot-and-dash line and it is obtained according to the data of a pressure in hydro-cylinders of powered support props in the experimental area. Actual data were measured along the length of an extraction panel at the interval of 610-830 m from installation chamber (Fig. 2a). Figure 2b demonstrates the scheme of the shifting process along the stope l at the distance of 548 m from the installation chamber.

Period and amplitude of a pressure arch thickness in a bottom-hole seam zone generally corresponds to the model of geo-mechanic process development in the enclosing rock [10]. Moreover, the sine curve amplitude that approximates actual values is always lower then the calculated one in about $\sqrt{2}$ times ($\sqrt{2}$ – a value of geo-environmental canonical parameter). The contrast of the period figures can be explained by roof rock anisotropy and neighboring of a working area with a goaf in an abandoned place.

4. Conclusions

Thus, the undulating character of geo-mechanic layers formation both along the length of the extracted panel (Figure 2a) and along the length of the stope (working face) (Figure 2b) is registered. The formation period for shifting arches of disintegrated massive in a goaf is divisible by the length of a stope. The following features are revealed:

- during radial stress relief ahead of a moving stope the period of forming zonal disintegration in a seam is close to divisibility by single square root of two;

- during more intensive vertical stress relief the period for forming mass shifting arches in a goaf is divisible by two. The peculiarities of forming the radial stress relief are in correspondence with shifting processes in development headings when shifting covers only rock layers that are in immediate neighboring to a mine working.

As for practical results it is necessary to note that while laying out of plan for a working area (block) the length of a stope *l* should be taken as divisible by $h_0 2^{n+1}$.

Following this condition helps:

- to soften negative influence of a ground pressure while geometrical measuring of a working area (block) and the position of holing chute (crosscut);
- to specify the parameters of the main roof control such as defining the caving step;
- to measure bearing support zone made in the form of stable pillars contouring the working area (block) for sustainable process of arch-formation. As it is seen in figure 2b (shifting arches goes to negative zone along the X- axis) a part of relief arch bases is behind the belt entry and the airway;
- to provide effective ventilation and outgassing due to specifying the height of shifting arches and the areas of relieving enclosing massif which initiates gas flows from adjacent seams to the working areas (blocks).

In whole, specifying geomechanical processes in enclosing massif improves the evaluating adequacy of a massif disintegrating characteristics, forecasting precision of geomechanical parameters, basic methane release both while planning the mining and during the mining process.

References

[1] Jacobi O 1987 *Controlling Ground Pressure* (Moscow: Nedra) p 566
[2] Reuter M, Kurfürst W, Mairhoffer K and Weksler U 2009 *J. of Mining Sci.* **2** 38–44
[3] Reuter M, Krach M, Kießling U and Veksler Y 2015 *Fiziko-Tekhnicheskie Problemy Razrabotki Poleznykh Iskopaemykh* **51(2)** 237–42
[4] Polevshchikov G Ya, Kozyreva Ye N and Shinkevich M V 2014 *Occupational Safety in Industry* **6** 50–4
[5] Polevshchikov G Ya, Kozyreva Ye N and Shinkevich M V 2012 *Bulletin Nts VostNII* **2** 20–6
[6] Polevshchikov G Ya, Shinkevich M V, Kozyreva Ye N and Bryuzgina O 2008 *GIAB* **2** 139–43
[7] Kozyreva Ye N, Shinkevich M V and Ryabkov N V 2010 *GIAB* **3** 356–9
[8] Kozyreva Ye N, Shinkevich M V and Rodin R I *2012 2nd Chinese-Russian Sci. Conf. on Non-linear Geo-mechanic and Geo-dynamic Processes During Deep Mining* pp 267–72
[9] Polevshchikov G Ya 2013 *Fiziko-Tekhnicheskie Problemy Razrabotki Poleznykh Iskopaemykh* **5** 50–60
[10] Kozyreva Ye N, Shinkevich M V and Leontyeva Ye V 2015 *Naukoyomkiye Technologii Razrabotki i Ispolzovaniya Mineralnykh Resursov* **2** 143–8
[11] Shinkevich M V and Leontyeva Ye V 2015 *Bulletin of KuzSTU* **3** 23–31
[12] Shemyakin Ye I, Kurlenya M V, Oparin N V, Reva V N Glushikhin F P and Rosenbaum M A 1992 *Discovery No 400 (Otkrytiye No 400) Phenomena of Rock Mass Zonal Disintegration around Underground Mine Working (Opening)* Bull. **1** p 3
[13] Rodin R I and Plaksin M C 2016 *Bulletin of NTs VostNII* **1** 42–8
[14] Cherdantsev N V, Pressler V T and Izakson V Yu 2010 *Fiziko-Tekhnicheskie Problemy Razrabotki Poleznykh Iskopaemykh* **2** 62 –8
[15] Klishin V I, Kokoulin D I, Kubanychbek B and Durnin K M 2010 *Ugol' (Coal)* **4** 40–2
[16] Tailakov O V, Tailakov V O, Makeev M P, Sokolov S V and Kormin A N 2013 *GIAB OB* **6** 160–5
[17] Alkov V I 2014 *Bulletin of NTs VOSTNII* **2** 9–12
[18] Kozyreva Ye N 2014 *Bulletin of NTs VOSTNII* **1** 109–13
[19] Plaksin M S, Rodin R I and Ryabtsev A A 2015 *Ugol' (Coal)* **2** 48–50
[20] Gerike B L, Filativ A P, Gerike P B and Klishin V I 2006 *Fiziko-Tekhnicheskie Problemy*

Razrabotki Poleznykh Iskopaemykh **6** 98–105
[21] Klishin V I, Nikolskiy A M, Opruk G Yu, Neverov A A and Neverov S A 2008 *Ugol' (Coal)* **11(991)** 12–17
[22] Klishin V I and Klishin S V 2010 *Fiziko-Tekhnicheskie Problemy Razrabotki Poleznykh Iskopaemykh* **2** 69–81
[23] Klishin V I, Shundulin I A, Yermakov A Yu and Solovyev A S 2013 *The Technology of Thick steep Seams Development with the Release of Coal* (Novosibirsk: Nauka) p 248
[24] Gerike B L, Gerike P B, Klishin S V and Filatov A P 2008 *Fiziko-Tekhnicheskie Problemy Razrabotki Poleznykh Iskopaemykh* **5** 81–8

IOP Conf. Series: Earth and Environmental Science **84** (2017) 012008 doi:10.1088/1755-1315/84/1/012008

Coal resources of southern Kuzbass (Novokuznetsk administrative district)

Ja M Gutak

Siberian State Industrial University, 42 Kirova Street, Novokuznetsk, 654007, Russia

E-mail: gutakjaroslav@yandex.ru

Abstract. The article summarizes data on geological structure of Novokuznetsk district of the Kemerovo region and its coal reserves. Survey is conducted consistently in regards to geological and economic areas of Kuzbass, included in the territory of Novokuznetsk district (Aralichevsky, Baidayevsky, Bunguro-Chumyshsky, Erunakovsky, Kondomsky, Mrassky, Osinovsky, Tersinsky, Tutuyas, Central, Uskatsky). Stating presence of significant coal reserves in this territory, the survey points out that in some geological and economic areas reserves are either depleted (Aralichevsky) or there are no prospects for reserves increase (Baidayevsky, Mrassky). The largest coal reserves explored are concentrated in Erunakovsky, Tersinsky and Osinovsky geological and economic areas. The main reserves of highly dmanded coking coals are located in Erunakovsky and Osinovsky areas. As for the Novokuznetsk administrative district, a significant increase in coal production level is possible, although it has to be linked to demand, since the extracted coal can not be stored for a long time. Today key point is not the increase in coal production, but qualitative deep processing of coal.

1. Introduction

Traditionally, review of the mineral resource base is conducted for a large geological region (Kuznetsk coal basin, geological and economic region) or for administrative region (the Kemerovo region) [3, 5]. For smaller areas of administrative district level, such reports are not available. In this sense, provided review of resources of the Novokuznetsk administrative district of the Kemerovo region is seen as the start of analysis of the mineral resources base of the Kemerovo region by districts. Such survey can be useful for municipal administrators, for a variety of economic and administrative structures involved in extraction and processing of minerals.

In terms of coal mineralization, the Novokuznetsk district holds a special position among the rest of the Kemerovo region. The Kemerovo region is always associated with the Kuznetsk coal basin with a unique coal deposit. In any place of this basin the researcher will meet a certain number of coal seams. In this sense, the term coal deposit in the Kuzbass lost its original meaning (Kuzbass is a unique coal deposit as a whole). It is often now, that coal deposits in our region are understood as coal mining sites and areas prosperous for its development. Thus, the Sibirginskoye field in Mrassky geological and economic area smoothly passes into the Uregolskoye field, and the latter to the Kyzakskoye one. From the geological point of view, these are the same coal seams and the deposit location is determined only by mining allotment. Sometimes several mining companies work at one site of coal mining and then the names – Taldinskoye – the first, Taldinskoye – the second etc., appear. This brings a certain confusion and uncertainty in concept of the deposit (as an accumulation of mineral matter on the surface or in the depth of the Earth, feasible for industrial use in terms of quantity, quality and

IOP Conf. Series: Earth and Environmental Science **84** (2017) 012008 doi:10.1088/1755-1315/84/1/012008

conditions of occurrence) [2]. It is for this reason that there is no clear list of specific coal deposits in the Kemerovo region, and when economical assessment the region's potential takes place, researchers operate with the concept of geological and economic area. Kuzbass is divided into 22 districts by geological and economic features. Eleven of them are fully or partially located within the territory of the Novokuznetsk administrative district. They are as follows: Aralichevsky, Baidayevsky, Bunguro-Chumyshsky, Erunakovskiy, Kondomsky, Mrassky, Osinovsky, Tersinsky, Tutuyas, Uskatsky and Central (Fig. 1). In these areas, the main coal reserves of various brands, including especially valuable grades of coking coal, are concentrated. Not all the eleven areas listed above within the Novokuznetsk district have the same value as for now. Some coal sites are operated from before the World War II, others have only been put into operation resently (Kiyzaksky open pit mine). As of January 1, 2013 there are 307 extraction licenses acting in the Kemerovo region (excluding gas, groundwater and common mineral resources). Of these, coal accounts for 85% - 261 licenses [4].

2. Methods of research

The Novokuznetsk municipal district geologically is a broad terrain of the east-west trending, that includes the structures of the western slope of the Kuznetsk Alatau (in the east), Kuznetsk coal basin (the main part), the Mountain Shoria and Salair, including the Nenya-Chumysh depression (the western part) that divides them. Geological exploration status of these territories is uneven. Kuznetsk coal basin and the Mountain Shoria are of the greatest degree of exploration. Low population areas of Kuznetsk Alatau and Salair are less explored.

The materials available to date allow us to interpret geological history of the territory with sufficient confidence. Mountain-folding structures of Kuznetsk Alatau and the Mountain Shoria are considered as structures of the Caledonian stage (the Lower Paleozoic) including its early Salair phase (the Cambrian period). The geological structures of Salair and the Kuznetsk basin adjacent to them are considered to be the structures of the Hercynian orogeny (the Late Paleozoic). The Kuznetsk depression has sharply asymmetric structure, in the east thickness of the section is sharply reduced. The greatest thickness of sediments, including coal-bearing deposits, occurs in the western pre-Salair part of the depression. Salair structures encroachment on the adjoining sediments of Kuzbass is observed. Time of the Kuznetsk depression formation for a long time was subjected to discussions. The author shares the view that the beginning of sedimentation processes in this geological structure started in the Givetian age of the Middle Devonian. According to palaeobotanic and paleontological data, this is the age of the deposits lying on a dislocated foundation of the Early Paleozoic and the Early Devonian rocks of Kuznetsk Alatau. Based on that Kuznetsk depression can be attributed to piedmont depression on the border with the Tom-Kolyvan folded zone. Nenya-Chumysh basin separating Salair and the Mountain Shoria, was formed in the Lower Cretaceous. In addition to these structures two discordant basins (Central and Tutuyas), composed of sediments of the Jurassic system, have been distinguished in the Novokuznetsk district in the area of coal deposits of the Permian system. Volcanic-sedimentary deposits of the Triassic are recorded at the base of the Central basin section.

Such a complex and long history of geological development of the territory predetermined multiplicity of the region's minerals, but the main wealth of the area is made up of coal.

3. Results and discussion

The Aralichesky geological and economic area is one of the oldest in the Novokuznetsk district (part of Novokuznetsk and Prokopyevsky district) and is located in the valley of the rivers Tom (Kondoma, Aba, Sharap tributaries). The first coal mining enterprises were the Ordzhonikidze mine (1930-1994) and the Dimitrova mine (1931-1996) in Novokuznetsk. 85mln tons of coal were produced during the operation time. Until now resource potential of the Aralichevsky area has been completely exhausted taking into account increase in reserves prosperous in terms of industrial development and establishment of large, economically efficient production [7].

Baidayevsky geological and economic area orographically occupies the right bank of the river Tom and the basins of the rivers Abashevaya and Esaulka, and is partially located within the city of Novokuznetsk. The first information about findings of coal in the area belongs to the middle of the XVIII century. In regard geology and structure, area is represented by a series of brachyfolds developed along the deposits of the Ilyinskaya and the Erunakovskaya subseries of the Permian system. Coals in the area are almost all coking, low-ash. The total coal reserves of A, B and C_1 categories (as of 01.01.2001) are 1240 million tons. Inferred resources are 1720 million tons. Almost the whole area is affected by mining operations in one form or another. Until recently, Abashevskaya mine (closed in 2014), Zyryanovskaya mine (closed in 1999), Novokuznetskaya mine (closed in 2001), Nagornaya mine (closed in 1996) had been operated here. Yubileynaya mine (Topprom JSC), Esaulskaya mine (Raspadskaya United Company), Polosukhinskaya mine, Antonovskaya mine, Bolshevik mine (Sibuglemet Group) continue operation. Large reserves and inferred resources of coking coals of especially valuable brands in the area are concentrated in low-density seams and are not processed because of absence high-performance machinery. There are no prospective areas for discovery of economically profitable coal reserves in this area. The time for existing reserves extraction in operating mines is about 50 years.

The Bunguro-Chumyshsky geological and economic area occupies the extreme western part of the Novokuznetsk district, including watershed of the rivers Tom (with Kondoma and Aba tributaries) and Chumysh (in the western part of the district). Over the long history of geological exploration of the area all promising coal plots have been prospected and explored. Carboniferous deposits belong to the Balakhonskaya series (the Upper Carboniferous – the Permian) and are faulted into the system of anticlinal and synclinal folds with steep incidence angles. The reserves of coal of A, B and C_1 categories are 3300 million tons, inferred resources are 14451 million tons. Despite significant reserves and inferred resources, development of coal mining is constrained by the absence of coking and technological grades of coal of high demand. Among factors of constraint are complex geological conditions of coal seams occurrence (steep incidence angles, fault tectonics). There are no huge areas suitable for highly effective underground mining in the area, development of open pit mining is handicapped by an increase of overburden ratio with depth. Currently, open pit mining is conducted here in small sections of lean coals and anthracites used for energy production. Several sections of "Listvyansky open pit mine" OJSC of "Kuzbassrazrezugol" company "Bungursky open pit mine" LLC (Bunguro-Listvyansky 4 and Mikhailovsky sites), "Kalininsky open pit mine" OJSC operate here.

Erunakovsky geological and economic area is located within the territory of Novokuznetsky, Belovsky and Prokopyevsky districts. Novokuznetsky district includes the part of the left bank of the river Tom from the Iliynskoe village to Ust-Naryk village. By its geological and structural features Erunakovsky area is subdivided into 11 perspective coal fields, recognized as deposits. Of these, Krasulinskoye, Tagaryshskoe, Kukshinskoye, parts of Taldinskoye, Zhernovskoye, Novokazanskoye coal fields belong to Novokuznetsk district. The carbon mineralization is associated with the deposits of the Kolchuginskaya series of the Middle-Upper Permian. Denoted coal fields, as a rule, correspond to axial parts of brachiantinclines and brachinsynclines. Each of these structures has its own name, which is transformed into the name of the deposit. Steep fold wings refer to areas challenging for coal mining. The southwestern part of Erunakovsky geological and economic area belongs to the structures of pre-Salair part of Kuzbass (elongated linear folds), the northeastern part belongs to the central zone of Kuzbass (flat brachiform folds). The total reserves of coal of A, B and C_1 categories are 25545 million tons, inferred resources are 83265 million tons. Coals here, excluding non-caking long-flame, are suitable for coking, a significant part of coal is suitable for semi-coking, coal-derived gas and liquid fuel production. By their territory "Erunakovsky open pit mine", part of "Taldinsky open pit mine" of "Kuzbassrazrezugol" company, OJSC "Iliynskaya Mine", "Uskovskaya Mine" OJSC (former Ulyanovskaya), OJSC "Otkrytchik open pit mine", "Zhernovskaya Mine" OJSC belong to Novokuznetsk administrative district. Erunakovsky geological and economic area of Kuzbass and its part, included in Novokuznetsk administrative district, has not yet been sufficiently developed by coal

IOP Conf. Series: Earth and Environmental Science **84** (2017) 012008 doi:10.1088/1755-1315/84/1/012008

mining industry and has a significant reserve of explored subsoil plots for construction of new mines and open pits. Here it is possible, first, to increase significantly energy coals production in operating open pit mines, and, second, to start selective extraction of particularly valuable grades of coal at Zhernovskaya and Uskatskaya mines. Further increase in reserves is possible due to exploration of Kukshysky and Naryksky coal fields.

Kondomsky geological and economic area occupies the southwestern part of Novokuznetsk administrative district, its southern border coincides with the outer contour of Kuzbass. This is the northern edge of the Mountain Shoria, confined to the valley of the river Condoma and its tributaries rivers Kinerka and the Malaya and the Bolshaya Tesh. Area is conditionally divided into 9 fields perspective for coal mining, which are recognized as deposits. These are as follows: Razvedchik, Targayskoye, Shushtalepskoye, Nikolaevskoye, Karachiyakskoye, Alardinskoye, Teshskoye, Chernokaltanskoye and Verkhnetneshskoye. Carbon mineralization of the area is associated with the Balakhonskaya series (C_3-P_1). Coal mining has been conducted sins 1949 (Shushtalepskaya mine, 1949-1998). The reserves of coal of A, B and C_1 categories in the Kondomsky area are 3695 million tons, inferred resources are 35149 million tons. Almost half of all prospected coal is coking coal, but not particularly valuable. Energy coals are classified as lean and are used mainly at enterprises of the Kemerovo Region (Kaltanskaya TPP). Reserved areas with prepared coal reserves are absent here. At the same time, operating mines of the Kondomsky area are provided with reserves for the near and distant future. Alardinskaya mine is provided with reserves for 200 years, the rest - for 40-50 years in average. A significant increase in production of lean energy coal is possible by means of open pit mining on thick seams. Currently, the following organizations are engaged in coal mining in the area: “Alarda Mine” LLC, “Kaltansky open pit mine” JSC (currently this company includes Osinovsky open pit mine). Targaisky open pit mine is being prepared for operation. Prospects for increase in coal mining in Kondomsky area are in developing open pit mining on thick seams of lean coal (“Razvedchik”, “Chernokaltansky”, “Teshsky”). In addition, thick seams of lean coals are found on sites explored for underground mining: Urzizakovsky, Targaya, Krasnogorsky, Taylensky, Korchakolsky, Teshsky, Chuazasky.

Mrassky geological and economic area is located in the south of the Novokuznetsk district in the right bank of low course of the river Mrassu. The southern border of the district is conventionally accepted along the Tom-Mrassu watershed, the border with the Kondomsky area is drawn between Chuazasky and Uregolsky coal fields. Carbon mineralization is associated with sediments of the Balahonskaya series (C_3-P_1). From the geological point of view, this territory is often referred to as the Mountain Shoria monocline (the entire section is represented by sedimentary rocks including carbonaceous rocks slightly pitching towards the central part of Kuzbass) [6]. Coal mining had started in 1963 in “Mezhdurechensky” open pit mine. Later on, Sibirginsky open pit mine had been launched. Total reserves of coal in the area amount to 3493 million tons of A, B and C_1 categories. Inferred resources are 13.504 million tons. Coals are mainly energy ones. In small quantities there are coking coals of abundant brands. Coal production in the area is mainly carried out by the open pit mining method in Sibirginsky open pit mine OJSC, partly by Krasnogrsky open pit mine of Yuzhny Kuzbass company and Mezhdurechye OJSC of Sibuglemet group. These enterprises have covered the entire range of industrial coal seams incrops of the Quaternary deposits. Uregolsky coal site was commissioned the latest (part of this plot is mined by Siberginsky open pit mine, the other part - by recently created “Kizassky open pit mine” OJSC) and there are no areas for new coal open pit mines in Mrassky geological and economic area. Sibirginsky open pit mine has already partially switched to underground coal mining on the deep horizons of Sibirginsky coal field.

Osinovsky geological and economic area occupies watershed of the Kondoma and the Tom in the basins of the rivers Chernaya, Rychagi, Bukanai, Kalandas. Two prosperous coal fields having status of deposits are distinguished in the area by their geology and structure: Osinovskoe (the main part of the area) and Vorobienskoye (a small syncline with thin coal seams). Carbon mineralization is associated with the Upper Permian deposits of the Kolchuginskaya series. Coals of the area are of high quality. The basis of reserves here are coals of acute deficiency Zh (fat) and KZh (coking fat) grades.

Their reserves of A, B and C categories are 1350 million tons, the inferred resources are 671 million tons. Osinovsky district is the main supplier of fat coals in Kuzbass supplied for high-quality coke production. Coal mining is carried out in underground mines which are as follows: "Shakhta Vysokaya" company, Osinnikovskaya mine (Raspadskaya United Coal Company), Talgina Mine. All operating enterprises are provided with long-term reserves, but reserve fund development is handicapped by complex geological and mining conditions (thin seams, complex tectonics, abundance of coal balls in coal seams, high gas saturation and presence of a powerful overburden of the Jurassic deposits).

Tersinsky geological and economic area occupies pre-Alatau part of Novokuznetsk administrative district. The western boundary of the district is along the river Northern Tom along the river Lower Ters, the southern border coincides with the border of the Jurassic deposits of the Tutuyassky basin, the eastern border coincides with the boundary of Kuznetsk Basin. The area is relatively poorly developed, economic activity is conducted only in the south-western part of the area adjacent to Baidaevsky geological and economic area. Kusheyakovskoe and Uvalnye coal fields are explored here. In the eastern part of Tersinsky district exploration works cover Chexinskoye, Tersinskoye and Makaryevskoye coal fields. In the same area the Tersinskoye mineral water deposit is located. Carbon mineralization is associated with the deposits of the Balakhonskaya and the Kolchuginskaya series. Coals are diverse in their properties due to the difference in stages of coal metamorphism and their petrographic characteristics. Coal reserves in the region of A, B and C_1 categories are 6.598 million tons, inferred resources are of 26.347 million tons. Coal is produced by Kusheyakovskaya Mine OJSC (Raspadskaya United Coal Company), in 2016 "Mine Uvalnaya" OJSC ("Sibirskaya Coal company") started operation. Coals of the Tersinsky geological and economic area are suitable for synthetic liquid fuels production by means of destructive hydrogenation. Based on evaluation criteria (ash content, fusinite content, vitrinite reflection index, carboxylic acid number, chemical composition of ash) of the degree of coals suitability for hydrogenation, seams of "Tersinsky-2", "Uvalny-9-10", "Kusheyakovsky-I" and "Kusheyakovskiy-II" geologic plots are of certain interest [1]. The outlines of Tersinsky coal field partially lay in state park of Kuznetsk Alatau, that seriously reduces chances of its development in the coming years. In 2015, three parts of Makaryevskoye coal field were put up for mining license auction.

Tutuyassky Central and Uskatsky geological and economic areas in terms of coal mineralization have no importance for the economy of the district because of either poor quality of coals (Tutuyassky and Central areas) or limited territory within the boundaries of Novokuznetsk administrative district (Uskatsky area).

4. Conclusions

Concluding review of coal resource base in Novokuznetsk district, it can be said that it is sufficient to plan economic activities for the nearest future and long term period. A significant increase in coal production level is possible in the district, although it has to be linked to demand, since the extracted coal can not be stored for a long time. Today key point is not the increase in coal production, but qualitative deep processing of coal. Deep processing of coals of Kuzbass and Novokuznetsk district in particular is still far from desired.

It should be acknowledged that development of coal resource base of Novokuznetsk administrative district is largely due to geologists of "Zapsibgeologiya" IGE, once the largest geological exploration organization in the south of Western Siberia. Its abolition in the early 1990s had negatively affected development of mineral resource base of the entire south of Western Siberia and the district in question in particular. Revival of a state geological exploration enterprise is a matter of time. Reality shows the need for a comprehensive geological exploration enterprise in a huge mining area.

References

[1] Anferov B A, Kuznetsova L V and Borisov I L 2014 *Bulletin KuzSTU* **5** 44–9
[2] *Geological dictionary* 1978 (Moscow: Nedra)

[3] *Geology of the USSR. Western Siberia (Kemerovo, Novosibirsk, Omsk, Tomsk regions, Altai Territory) 1982 Minerals vol XIV* (Moscow: Nedra) p 319
[4] Gukov S V 2013 *Proc. Sci. Conf. "70th Anniversary of Kemerovo Region, the 150th Anniversary of V.A. Obruchev and the 130th Anniversary of M.A. Usov"* (Novokuznetsk) pp 7–13
[5] Gutak Ja M 2003 *Izvestiya. Ferrous Metallurgy* **6** 61–5
[6] Gutak Ja M, Kauchakova E E and Ruban D A 2009 *Problems of Regional Ecology* **5** 178–84
[7] *Coal Base of Russia. Coal Basins and Deposits of Western Siberia (Kuznetsky, Gorlovsky, West-Siberian Basins, Deposits of Altai Territory and the Altai Republic)* 2003 vol II (Moscow: Geoinformcentre) p 604

IOP Conf. Series: Earth and Environmental Science **84** (2017) 012009 doi:10.1088/1755-1315/84/1/012009

Mineral resources of Novokuznetsk administrative district of Kemerovo region (metallic and non-metallic minerals)

Ja M Gutak

Siberian State Industrial University, 42 Kirova Street, Novokuznetsk, 654007, Russia

E-mail: gutakjaroslav@yandex.ru

Abstract. The article summarizes data on metallic and non-metallic minerals of Novokuznetsk district of Kemerovo region. Consistently reviewed are iron deposits (Tersinskaya group of deposits), gold deposits (placer accumulations and vein gold deposits), mineral water deposits (Tersinskoe deposit), deposit of refractory clay (Barkinskoe) and wide spread mineral deposits such as brick clay, keramzite materials, sand and gravel, building stones, ornamental stones, facing stones, peat, materials for lime production. It is indicated that resource base of metallic and nonmetallic minerals is inferior to that of mineral coal. At the same time it can be of considerable interest to small and medium-size businesses as objects with quick return of investment (facing and ornamental stones). For a number of wide spread mineral resources (brick clay, keramzite materials, sand and gravel) it is an important component of local industry.

1. Introduction

The resource base for metallic and nonmetallic minerals in Novokuznetsk district is much inferior to coal resources (only deposits of common non-metallic minerals such as gravel-sand mixtures, brick clays, mineral water in small amounts) are under operation. At the same time, a number of promising deposits of various minerals are known in the district, development of which would diversify local mining industry and prevent its dependence on a single type of minerals. Traditionally, review is started with metal minerals, of which iron and gold are presented in Novokuznetsk district.

2. Results and discussion

Iron. Resource base of the Novokuznetsk administrative district for this type of raw material, which is very important for economy in Kemerovo region, includes Lavrenovskoye and Zapovednoe deposits and a number of ore occurrences (Levoberezhnoye, Zaozernoye, Podhrebetnoye, Savelievskoe, Chernoyiusskoye, Goltsovoe, Shatskoye) of Tersinskaya group located in the watershed of Kuznetsk Alatau near the border with the Republic of Khakassia. Economically, area is not developed. It is a part of the "Kuznetsk Alatau" state park, any economic activity is prohibited within its borders. Thus, iron ore area of the Upper Ters River currently is only of cognitive significance. The total balance reserves of iron ore here are estimated as 68.1 million tons, including 17.9 million tons of A + B + C category reserves. Deposits are located in the upper flow of the rivers Verhnyaya Ters and Kibras, 110km to the NE from Novokuznetsk. Ores are represented by magnetite scarns. Iron content in ore ranges from 43.2% to 55%. The inferred iron ore resources are estimated at several hundred million tons. Length of ore strip is about 16km. It is confined to the Kanym graben-syncline (volcanic depression, filled with the Middle Cambrian volcanic-sedimentary deposits of the Kanymskaya suite)

IOP Conf. Series: Earth and Environmental Science **84** (2017) 012009 doi:10.1088/1755-1315/84/1/012009

on its contact with breakthrough intrusions of diorites, gabbro diorites and plagiogranites. Magnetite ore deposits are located conformable to rocks of ore-bearing strata (scarned marble limestones, and metamorphosed tuffaceous sandstones). Mineral composition of ore is of the same type: magnetite (30-90%), amphibole (up to 40%), garnet (up to 18%). Permanent impurities: pyrite, pyroxene, chlorite, carbonates, feldspar, quartz [4, 7].

Gold. Gold mineralization in Novokuznetsk district is represented by lode and placer occurances. Placer gold mineralization was established in the district in the middle of the nineteenth century. For a long time this was the only source of metal mining in here. By the beginning of the 20th century, alluvial placers of the rivers in Kemerovo region were largely depleted, gold mining declined [3]. Introduction of new methods of extraction and its intensification, made a number of valleys of the Kuznetsk Alatau rivers of certain interest for goldmining operation. Among them, the valley of the river Tutuyas (right tributary of the Tom River above the Tomusinskaya power plant) and the valley of the river Middle Ters (right tributary of the Tom River lower Osinovoe Pleso village) belong to the territory of Novokuznetsk district.

Tutuyas alluvial valley placer. Thickness of loose sediments is 3 to 4m in average, slopes of the valley are no more than 0.006. Alluvium stone content is less than 10%, good washability. Hydrogeological and technical mining conditions enable operation of 80-liter dredger. The plot of placer proposed for further study and development has length of 30km (beginning of site is 19km upstream from the river mouth) with a width of inferred resources contour of 40m. Gold is fine with predominance of 0.2mm size. Inferred resources are calculated by estimated parameters of the Taidon-Tersinsky area and amount to 416kg of uncleaned gold. In individual samples from the placer, zircon content of 70-80g/m^3, ilmenite - of 150g/m^3 is detected. Researchers believe that conglomerates of the Tarbagan series of the Lower Jurassic are the main sources of gold of this placer [3].

Placer of the Middle Ters river. For this placer, alluvial gold reserves were approved for open mining at a site 10km from the mouth of the Middle Ters river to confluence of its right tributary the Alexandrovka river. They include in C_1 category: sand 4112.3 thousand m^3; metal 2679.43kg with an average content of 797 mg/m^3 of chemically pure gold; in C_2 category - sand 889.96 thousand m^3, gold 646.2kg, with content of 825mg/m^3. By complexity of geological structure the placer is attributed to the third group of deposits. The main limit for mining operation, is proximity of the Kuznetsk Alatau state park and need to install a water purification system to prevent pollution of the river Tom [3].

The hardrock gold mineralization in Novokuznetsk district is represented by the Fedorovsko-Talanovskoye deposit [7, 10]. It is located in the left-bank part of the river Sayansas (left tributary of the Tylon river, Kuznetsk Alatau). Distance to Novokuznetsk is 135km through villages Osinove Plesso, Mutny and Pesas. It was discovered in 1968 during geochemical prospecting. The deposit is localized in scarns of the Early Devonian age. Gold-beresitic mineralization is controlled by a series of dykes. Tectonic zones in the south of ore field contain gold-listvenite and gold-sulfide mineralization, accompanied by the exposure of cinnabar, fluorite, barite, and fahl ores. It is believed that high-temperature scarns, medium-temperature gold-bearing beresites and low-temperature gold-listvenitic and gold-sulfide mineralization represent stages of development of single orogenetic system. Gold content reaches tens of g/t. The veins contain variations of hackly gold (1-5mm) and dust gold. Fineness is 820-840‰. Thickness of ore metasomatites with gold content in the first g/t (fineness 701-776‰, 725 ‰ on an average) varies from 10-40m to 300m. The Cretaceous kaolinite and the Cenozoic hydromica-kaolinite gold-bearing weathering crusts (thickness of 90m to 200m, Au content of 1.5g/t) remain in the deposit. Gold from crusts have dimension of up to 0.5mm. Their fineness is 530-890‰, 730‰ on an average. Almost none impurities, apart from silver. The main potential of the Fedorovsko-Talanovsky ore field is associated with large and extended mineralization zones, suitable for open and underground mining. Some researchers have estimated the total potential of the Fedorovsko-Talanovsky ore field (26km^2) to be at least 150 tons of gold [2]. The deposit is located on the territory of the Kuznetsk Alatau state park and for this reason there is no prospect of its industrial development in a short term.

The list of non-metallic mineral deposits in the district is much wider, but few are involved in industrial development.

Mineral water. Tersinskoye field. It is located in the valley of the river Upper Ters to the NE from Novokuznetsk. It was discovered when drilling a well for exploration of the Makarevskoye coal deposit in 1967. In addition to production well, there are three monitoring wells on the field. The deposit is known to consumers under the name “Tersinka”. Water is certified by the GOST “Mineral water drinking medical and therapeutic” as a separate “Tersinsky” type - therapeutic and drinking hydrocarbonate calcium-sodium carbon dioxide water. Commercial reserves of the field are approved for operating well № 1011 for a 10-year service life as 90m^3 per day. Water-bearing rocks are carboniferous sediments of the Carboniferous and the Permian (sandstones, siltstones, mudstones, and coal seams) forming the Tersinskaya anticline. Basalts and diabases are introduced into sedimentary rocks. Composition of “Tersinka” mineral water is close to “Borjomi” and can be used as therapeutic and drinking water[5]. According to Department of Subsoil Management in Kemerovo region (Kuzbassnedra), the Tersinskoye field is industrially equipped, a road and a power line are laid to it. “MVM” LLC extracts water, and another Tersinka LLC company bottles it. In regard to anticipated start of mining operations at the Makaryevskoye coal field, future of the mineral stream causes certain concerns.

Refractory clay. *Barkinskoye deposit.* It is located 3.5 km from Kuzedeevo village. Refractory clays form a series of plastoid bodies confined to the upper part of cyclically constructed clayey-sandy stratum of lake genesis, lying on the weathering crust of the Cretaceous-Paleogene age, overlapping with quaternary loams. The deposit was explored by wells and pits. The reserves of refractory clays in $A + B + C_1$ categories are 9 million tons. It is classified as a state reserve [5, 8].

Brick clay. Deposits serve as a base for brick factories operation. Demand for these products in the region is growing from year to year along with the growth of construction works. Bricks were produced at brick factories located in large settlements. Factories had used resources of small and medium-sized deposits of brick clay in the alluvial plain of the Tom and Kondoma rivers. The period of operation of such enterprises, as a rule, is small and is estimated as the first ten years [5, 8]. Thus, on basis of Abashevskoye field discovered in 1960 (located in the right bank of the Tom river 1-1.5 km from Staro-Abashevo village, 25km from the center of Novokuznetsk), from 1986 to December 2006, “Abashevsky brick plant” was operating. By now, Novokuznetsk brickworks No21 deposit, located in the south-western part of the city, has been developed by Novokuznetsk brick factory. In 2008, Kaltansky brick plant completed development of the Kaltanskoye deposit (the bank of the Kondoma river in the northeastern part of Kaltan). Myskovsky brick factory completed development of the Myskovskoye deposit.

At present, only two fields are being exploited, one in Novokuznetsk, one in Osinniki.

Baidayevsky brick plant is developing an average size *Baydaevskoe 1 and 2 deposits* of expanded clay and brick clay. Clays are suitable for bricks and keramzit production. For the second plot, commercial reserves of brick clays have been approved for production of: an ordinary construction brick of “100” – “150” grades; ceramic brick with 32 cells of the “125” grade; ceramic brick with 18 cells of the “100” grade; ceramic stones with 28 cells of the “150” grade; keramzit gravel of the “700” grade by bulk density as a filler for lightweight concrete. Total reserves of raw materials of A category estimated as 1861.1 thousand m^3; B category -2860.1 thousand m^3; C_1 -1346 thousand m^3. This amount of reserves corresponds to 204.4 thousand m^3 of overburden. Preliminary reserves of raw materials of C_2 category on the site are 2262.9 thousand m^3.

Osinnikovskoye deposit is located on the southeastern outskirts of the town of Osinniki, and includes three plots - Osinnikovsky, Osinnikovsky Eastern and Osinikovsky Southern. It is being developed by “Osinnikovskiy brick factory” JSC. To date, work in the southern plot has been stopped, the eastern plot with reserves of brick clays of $A + B + C_1$ category of 2,438 thousand m^3 is in reserve. Mining operations are carried out at the Osinnikovskoye deposit where the clay reserves of $A + B + C_1$ categories of 1007 thousand m^3 are concentrated. The loams of the deposit are suitable for the "50" grade brick production.

IOP Conf. Series: Earth and Environmental Science **84** (2017) 012009 doi:10.1088/1755-1315/84/1/012009

Undistributed fund currently includes 4 deposits. Among them are the medium-size Abagur deposit of brick clays (deluvial loams of the slope of the Tom river valley). The raw material is suitable for production of the “125” grade brick. The total reserves of A + B + C_1 categories amount to 8756 thousand m^3.

Nekrasovskoye deposit is located 0.5km from the town of Kaltan. The quality of clays meets requirements of GOST 530-80 “Brick and ceramic stones”. Commercial reserves of the Nekrasovskoye deposit clays for production of common brick Kr 125/1450/25, subjected to introduction of 4-5% of sawdust and 3-4% of coal as a leaner, are approved by the Protocol of the TCNR in Kemerovo region No. 631 of 01.01.1993 in the amount and by categories (thousand m^3): B - 49.2, C_1 - 169.0, total B + C_1 - 218.2. By conditions of occurrence, consistency, quality of clays, the Nekrasovskoye deposit is assigned to the first group of medium-sized blanket deposits, simple in structure, thickness and quality of raw materials.

Alardinskoye deposit is located in one kilometer from Malinovka village. Loams of the deposits are suitable for production of building bricks of the “100” and the “150” grades, if subjected to introduction of 12-15% of leaner (sawdust). Approved reserves of A + B + C_1 categories are 3845 thousand m^3.

Shushtalekskoe deposit located 5 km to the west from the town of Kaltan with loam resources suitable for production of brick of the “100” grade of A + B categories estimated as 5105 thousand m^3 is in reserve. A condition for development of this deposit is mandatory introduction of large amount of leaner (sawdust) into the batch for bricks production.

Keramzit raw materials in Novokuznetsk district are represented by Myskovskoye expanded clay deposit which is the resource base for Myskovsky keramzit gravel plant OJSC operation. Approved commercial reserves of loams for production of keramzit gravel of the “400” – “500” grades with introduction of 1% of fuel oil or 1% of used oils are: of A category - 434.8 thousand m^3; of B category - 1052.6 thousand m^3; C_1 category -1209.3 thousand m^3. By complexity of geological structure, the deposit is assigned to the 1st group.

Sand and gravel materials. In Novokuznetsk administrative district, the main reserves were concentrated in the fields near major cities (Novokuznetsk, Kaltan, Myski, Osinniki). Some of them have already been exploited by now (Karchitskoye, Topolniki-Feski, Ostrovskoye, Antonovskoe). The remaining reserves are concentrated in alluvial plains of the rivers Tom and Kondoma. Among them are Abaguro-Atamanovskoye, Kuznetskoye, Borodinskoye, Sidorovskoye. Sand-gravel mixture of these deposits is suitable as large and fine filler in concrete of the “50” to “500” grades, as well as for railroads and highways gravelling [8]. Development of deposits in alluvial plain of the river Kondoma (Ashmarinskoye, Kaltanskoe, Osmanskoye 1 and 2) is now considered unprofitable due to poor quality of raw materials, or because of location in water protection zone of the river [8]. Three small-size deposits of construction sand (Sredinnoye, Osmanovskoye and Kuzedeevskoye) located in alluvial plains of the rivers Tom and Kondoma are of definite importance for local construction industry. Sands of the Kuzedeevskoye deposit are suitable as molding compound. Reserves of the deposit are small and amount to 503 thousand tons for the Sredinnoye deposit, 583.000 tons for the Osmanovskoye deposit, and about 300.000 tons for the Kuzedeevskoye deposit.

Building stones. Gravelstone and sandstone of coal-free parts of Kuzbass (for example, the Kuznetsk suite) are of certain interest as quarry stone. In the post-war period these sandstones were developed by small quarries in outskirts of Novokuznetsk. However, quality of this material by mechanical strength and consistency of physical and mechanical properties left much to be desired. At present, almost every operating local coal open pit mine has subsidiary production units producing sandstone-gravel mixtures from overburden sandstones. Quarry stone is extracted by small quarries in a number of plots in Novokuznetsk district. A quarry at Kuzedeevo village entrance may serve as an example. Coarse-grained sandstone of the Upper Devonian appeared to be quarry stone. The author had studied this quarry in 2008 [1], since these sandstones contain imprints of the Late Devonian plants. Local residents evidence small-scale mining operations conducting there in summer. As a rule, such objects mined without carrying out geological prospecting and reserves estimation. The latest report on the deposits of nonmetallic resources in Kemerovo region [8] provides data on the Shushtalepskoye deposit as the only quarry in the district (southeastern outskirts of the Malyshev Log settlement), which was discovered in 1950 by the Shushtalepskaya PE. Reserves of quarry sandstone here are extimated by A + B categories in the amount of 1.102 thousand tons, low quality of raw materials is noted. Sandstones of the deposit lie directly on coal seams. Their extraction is only possible in conjunction with coal seams development.

Ornamental stones (agates). Tersyukskoe deposit. It is located in the northern part of Novokuznetsk district, in the upper flow of the river Middle Tersyuk, right tributary of the river Tom, with a mouth in Yachmenyukha village. The nearest settlements are the villages of Yachmenyuha (about 15km) and Osinovoe Pleso (about 50km), with which the deposit is connected by gravel roads. Agates are formed with siliceous, zone-colored amygdales in the basalts of the Yamino suit of the Lower Triassic. Amygdales aggregates form a layer-like productive stratum, lying at angles of 7-25° falling to the northeast, with thickness of 0.7-2m. The amygdales in productive strata range insignificantly and are of 36 kg/m^3 on an average. The size of amygdales is 5 × 5 × 2cm on an average. The reserves of amygdales are 140.000 tons. Commercial yield is 25% (conclusion of the "Baikalvartsamotsvetsy" IGE). Standard reserves of conditioned agates are 5600 tons. Experimental quarries are in operation on the deposit, about 42 tons of conditioned agates were mined [6, 8].

Facing stones. Description of deposits of this type in Kemerovo region has many pages in the last report on nonmetallic deposits of Kemerovo region [8]. The illusion of significant resource potential of the region, including the Novokuznetsk district, is created. The real situation is quite different. In the city of Novokuznetsk there is no a single building and construction that would have been trimmed with local facing stone. The data on deposits and reserves of the facing and finishing stone given in the mentioned above monograph should at best be considered as inferred resources, since no experimental quarries have been created anywhere, there is no data on the blocky structure of rock massif and other parameters necessary for this type of raw material. A number of occurrences are located in places located far from human habitats and there are no roads, no electricity, no shelter, or located in water protection zone of rivers. Nevertheless, this type of raw materials should be considered as potential basis for local industry development, including small businesses. The author saw examples of such private small stone cutting enterprises in Serbia during geological expedition organized by NIS "Naftogas", Serbia in 2005. Small business can successfully be engaged in development of such deposits and meet local needs in natural facing raw materials. Objects of extraction and processing of facing stone can be as follows:

- *Ust-Talovskoye deposit of labradorite porphyrites* - located opposite to Osman village on the left bank of the river Kondoma. It is localized in volcanic-sedimentary stratum of the Lower Devonian (the Tel'bes series), is an extensive coastal cliff of the river Kondoma, inferred resources of labradorite porphyrites are about 2 million tons, it is necessary to take into account the position of the deposit in protected zone of the river;
- *Atlasskoye deposit of decorative limestone* – the right side of the valley of the Lower Ters river in the area of the former Pesas settlement, cherry-red and archeocyanate limestones of the Early Cambrian. Certain decorative properties have cherry-red listvenites, sealing wax lava breccia, conglomerate breccia of the deposit. Inferred resources of raw materials are 112 million tons with a pit depth of 30 meters.
- *Tachelginskoye deposit of marble* - the valley of the river Tashelga (the right tributary of the river Mrassu, explored by the Tashkelginskaya prospecting party in 1975 with the organization of experimental quarries, marble from white to green and pink color, marbled breccia (right bank plot) are decorative. Reserves of $B+C_1$ categories are estimated as 2728 thousand m^3, of C_2 category - 14.6 million m^3. On the right-bank plot, reserves of C_2 category are 8.8 million m^3.
- *Porozhinskoye granite deposit* - the valley of the river Mrassu between the estuaries of the Tashelga and Kizes rivers, the Porozhinskiy massif of subalkaline granites and leucogranites of the Late Triassic age [11] (the Middle Riphean age is erroneously indicated in the summary

[8]). Granites are suitable for building panels and flooring coating production. Reserves of C_2 category estimated as 5.4 million m^3. The deposit is located in a remote area.

Peat. Summary report on nonmetallic deposits in Kemerovo region [8] indicates that there are 4 explored peat deposits in Novokuznetsk district (Atamanovskoye, Kurgany, Borovskoye, Akushkinskoe, Tailepskoye). Peat is suitable for introduction into the soil as a disintegrant, as well as for peat mixtures preparation. Phosphorus-containing peat can be used as a fertilizer. Currently, peat extraction is not carried out at these locations, except for small amount used on private farms by local residents.

Akushkinskoye deposit, located in alluvial plain of the river Kondoma 6 km from village Sarbala, was studied in the most detailed way. Peat reserves of A category estimated as 214 thousand tons with its ash content of 22%.

Peat of the Kurgany deposit on the left bank of the river Tom is enriched with phosphorus boggy phosphates) with P_2O_5 content of 1.81%.

At the Tilepskoye deposit, as a result of preliminary prospecting, inferred peat reserves of C_2 category in the amount of 772 thousand tons were calculated in an area of 334 hectares.

Raw materials for lime production. In Novokuznetsk district, one deposit of limestone feasible for lime production - *Abramovskoe* is known. It is located on the right bank of the river Kondoma near the Osman railway station (former Ail Abramovsky) in 9km from Kuzedeevo village. The deposit is confined to the sediments of the Vassinsky horizon (the Frasnian stage) of the Upper Devonian and is composed of monoclinically falling sequence of organogenic detritus limestones (incidence angles of 10-200). Thickness of limestone sequence is 10-14m. It was studied and developed in the 1940s, as is evidenced by numerous small open pit mines on limestone outcrops. Geological section of the Devonian in the area of the Abramovskoye deposit is one of reference sections of the Devonian of the south of the Kuznetsk Basin, containing locations of the Upper Devonian fossils and the Middle Devonian vegetation [9]. Reserves of limestone suitable for lime production according to prospect data of 1941 amount to 734 thousand tons.

3. Conclusions

The above review shows significant potential of Novokuznetsk district in a wide range of minerals. As for metal minerals, some at the moment have only cognitive significance (iron ore deposits) because of their location within state park, others (alluvial gold) may be involved industrial development in the near future. It is necessary to take into account that the resources of alluvial gold are significantly inferior to coal resources and for this reason large business can hardly be of interest in it. At the same time, for small and medium-size enterprises, they can be of considerable interest as objects with a rapid return of investments.

Common mineral resources (sand and gravel materials, expanded clay, brick clays, mineral water) are of significant importance for local industry. Some of the deposits of this group known in the district have only theoretical value and are unlikely to ever be developed (peat, raw materials for lime production). In the foreseeable future, deposits of ornamental (agate) and facing (marbled porphyrites, decorative limestones, marbles, granites) stones may be in demand.

References

[1] Antonova V A, Gutak Ja M and Nadler Yu S 2009 *Regional Geology, Stratigraphy and Paleontology of the Phanerozoic of Siberia* (Novosibirsk: SNIIGGiMS) p 105

[2] Barmotin V K 2011 *Proc. Sci. Conf. "90 Years of the State Geological Service of Western Siberia"* (Novokuznetsk) pp 39–40

[3] Butvilovsky V V, Avvakumov A E and Gutak O Ya 2011 *Placer Gold Mineralization of the South of Western Siberia. Historical and Geological Survey and Assessment of Opportunities* (Novokuznetsk: KuzSPA) p 241

[4] *Geology of the USSR. Western Siberia (Kemerovo, Novosibirsk, Omsk, Tomsk regions, Altai Territory)* 1982 vol XIV book 1 (Moscow: Nedra) p 319

[5] *Geology of the USSR. Western Siberia* (Kemerovo, Novosibirsk, Omsk, Tomsk regions, Altai Territory) 1982 vol XIV book 2 (Moscow: Nedra) p 196
[6] Lazarev I V anf Voznaya A A 2015 *Proc. VIIth Sci. Conf. "Young Russia"* (Kemerovo: KuzSTU)
[7] Kondakov A N and Voznaya A A 2013 *Mineral Resources of Kemerovo Region* book 1 (Kemerovo: KuzSTU) p 290
[8] Kondakov A N and Vvoznaya A A 2016 *Mineral Resources of Kemerovo Region* book 2 (Kemerovo: KuzSTU) p 496
[9] Udodov V P, Gutak Ja M and Mezentseva O P et al 1987 *Nature and Economics of Kuzbass: Theses of Reports for the Upcoming Conf.* (Novokuznetsk: NGPI) pp 51–3
[10] Shevyrev L T and Savko A D 2012 *Ore Deposits of Russia and the World. Handbook* (Voronezh: VorSU) p 284
[11] Shokalsky S P, Babin G A and Vladimirov A G et al 2000 *Correlation of Igneous and Metamorphic Complexes of the Western Part of the Altai-Sayan Folded Area* (Novosibirsk) p 187

IOP Conf. Series: Earth and Environmental Science **84** (2017) 012010 doi:10.1088/1755-1315/84/1/012010

Paradigm oil and gas complex of Russia at the present stage

A E Kontorovich, L V Eder and I V Filimonova

Trofimuk Institute of Petroleum Geology and Geophysics, Siberian Branch of the Russian Academy of Sciences, 3 Koptyug Ave., Novosibirsk, 630090, Russia
Federal Research Center of Coal and Coal Chemistry, Siberian Branch of the Russian Academy of Sciences, 18, Sovetskiy Ave., Kemerovo, 650000, Russia

E-mail: KontorovichAE@ipgg.sbras.ru

Abstract. The paper presents the features of the Russian oil complex development paradigm in the 20th century, including the priority search and development of unique and major fields; expansion of the oil industry geographical location; priority input into the development of reserves that are characterized by high quality characteristics. It is substantiated that at the present stage there is a need to change the development paradigm as well as the transition from extensive to intensive development, including the development of sedimentary basins of the Russian shelf of the Arctic Ocean; poorly studied petroleum provinces in the east of the country; unique oil resources of nonconventional sources; complex and rational use of residual reserves of unique and major fields.

1. Introduction

So far, the bowels use paradigm in the fuel and energy sector of Russia has been predominantly extensive, characterized primarily by a quantitative increase in the production of raw materials with a relatively unchanged technical potential.

The extensive development path provided for the expansion of the geography of the oil-and- gas production industry at the expense of new provinces on the continental part of Russia; priority search and development of unique and major fields; priority input into the development of reserves characterized by high quality characteristics; increasing the processing of raw materials with the production of relatively low quality products. The extensive development path, on the one hand, creates opportunities for growth in the economy and its individual sectors because it increases production resources, and, on the other hand, it limits the opportunities for qualitative growth.

At the present time, a paradigm shift in the development of Russia's resource and raw materials complex should occur. The paradigm shift is connected with the transition from an extensive to an intensive trajectory of development. A qualitative leap and a sharp increase in the innovative character of the raw material industries development must occur for sustainable development of the economy as a whole, taking into account both the importance of the raw material complex in its structure, and, separately of the raw materials industries and resource regions. Such a change should occur regarding the integrated development of Russia's resource regions and enhance the efficiency of Russia's bowels use system throughout the entire production chain, starting from geological exploration and ending with the formation of processing capacities and the transport system for raw materials and products of their processing.

2. Methods of research

IOP Conf. Series: Earth and Environmental Science **84** (2017) 012010 doi:10.1088/1755-1315/84/1/012010

The methods of economic, normative, comparative and expert analysis have been used in the work. To justify the parameters of the paradigm shift in the development of Russia's oil and gas complex, a combined approach has been used which involves a combination of macro-forecasting methods and a method of comparing the main targets of subsoil use policy and the associated complex institutional effects.

3. Results and discussion

The paradigm of the Russian oil complex development in the twentieth century

The paradigm of the Russian oil complex development strategy in the twentieth century was focused on a predominantly extensive development path, which includes the following main features:

1) priority search and development of unique and major fields;

2) expansion of the geography of the oil industry at the expense of new provinces on the continental part of Russia (moving from West to East) [1];

3) Priority input into the development of reserves characterized by high quality characteristics (porosity, permeability, viscosity and density).

The oil industry at the present stage; the need to change the development paradigm and transition from an extensive to an intensive development

At present, the significant structural changes are taking place in the structure of the mineral and raw materials base, which in essence will determine a new development paradigm of the Russian oil industry for the long term.

Firstly, a qualitative change in the structure of the fields being developed, in terms of developing unique and major fields and increasing the share of medium-sized and small fields in the structure of reserves is taking place.

Secondly, the geography of oil production in Russia is significantly changing with the increasing role of the eastern regions of Russia and its sea shelves [3].

Thirdly, there is a significant reduction in the amount of recoverable reserves in the fields prepared for commercial production, while increasing the number of such fields;

Fourth, there is a significant quality loss of oil reserves, due to the increase in the share of hard-to-recover oil reserves.

Fifth, there are changes in the structure of oil reserves by categories with an increase in the share of reserves of highly reliable categories already involved in development.

The role of unique and major fields at the present stage

At present, the raw materials base structure of the oil industry in Russia has changed significantly. The role of unique and major fields in the oil production has decreased. As of January 1, 2016, 11 unique and 87 major oil fields remained in the Russian Federation in terms of residual oil reserves. About 3.7% of all unique and major oil, gas and oil-and-gas fields account for about 56% of residual oil reserves (Figure 1). At the same time, there are still a number of problems related to the development of unique and major fields. First of all, these are high depletion, water cut and relatively low production rates.

Small and smallest fields are the main category of discovered fields at the present moment and for the long term

Small and smallest fields are the main category of newly discovered fields at present

As a result, the role of medium-sized, small, and in the medium term the smallest fields is sharply increasing. In the 21st century only small fields were discovered in the main oil producing regions of Russia. In the Orenburg region, the Republic of Tatarstan and the Republic of Bashkortostan, the share of discovered fields with the reserves of less than 1 million tons makes up almost 70-90%. In Khanty-Mansiysk Autonomous Okrug (KhMAO) and Tomsk region this figure is just under half. At the same time, the average level of the discovered field in the KhMAO, which is the largest oil region, does not exceed 3 million tons. As a result, according to a new scale classification of fields on average, the smallest fields are discovered, with the recoverable oil reserves no more than 5 million tons.

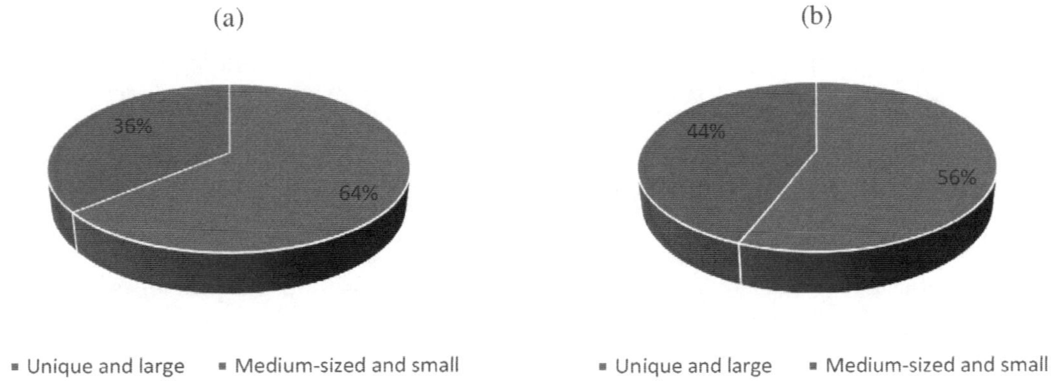

Figure 1. The share of unique and major fields in Russia in (a) initial recoverable oil reserves, (b) current recoverable oil reserves.

Structural shifts in the production geography

Western Siberia is the main center of the Russian oil industry, where more than 56% of Russian oil is produced. However, over the last decade, the region's share in the production structure has decreased by 12%. In 2016, oil production in Western Siberia amounted to 309.6 million tons. Eastern Siberia and the Far East are regions with the most dynamically developing oil production. So, while in 2008 their share in the oil production structure was less than 3%, in 2016 they produced12%. Since 2012 a gradual increase in oil production has been taking place in the European part of Russia. Thus, production increased from 151.4 million tons in 2012 to 169.3 million tons in 2016 (figure 2).

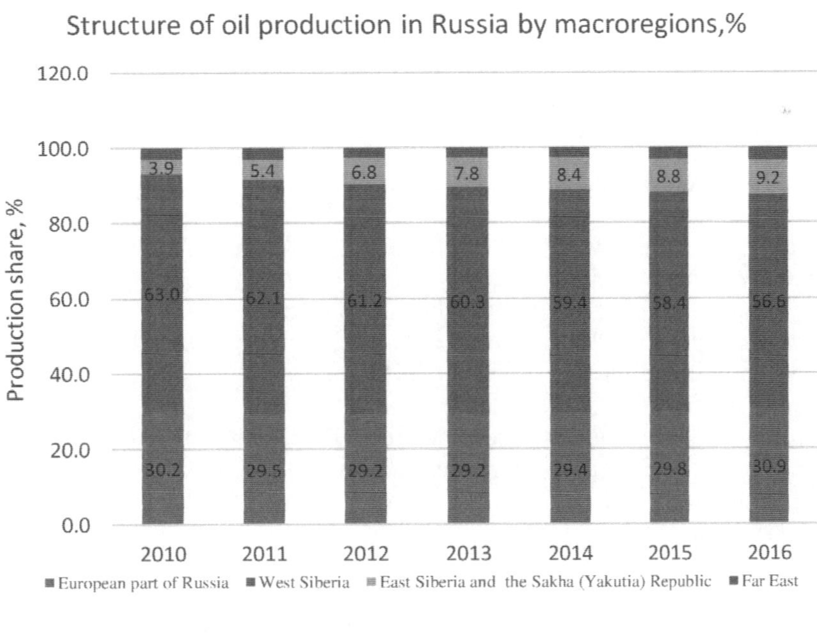

Figure 2. Structure of oil production in Russia by macroregions, %.

IOP Conf. Series: Earth and Environmental Science **84** (2017) 012010 doi:10.1088/1755-1315/84/1/012010

The deterioration of the oil reserves structure, associated with the increase in the share of hard-to-recover oil reserves

The structure of the explored oil reserves in Russia is continuously deteriorating. The share of hard-to-recover reserves in the overall structure of oil reserves is growing rapidly. Under the hard-to-recover reserves, in this case we mean the accumulation of oil in low-permeability reservoirs, in subgas and water cut zones, multilayer reservoirs with relatively low thickness of oil-saturated reservoirs, viscous and dense oil (Table 1). To date, the actual reserves have been developed on average by 75%, while the hard-to-recover reserves have been developed only by 35%.

Table 1. The structure of oil reserves by regions with differentiation in terms of the oil quality characteristics.

Federal Districts	Share of oil reserves with high density by region,%		Share of oil reserves with high viscosity by region,%	Share of oil reserves with low permeability,%
	0.87-0.9 g/cm^3	More than 0.9 g/cm^3	More than 30 MPa*s	Less than 0.05 mm^2
Northwestern	11.9	39.8	31.5	16.7
Southern	4.7	21.7	12.1	15.8
North-Caucasian	1.2	2.1	2.1	48.1
Privolzhsky	30.9	26.7	25.2	27.9
The Urals	17.1	10.3	8.1	58.2
Siberian	7.6	15.7	0.2	30.3
Far Eastern	17.2	4.4	1.9	2.1
Shelf of the Russian Federation	16.7	21.6	4.5	11.6
Russian Federation	18.4	16.4	12.2	44.0

The new development strategy paradigm of raw materials and oil production in the Russian Federation

Major structural shifts in the resource and raw materials base and oil production in Russia determine the need for a new paradigm for the oil-and-gas industry, since the old one has largely been realized and exhausted.

In the 21st century, the main new objects of prospecting, exploration and development of oil fields, as well as the priority tasks of the oil-and-gas industry of the Russian Federation will be:

1) sedimentary basins of the Russian shelf in the Arctic Ocean;

2) poorly investigated provinces on land (Leno-Tunguskaya province) [4];

3) large objects not entered into development in the northwest of Western Siberia in the Yamal-Nenets Autonomous District [5];

4) unique oil resources from nonconventional sources (Bazhenov suite, etc.);

5) rational use of residual reserves of unique and major fields [6, 7, 8, 9];

6) prospecting, exploration and development of small, smallest and medium-sized fields in mature petroleum basins. [10, 11].

4. Conclusions

At the present time, a paradigm shift in the development of Russia's resource and raw materials complex should take place. The paradigm shift is connected with the transition from an extensive to an intensive trajectory of development. The intensive path of development is connected, first of all with the innovative approach to extraction and use of mineral raw materials. In the oil and gas complex in the sphere of production, such key areas are the development of hard-to-recover and nonconventional hydrocarbons, the shelf of the Arctic Ocean; poorly studied provinces in hard-to-reach arctic regions; rational use of residual stocks of unique and major deposits; prospecting, exploration and development of small, smallest and medium-sized deposits in mature oil and gas provinces.

Acknowledgements
The authors are grateful for financial support of the Russian Science Foundation No. 16-18-10182, as well as to the Russian Foundation for Basic Research No. 15-06-090994, 16-36-00369, 17-06-00537.

References

[1] Kontorovich A E, Eder L V and Nemov V Yu 2013 *Neftyanoe Khozyaystvo (Oil Industry)* **1** 4–8
[2] Kontorovich A E, Eder L V, Filimonova I V, Mishenin M V and Nemov V Y 2016 *Russian Geology and Geophysics* **57(12)** 1653–67
[3] Korzhubaev A G, Filimonova I V, Eder L V and Sokolova I A 2009 *Neftyanoe Khozyaystvo (Oil Industry)* **3** 14–7
[4] Kontorovich A E, Korzhubaev A G, Filimonova I V and Eder L V 2008 *Neftyanoe Khozyaystvo (Oil Industry)* **5** 24–7
[5] Kontorovich A E, Eder L V and Nemov V Yu 2012 *Neftyanoe Khozyaystvo (Oil Industry)* **7** 66–70
[6] Tagaeva T O, Gilmundinov V M and Kazantseva L K 2017 *Economy of Region* **14(7)** 165–77
[7] Tagaeva T O 2011 *Source of the Document Studies on Russian Economic Development* **22(3)** 331–8
[8] Kryukov V A, Bozo N V and Malysheva Y V 2005 *9th Russian-Korean Int. Symp. on Sci. and Tech., KORUS-2005* 898–903
[9] Nikitenko S M and Goosen E V *2017 IOP Conf. Series: Earth and Environmental Science* **53** 012018
[10] Sharf I, Filjushin V, Shenderova I and Kochetkova O 2015 *IOP Conference Series: Earth and Environmental Science* **27(1)**
[11] Nikitenko S M, Goosen E V and Sablin K S 2016 *IOP Conference Series: Earth and Environmental Science* **45** 012001

IOP Conf. Series: Earth and Environmental Science **84** (2017) 012011 doi:10.1088/1755-1315/84/1/012011

Current state and problems of integrated development of mineral resources base in Russia

I V Filimonova[1,2], L V Eder[1,3], M V Mishenin[1] and T M Mamakhatov[1]

[1]Trofimuk Institute of Petroleum Geology and Geophysics, Siberian Branch of the Russian Academy of Sciences, 3 Koptyug Ave., Novosibirsk, 630090, Russia
[2]Novosibirsk State University, 2 Pirogova Street, Novosibirsk, 630090, Russia
[3] Federal Research Center of Coal and Coal Chemistry, Siberian Branch of the Russian Academy of Sciences, 18 Sovetskiy Ave., Kemerovo, 650000, Russia

E-mail: FilimonovaIV@list.ru

Abstract. The article deals with the issues of integrated development of subsoil resources taking into account the actual problems facing the Russian oil and gas complex. The key factors determining the need for integrated development of subsoil resources have been systematized and investigated. These factors are the change of the hydrocarbon resource base quality, the improvement of the depletion degree of basic (unique and major) oil fields, the increase in the number of small and smallest oil fields discovered and introduced into development, the increased capital intensity and the riskiness of geological exploration, and the territorial location of new subsoil use facilities.

1. Introduction

The oil and gas complex has a powerful influence on the Russian economy. Over the past twenty years, the oil and gas industry have been the guarantors of economic growth and contributed to the accumulation of state revenues [1, 2].

Long-term sustainable development of Russia's oil and gas complex is primarily related to the integrated development of mineral resources (IDMR). There are a number of important prerequisites that necessitate the transition to integrated development of subsoil resources including changing the quality of the hydrocarbon resource base, increasing the depletion degree of the base (unique and major) oil fields, increasing the number of small and smallest oil fields being discovered and introduced into development, the high capital intensity and riskiness of geological exploration, and the territorial location of new subsoil use facilities.

The transition to IDMR will allow to diversify the directions of economic growth and create new centers for the generation of value added (the cost created by processing industries), to reduce the dependence of the economy on the export of hydrocarbon raw materials.

2. Methods of research

As a methodological basis in the article, the system approach is implemented by way of a general scientific method of cognition, as well as applied methods of synthesis, classification, generalization and grouping, comparative and structural analysis. Based on the methods of system analysis, various aspects of the integrated development of mineral resources in Russia have been analyzed, and groups of factors that determine the complexity of development have been classified.

IOP Conf. Series: Earth and Environmental Science **84** (2017) 012011 doi:10.1088/1755-1315/84/1/012011

3. Results and discussion

3.1. Factors of integrated development of oil and gas industry in Russia

When formulating an effective policy for the integrated development of promising oil and gas areas, it is first of all necessary to take into account the strategic interests of the Russian Federation and the priority tasks of the social and economic development of these regions of the country. The need for integrated development of oil and gas reserves and resources is conditioned by the need to address the industry's priority tasks related to:

- changes in the quality of the hydrocarbon resource base;
- increasing the depletion degree of basic (unique and major) oil fields;
- an increase in the number of small and smallest oil fields discovered and put into development;
- high capital intensity and risky geological exploration;
- the territorial location of new subsoil use facilities, their distance from the traditional centers of production, processing and consumption.

The integrated development of oil and gas bearing areas should focus on the creation of new large centers, processing petroleum chemistry products. The core of integrated development is implementing a resource-innovative approach to the formation of the petroleum industry in Russia and resource-saving. *The resource-innovative approach* is understood as the restriction on the export of unprocessed raw materials as a complex mineral raw material and the creation in Russia of powerful centers to process the products of oil and gas chemistry, where high-tech products with high added value will be made. *The resource saving approach* means, firstly, minimizing the losses and waste of the raw materials complex due to the complex utilization of its products or conservation for future generations, if the production capacity exceeds the market needs, and secondly, the mandatory requirement of a qualified (giving the greatest value added) use of hydrocarbon raw materials.

3.2. Ensuring expanded reproduction of the mineral resource base

The reproduction of the oil and gas resource base ensures sustainable development and operation of the entire petroleum complex in Russia. The development of systems for the search, assessment and exploration of hydrocarbon reserves is one of the important elements of the complex technological re-equipment of the petroleum complex as a whole, its transfer to domestic equipment, technologies and software.

However, in recent years, the paradigm of reproduction of the resource base of oil and gas has changed. In mature oil and gas provinces, newly discovered fields and structures are represented by small and smallest objects in oil reserves, which in recent decades have been the main growth of reserves in Russia (table 1).

Table 1. Production and increment of oil reserves in Russia.

Indicator / Year	1991	1995	2000	2005	2010	2011	2012	2013	2014	2015	2016
Oil production, million tons	462	307	323	470	505	511	518	523	527	534	534
Increment in oil reserves, million tons	931	182	295	330	773	700	681	635	750	730	575
Multiplicity of oil reserves reproduction, units	2.0	0.6	0.9	0.7	1.5	1.4	1.3	1.2	1.4	1.4	1.1
"Eating" / expansion of oil reserves (increment minus production), million tons	469	-125	-28	-140	268	189	163	112	223	196	41

3.3. Ensuring the complex utilization of associated petroleum gas

The level of utilization of associated petroleum gas (APG) largely determines the degree of development efficiency of the entire oil and gas complex of the country. In this connection, the activities related to the extraction and use of APG in Russia are one of the topical and priority tasks for the sustainable development of Russia's oil and gas industry [3, 4].

The production of associated gas in Russia in 2015 amounted to 78.2 billion cubic meters, including 67.8 billion cubic meters (86.7%) extracted and used, and 10.4 billion cubic meters (13.3%) taken into account as gas burned in gas flares.

3.4. Need for processing of condensate gas

An important trend in the development of the Russian gas industry is the change in the structure of the raw materials base. The majority of hydrocarbon fields in East Siberia are of the oil-and-gas or gas-and-oil type, i.e. they contain oil, gas, condensate; and natural gas has a high content of homologues (methane, ethane, propane, butanes), as well as condensate and helium [5]. This is the so-called "fat gas", that requires the processing and separation of valuable components, which are raw materials for oil and gas manufactures (table 2).

Table 2. Comparative characteristics of natural (free) gas of East Siberia, the Republic of Sakha (Yakutia) and West Siberia.

Subject of the Russian Federation	Condensate content, g/m^3			Ethane content, g/m^3			Propane content, g/m^3			Content of butanes, g/m^3		
	weighted average	max	min	weighteda verage	max	min	weighted average	max	min	weighted average	max	min
Irkutsk region	132.9	200.0	11.0	70.0	161.9	60.6	32.2	76.3	19.9	21.3	51.1	14.0
Krasnoyarsk region	155.5	354.0	16.8	68.8	138.8	37.6	41.7	61.7	13.6	26.4	42.0	9.5
the Republic of Sakha (Yakutia)	39.4	139.0	1.5	60.2	108.8	25.0	27.9	63.2	3.4	13.9	23.7	3.0
East Siberia and the Republic of Sakha (Yakutia)	117.1	354.0	1.5	66.2	161.9	25.0	33.8	76.3	3.4	20.7	51.1	3.0

Therefore, when creating future oil and gas production centers in the east of the country, it is necessary to plan the establishment of facilities for processing and underground storage of hydrocarbons and associated components and synchronize the development programs for the oil and gas industries.

3.5. Providing separation and storage of associated components, for example helium

The main helium provinces of Russia are connected with the oil and gas basins of the East European and Siberian platforms. Helium is taken into account in dissolved gas at a content of not less than 0.035%, in free (including gas of gas caps) - not less than 0.05% [6].

The concentration of helium in the only developed Orenburg field in Russia is 0.053-0.055%. However, at the fields of East Siberia and the Republic of Sakha, which is the largest undeveloped helium province in the world, this indicator is significantly higher (on average 033-0.42%) and reaches 0.65% (table 3).

Table 3. Weighted average values of helium content in natural gas of East Siberia and the Republic of Sakha (Yakutia).

Subject of the Russian Federation	Structure of ABC1C2 reserves		Helium content,%	
	Helium-containing gas,%	Helium,%	max	min
Irkutsk region	47	34	0.32	0.04
Krasnoyarsk region	14	12	0.65	0.18

the Republic of Sakha (Yakutia)	39	54	0.65	0.04
East Siberia and the Republic of Sakha (Yakutia)	**100**	**100**	**0.65**	**0.04**

The start of large-scale natural gas production in the east of the country can make the regions of East Siberia the largest helium suppliers on the world market.

3.6. Ensuring the effective development of unique and major deposits

The strategy for the development of the Russian oil complex in the twentieth century was geared towards expanding the geography of the oil industry at the expense of new provinces on the continental part of Russia. At the same time, priority was given to the search and introduction into development of unique and major deposits. The expansion of oil production geography in Russia occurred from the western to the eastern and northern regions of Russia [7].

During the entire period of the petroleum industry development in Russia, 21 unique and 113major oil fields were discovered.

3.7. Ensuring the organizational and legislative environment for the development of small and smallest fields

Economic incentives for the development of oil and gas fields further strengthen the "biased" nature of the fields' choice. Naturally, giant and major fields are the first of the opened to be introduced in the development [8, 9]. At the same time, as the basins pass through the peak of their production, it is necessary to intensify the search and introduce small and smallest fields into development to maintain the oil production or to reduce its rate (table 4).

Table 4. Comparative characteristics of the resource base and the state of oil production in small and smallest fields in the Republics of Tatarstan and Bashkortostan by majors and independent companies.

Parameters	Tatarstan			Bashkortostan		
	majors	independ.	total	majors	independ.	total
Number of fields	89	89	178	160	1	161
Number of discovered fields, 1991-2013	46	38	84	35	1	36
Initial recoverable reserves, million tons	287.2	309.0	596.2	438.1	1.4	439.5
Cumulative production, million tons	88.0	80.2	168.2	267.5	0	267.5
Current production, 2013, million tons	3.5	5.3	8.8	5.8	0	5.8

4. Conclusions

Thus, the integrated development of subsoil is closely connected with the current problems of transformation of the oil and gas complex as a whole. At present, the issues of integrated development of subsoil resources are associated with the depletion of major fields and the increasing role of small and smallest fields and the need to change the institutional, legislative and organizational environment for subsoil use. Also, an important issue of ensuring the complexity of subsoil development is the qualified utilization of all associated components (APG, ethane, propane, butane, helium, condensate, etc.) [10, 11].

Ensuring the integrated nature of the subsoil resources development will allow us to diversify the economy, create new industrial centers, master remote areas and improve the socio-economic level of regional development.

Acknowledgements

The authors are grateful for the financial support to the Russian Science Foundation No. 16-18-10182, as well as to the Russian Foundation for Basic Research No. 15-06-090994, 16-36-00369, 17-06-00537.

References

IOP Conf. Series: Earth and Environmental Science **84** (2017) 012011 doi:10.1088/1755-1315/84/1/012011

[1] Kontorovich A E, Eder L V and Nemov V Yu 2013 *Neftyanoe Khozyaystvo (Oil Industry)* **1** 4–8
[2] Kontorovich A E, Eder L V, and Nemov V Yu 2012 *Neftyanoe khozyaystvo (Oil Industry)* **7** 66–70
[3] Tagaeva T O, Gilmundinov V M and Kazantseva L K 2017 *Economy of Region* **14(7)** 165–77
[4] Tagaeva TO 2011 *Source of the Document Studies on Russian Economic Development* **22(3)** 331–8
[5] Korzhubaev A G, Filimonova I V, Eder L V and Sokolova I A 2009 *Neftyanoe Khozyaystvo (Oil Industry)* **3** 14–7
[6] Kontorovich A E, Korzhubaev A G, Filimonova I V and Eder L V 2008 *Neftyanoe Khozyaystvo (Oil Industry)* **5** 24–7
[7] Kryukov V A, Bozo N V and Malysheva Y V 2005 *9th Russian-Korean Int. Symp. on Sci. and Tech., KORUS-2005* pp 898–903
[8] Kontorovich A E, Eder L V, Filimonova I V, Mishenin M V and Nemov V Y *2016 Russian Geology and Geophysics* **57(12)** pp 1653–67
[9] Sharf I, Filjushin V, Shenderova I and Kochetkova O 2015 *IOP Conference Series: Earth and Environmental Science* **27**
[10] Nikitenko S M, Goosen E V 2017 *IOP Conf Series: Earth and Environmental Science* **53** 012018
[11] Nikitenko S M, Goosen E V, Sablin K S 2016 *IOP Conference Series: Earth and Environmental Science* **45** 012001

IOP Conf. Series: Earth and Environmental Science **84** (2017) 012012 doi:10.1088/1755-1315/84/1/012012

The current state of the petroleum industry and the problems of the development of the Russian economy

L V Eder[1,2], I V Filimonova[1,2], I V Provornaya[1] and V Yu Nemov[1,3]

[1]Trofimuk Institute of Petroleum Geology and Geophysics, Siberian Branch of the Russian Academy of Sciences, 3 Koptyug Ave., Novosibirsk, 630090, Russia
[2]Federal Research Center of Coal and Coal Chemistry, Siberian Branch of the Russian Academy of Sciences, 18 Sovetskiy Ave., Kemerovo, 650000, Russia
[3]Novosibirsk State University, 2 Pirogova Street, Novosibirsk, 630090, Russia

E-mail: Ederlv@yandex.ru.

Abstract. In the article features of resource-raw development of the Russian economy for the period between 2000 and 2014 have been considered. The main features of the resource-raw development of the Russian economy at the present stage have been analyzed. The basic principles of the transition of the Russian economy to a resource-innovative development trajectory have been formulated.

1. Introduction

Favorable conjuncture of the world energy markets combined with the presence of an internal potential for the development of the oil and gas complex in the 2000s and the early 2010s provided a significant increase in the role of oil and gas revenues in the Russian economy. The growth in revenues from oil and gas exports with a simultaneous strengthening domestic currency discouraged structural reforms that could diversify the economy, which led to the formation of a resource-inertial model for the development of the Russian economy. Such a development model has a complex of both positive and negative consequences. Generally, positive impact, as a rule, has a more short-term manifestation, whereas negative factors have an increasing influence over time. The main result of the formation of the resource-inertial model was the development of resource- producing and export-oriented industries and the hypertrophied consumer sphere against the background of a significant inflow of income from hydrocarbon exports. At the same time, the high-tech sector, engineering and processing industries did not receive sufficient development. In fact, in the period of high oil prices, there was no incentive to structural reforms in the economy, capable of carrying out its diversification. To a considerable extent, there was no significant investment process in the economy at an essential level of imports of final products, the share of which reached about 70 % in the structure of final consumption of goods and services.

Since 2014, the period of a new economic reality has come, a period of low oil prices and an active reduction of oil and gas revenues in the economy. In these conditions, the issue of the budget deficit, the shrinking balance of payments, and the reduction in the volume of sovereign funds is acute. The reduction of real incomes of the population leads to a narrowing of the consumer sector, which causes a drop in basic macroeconomic indicators.

2. Methods of research

IOP Conf. Series: Earth and Environmental Science **84** (2017) 012012 doi:10.1088/1755-1315/84/1/012012

The methodological basis of the study is a set of methods of comparative historical analysis of the economic dynamics of the country's economic development. The article considers the current economic situation in Russia and the peculiarities of the impact on it of the oil and gas complex using general (analysis, synthesis, deduction, induction) and private methods (statistical analysis, economic evaluation, retrospective-perspective studies) of scientific research.

As the basic indicators of the modern economic system in Russia, the parameters of the Russian budget system, the indicators of the state's balance of revenues, including export revenues and savings in specialized funds, have been selected for the study.

3. Results and discussion

3.1. Features of resource and raw materials development of the Russian economy over the period between 2000 and 2014

The emergence of resource-inertial development. In the 2000s, in the first half of the 2010s, two main factors contributed to the emergence of a resource-raw material trajectory for the development of the Russian economy. They are an external one that means a favorable conjuncture of the world's raw material markets and an internal one, namely, the availability of resource and raw materials for increasing production and export of raw materials (figure 1).

Since the early 2000s the improvement of the conjuncture in the international energy markets has become a powerful stimulus for the development of the internal factor for the emergence of resource and raw materials development, i.e., the restoration of production and export of hydrocarbons from Russia. As a result, since the early 2000s the production and export of oil in Russia has almost doubled. During the same period, exports of petroleum products more than tripled (table 1) [3 - 4].

Table 1. Oil and gas extraction and export from Russia 1995-2013.

	1995	2000	2005	2010	2011	2012	2013
Extraction							
Oil, million tons	307	323	470	505	511,4	518	523
Petroleum products, million tons	131	125	149	175	181	182	188
Gas, billion cubic meters	595	584	641	650	671	654	668
Export							
Oil, million tons	122	144	253	244	241	240	237
Petroleum products, million tons	45	63	97	131	125	138	151
Gas, billion cubic meters	192	194	209	178	190	186	203

*Sources: Fuel and Energy Complex of Russia // Fuel and Energy Complex of Russia, No.1, 2000-2013; Summary indicators of energy resources production in the Russian Federation / / Info Energy, No.1, 2000-2013.

The bulk of the increase in hydrocarbon production was exported. The growth in the volume of hydrocarbon supplies to world markets stimulated the rapid growth in the construction of transport infrastructure. Large oil (BPS-1, BPS-2, ESPO-1, ESPO-2, etc.) and gas (Blue Stream, Nord Stream, South Stream, etc.) export channels were formed [5].

In conditions of growth in extraction and export of raw materials, issues related to the reproduction of the raw material base, efficient processing and the skilled use of raw materials (crude oil, associated petroleum gas, high-condensate natural gas) were unsatisfactorily solved, which points to the resource-inertial trajectory of development in the raw materials sectors themselves.

Formation of resource-inertial development. The increase in extraction and export of raw materials against the backdrop of high world prices for hydrocarbons, which is the basis of resource and raw materials development, has led to several major multidirectional trends:

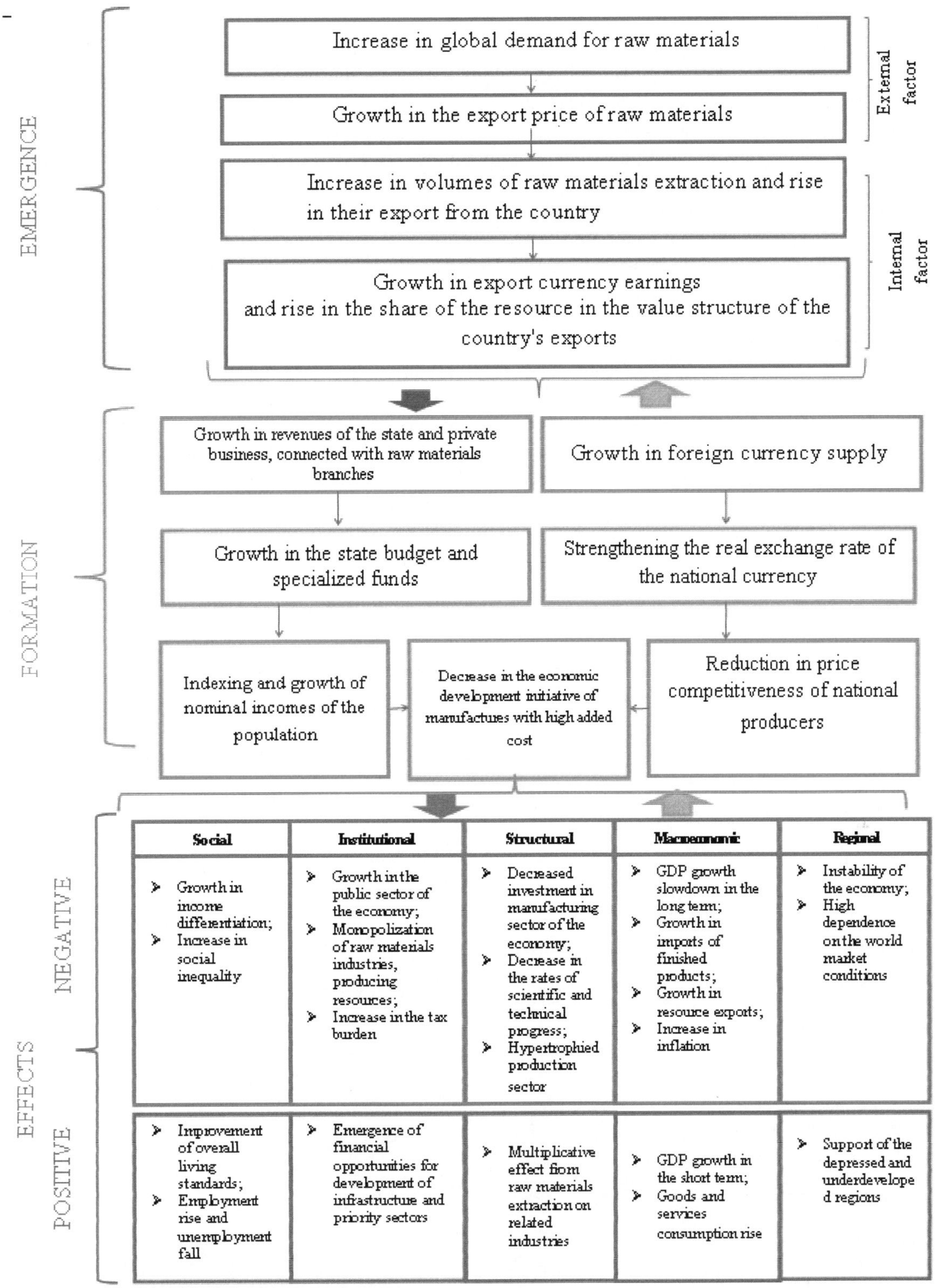

Figure 1. Scheme of the stages of emergence, formation and effects of resource-inertial development under favorable conditions of international energy markets.

- growth in state revenues, replenishment of the federal budget and specialized funds, which potentially creates favorable conditions for carrying out a set of measures that have a positive impact on economic development (table 2);
- an increase in the supply of foreign currency on the domestic market led to a strengthening of the real exchange rate of the national currency. As a consequence, in conditions of an overvalued national currency, domestic goods “lost” in price competition to foreign counterparts, both on domestic and international markets; this led to a number of negative effects destroying the country's economic potential;
- rapid growth in revenues from raw materials exports has another effect; there is a “sense of well-being”, leading to a reduction in the economic initiative to differentiate the structure of the economy, and to develop processing industries characterized by the release of products with high added value.

Effects of resource-inertial development. The resource and raw materials development with a favorable conjuncture in the world energy markets and the availability of opportunities for a significant increase in the volume of raw materials for export leads to an increase in state and business incomes related to extraction and export, as well as related industries, consumer and service sectors. This undoubtedly gives a powerful positive effect in the area of raising the living standards of the population, increasing employment and reducing unemployment, which causes an increase in consumer demand, the volume of the domestic market for goods and services. In the short term, this leads to an increase in GDP and industrial production and contributes to the accumulation of funds and reserves that can potentially be used to carry out structural reforms in the economy, supporting priority economic and social programs [6, 7, 8, 9].

Table 2. Indicators characterizing the role of hydrocarbon resources in the Russian economy, 2000-2016.

Indicator	2000	2005	2008	2011	2012	2013	2014	2015	2016
The share of petroleum industry in the gross added value of industries (constant prices)	no data	no data	no data	32	31	31	31	31	31
The share of petroleum industry in the gross added value of industries (cost related prices)	no data	34	31	37	38	39	37	34	33
Share of oil and gas revenues in the federal budget,%	no data	42	47	50	50	50	51	44	36
Share of hydrocarbons in the total value structure of exports,%	50	62	64	69	67	67	66	59	53
Share of investments in petroleum industry in total investment,%	no data	13	13	14	13	14	15	18	20
The volume of the Stabilization, Reserve and National Welfare Fund, billion rubles.	no data	1237	6612	3606	4576	5760	9334	8868	5331
The volume of international reserves, billion dollars.	35	125	479	479	499	532	510	368	378

*Source: Rosstat, Treasury of the Russian Federation

3.2. Features of resource and raw materials development of the Russian economy at the present stage
The fall in oil prices in 2014 and up to the present time has led to a significant deterioration in the main macroeconomic indicators in the country. It is worth noting the change in income coming from the oil and gas complex during the crises from 2015 to 2016. Over that period the share of oil and gas revenues in the structure of the federal budget fell by 30 % to 36 %, thereby causing a significant deficit. At the same time, the exchange rate and the increase in the rate of oil and gas taxes provided a significant compensating effect on budget revenues in rubles.

The main problematic area of the Russian economy is a tremendously low volume of investments. If the share of investment in the Chinese economy is about 50 %, then half of the total surplus produced is returned to the economy in the form of investment. In Russia in recent years this indicator has varied in the range of 19-24 % in the GDP structure. The lack of modernization of the existing industries, not to mention the large-scale extended reproduction, leads to the stagnation of the mono economy.

4. Conclusions
At present, the main problem area of the Russian economy is a tremendously low volume of investments. The lack of modernization of existing industries, not to mention the large-scale extended reproduction, leads to the stagnation of the resource mono economy. Structural reforms in the economy should concern, first of all, the investment sphere. Modernization of economic sectors should be carried out first of all in raw materials and related industries. It is necessary to change the paradigm of subsoil development and implement the transition from an extensive development path to an intensive one to ensure sustainable economic development on the basis of increasing technological development while building production and technological chains with the production of high added value products. In this connection, it is necessary to identify the opportunities for the formation of extensional specialization in resource-type regions, to develop the basic principles when choosing the innovative technological extensional specialization of resource regions.

Acknowledgements
The authors are grateful for financial support to the Russian Science Foundation No. 16-18-10182, as well as to the Russian Foundation for Basic Research No. 15-06-090994, 16-36-00369, 17-06-00537.

References
[1] Kontorovich A E, Eder L V and Nemov V Yu 2013 *Oil Industry* **1** 4–8
[2] Kontorovich A E, Eder L V et al 2016 *Russian Geology and Geophysics* **57(12)** 1653–67
[3] Korzhubaev A G, Filimonova I V, Eder L V and Sokolova I A 2009 *Neftyanoe Khozyaystvo (Oil Industry)* **3** 14–7
[4] Kontorovich A E, Korzhubaev A G, Filimonova I V and Eder L V 2008 *Neftyanoe Khozyaystvo (Oil Industry)* **5** 24–7
[5] Kontorovich A E, Eder L V and Nemov V Yu 2012 *Neftyanoe Khozyaystvo (Oil Industry)* **7** 66–70
[6] Tagaeva T O, Gilmundinov V M and Kazantseva L K 2017 *Economy of Region* **14(7)** 165–77
[7] Tagaeva T O 2011 *Source of the Document Studies on Russian Economic Development* **22(3)** 331–8
[8] Sharf I, Filjushin V, Shenderova I and Kochetkova O 2015 *IOP Conf. Series: Earth and Environmental Science* **27**
[9] Kryukov V A, Bozo N V and Malysheva Y V 2005 *9th Russian-Korean Int. Symp. on Sci. and Tech., KORUS-2005* 898 –903
[10] Nikitenko S M and Goosen E V 2017 *IOP Conf. Series: Earth and Environmental Science* 53012018
[11] Nikitenko S M, Goosen E V and Sablin K S 2016 *IOP Conf. Series: Earth and Environmental Science* **45** 012001

IOP Conf. Series: Earth and Environmental Science **84** (2017) 012013 doi:10.1088/1755-1315/84/1/012013

Socio-economic development of territories based on the principles of public-private partnership in the sphere of comprehensive mineral exploration

S M Nikitenko and E V Goosen

Federal Research Center of Coal and Coal Chemistry, Siberian Branch of the Russian Academy of Sciences, 18 Sovetskiy Ave., Kemerovo, 650000, Russia

E-mail: nsm.nis@mail.ru

Abstract. The article explores the possibility of using instruments of public-private partnership for a paradigm shift in subsoil use in the fuel and energy complex of Russia. The modern Russian fuel and energy complex (FEC) is characterized by high depreciation of production assets, technological inferiority compared to the developed countries, etc. The solution to all these problems seems to be closely connected with the transition from extensive use of natural resources to comprehensive mineral exploration (CME), with a stable socio-economic development of territories and mutually beneficial partnership between science, business and government based on the principles of public-private partnership (PPP). The article discussed the three main directions of PPP projects development in subsoil use. The first direction comprises the projects aimed at the establishment of core mineral resource businesses on the basis of concession agreements and production sharing contracts. The second direction concerns the projects focused on the development of territories and objects of industrial and social infrastructure in resource regions. The third direction is formed by the projects aimed at the development of new industries, focused on the creation of centers of innovative development, formation of markets for innovative products and innovative clusters in the energy industry.

1. Introduction

The modern Russian fuel and energy complex (FEC) is characterized by high depreciation of production assets, technological inferiority compared to the developed countries, high level of dependence on import of equipment, materials and services, excessive dependence on unstable foreign energy markets. The solution to all these problems seems to be closely connected with the transition from extensive use of natural resources to comprehensive mineral exploration (CME), with a stable socio-economic development of territories and mutually beneficial partnership between science, business and government based on the principles of public-private partnership (PPP) [1].

2. Methods of research

The research methodology consisted in the methods of analysis, comparison, generalization and expert assessment.

3. Results and discussion

IOP Conf. Series: Earth and Environmental Science **84** (2017) 012013 doi:10.1088/1755-1315/84/1/012013

Within the FEC, PPP can focus on such objects as deposits of minerals, industrial (energy, railway transport, highway facilities, ports, airports, utility systems, etc.) and social (housing, public buildings, education and health institutions, etc.) infrastructure in the territories where active mining and processing of minerals take place [2, 3].

The market for PPP projects in Russia began to formally develop after the adoption of the Federal Law of 21.07.2005 No 115-FZ "On concession agreements". Despite the fact that most of the Russian projects are concession agreements in the infrastructure and social sector, there still is some positive experience of PPP projects implementation and in industries involved in mining, processing and transportation of mineral resources. The share of such projects is not large. They are implemented through major programs of development of branches and regions, and financed through the Investment Fund of the Russian Federation, and are not officially considered PPP projects.

Two official Russian databases of PPP projects (the PPP Information portal and the official website of the Ministry of Economic Development where the projects cofunded by Investment Fund of the Russian Federation are registered) contain the information about only 12 PPP projects related to subsoil use, which makes less than 1% of all PPP projects registered on both sites. Only 6 projects can be directly attributed to the fuel and energy complex, namely: Development of project documentation for the implementation of the investment project "Comprehensive development of South Yakutia" (coal mining, enrichment, energy), Development of the transport infrastructure for the exploration of mineral resources in the south-east of the Transbaikal Kray (enrichment, transportation), Complex of refineries and petrochemical plants in the city of Nizhnekamsk (oil refining), Complex development of the Lower Angara region (transport, power industry, nonferrous metallurgy), Development of a complex for processing of Northern Caspian gas into ethylene, polyethylene and polypropylene (I phase) (gas processing, energy and transport infrastructure), Construction of the Kyzyl – Kuragino railway line in the context of the exploration of the mineral resources of the Tyva Republic (transport) [4, 5, 6, 7].

The small number of projects in the FEC sphere in Russia is largely due to legislative restrictions. Article 3 Part 1 of the Federal Law of 13.07.2015 No 224-FZ "On public-private partnership, municipal-private partnership in the Russian Federation and amendments to certain legislative acts of the Russian Federation" prohibits the establishment of PPP agreements in subsoil use. The licensing nature of subsoil use also restricts the establishment of PPP projects in fuel and energy complex in the framework of the Law on Concessions.

Another legislative barrier to PPP development in the energy sector is the definition of the private partner in PPP projects given in Article 3 Part 3of the Law on PPP, under which state and municipal unitary enterprises, state-owned companies and non-profit organizations and subsidiaries of state companies can not be private partners in PPP projects. This restriction automatically makes it impossible for the most part of major oil and gas companies, where a significant share is owned by the state, to participate in PPP projects.

The organizational structure of the companies constituting the FEC is another serious constraint to PPP development in that sphere. Most regional energy companies are not independent, as they make part of large vertically integrated groups that are registered and have profit centres in Moscow and Saint-Petersburg; thus the regional companies are limited in their ability to decide the issues about the volumes and directions of investments on their own.

Despite the problems listed above, PPP projects in the sphere of subsoil use and the FEC can be created in Russia in three main areas.

The first direction comprises the projects aimed at the establishment of core mineral resource businesses on the basis of concession agreements and production sharing contracts. Examples of this kind of PPP in Russia are the projects of the exploration of Elga coal deposit (Republic of Sakha, Neryungri district), oil and gas fields in Evenkia (Yurubcheno-Tokhomskoe field, Kuyumbinskoe field, Nizhneangarsk group, Sobinsk-Teterinsky group). The examples of PPP projects in the energy industry with participation of foreign capital are (with certain reservations) the projects of hydrocarbons exploration in the form of the sharing agreement: the Sakhalin-2 project (the Piltun-

IOP Conf. Series: Earth and Environmental Science **84** (2017) 012013 doi:10.1088/1755-1315/84/1/012013

Astokhskoye oil and gas field and the Lunskoye gas-condensate field); the Sakhalin-1 project (Chaivinskoye, Arkutun-Daginskoye and Odoptinskoye oil and gas fields); 'Kharyaga' project; Samotlor oil and gas condensate field development project [8].

It is important to note that such PPP projects do not make serious changes to the paradigm of subsoil use and retain a predominantly intensive character of FEC branches development, fixing the dependence of 'host regions' on resource extraction and limiting their development.

The second promising direction of PPP development in the energy industry are projects aimed at the development of territories and creation of the objects of industrial and social infrastructure in resource regions. That is where the terms of subsoil use in Russia are formed. The examples of kind of projects are the large investment and infrastructure projects with the participation of business and government, focused on integrated development of territories, creation of industrial production oriented at consumption and processing of energy resources and the development of transport and energy infrastructure. Those are the quasi-PPP projects in the sphere of comprehensive mineral exploration, namely the integrated mega-projects "Ural Industrial – Ural Polar", "Complex development of the Lower Angara region", "Comprehensive development of South Yakutia", and others [5].

The third direction is formed by the projects aimed at the development of new industries, focused on the creation of centers of innovative development, formation of markets for innovative products and innovative clusters in the energy industry. Only those projects are directly aimed at creating conditions for a paradigm shift of subsoil use. Such projects are very few all over the world. According to the IPP Journal, in 2016, 68 innovative PPP projects using the PPP mechanisms in the energy industry were implemented all over the world, of which were registered 15 in the USA, 4 in the UK, and 3 in Australia[8]. Most of the projects are undertaken in the energy sector (the construction of modern power plants), while in the oil and coal chemistry they are much fewer.

The examples of the attempts at implementing the third group of the PPP projects in the FEC in Russia to serve are three technological platforms for the extraction of natural resources and oil and gas processing, approved by the Government Commission on high technologies and innovations in 2011: the 22nd technology platform for solid minerals focused on deep processing and upgrading of Russian processing enterprises with high technologies; the 23rd technology platform for the extraction and use of hydrocarbons, the result of which should be the development and implementation of new technologies for the extraction, preparation, refining and transportation, drilling, etc.; the 24th platform for deep processing of hydrocarbon resources, involving the creation of conditions for technological modernization and increase of competitiveness of oil and gas processing, petrochemical industry and organic synthesis using the foresight procedures. The main aim of those technology platforms is the selection of ideas in basic research and using them as the basis for exploratory research, scientific research and the development of innovative business concepts [9, 10].

The main difference between PPP projects of the third group is that the recipients of investment are not the producing and processing companies but the companies operating in the FEC, forming the internal and external selling markets, developing the resource, production and scientific bases and the human capital. This limits the ability of the large vertically integrated energy companies to accumulate all the resources and leads to a synergistic effect: a larger number of competing highly- specialized small and medium-sized innovative companies are created around the basic PPP project, and the conditions for the development of scientific and social infrastructure are formed. Ultimately, this creates the conditions for the transition to sustainable model of subsoil use by the FEC companies, changing the vector of resource territories development and breaking the vicious circle of the "resource curse" [11, 12].

4. Conclusions

Russia has enormous energy resources which are affordable, competitive and highly demanded. Of all the existing forms of PPP projects implementation, the projects focusing on the innovative model of comprehensive mineral exploitation, are the most promising. They are able to fundamentally solve the problems of the FEC and the socio-economic development of territories. The attempt to integrate the government and business by means of the science can be a successful experience and a platform for further development of PPP projects in the third direction.

Acknowledgements
This article was prepared with the financial support of Russian Science Foundation (Agreement No. 16-18-10182, the project "Formation of organizational and economic mechanisms for comprehensive exploration of mineral resources in resource type regions on the basis of partnership between science, business and government").

References

[1] *State Report "On the state and use of the Russian Federation's mineral resources in 2014"* http://www.mnr.gov.ru
[2] Nizhegorodtsev R M, Nikitenko S M and Goosen E V 2012 *Public-private partnership in the sphere of innovations: the global experience and the prospects for Russia* (Kemerovo: Trapeznikov Institute of Control Sciences of the Russian Academy of Sciences) p 482
[3] Kontorovich A E et al 2015 *State and Prospects of Public-Private Partnership Projects Development in the Context of Comprehensive Exploration of Mineral Resources* (Moscow-Kemerovo) p 331
[4] *Reference Materials* of http://www.pppi.ru on the Projects Implemented with the Support of the Investment Fund of the Russian Federation 2015 (Moscow) http://economygov.ru
[5] *Ural Industrial – Ural Polar* 2009 Passport of the Program http://uploadrb.ru
[6] *Passport of the Investment Project "Development of the Transport Infrastructure for the Exploration of Mineral Resources in the Routh-east of the Transbaikal Kray"* http://passport.zab-investportal.ru
[7] *Project "Construction of the Kyzyl – Kuragino Railway Line in the Context of the Exploration of the Mineral Resources of the Tyva Republic* http://www.mert.tuva.ru
[8] *Public-Private Partnerships Reference Guide Version 2.0 2016 (Washington)* URL: http://documents.worldbank.org
[9] Polianskaia I G 2009 *Business, Management and Law* 1(18) 46 http://documents.worldbank.org/curated/en/600511468336720455/pdf/903840PPP0Refe0Box385311B000PUBLIC0.pdf 50
[10] Nikitenko S M, Goosen E V and Sablin K S 2016 *IOP Conference Series: Earth and Environmental Science* **45**
[11] Kontorovich A E, Korzhubaev A G, Filimonova I V and Eder LV 2008 *Neftyanoe Khozyaystvo (Oil Industry)* **5** 24–7
[12] Korzhubaev A G, Filimonova I V, Eder L V and Sokolova I A 2009 *Neftyanoe Khozyaystvo (Oil Industry)* **3** 14–7

IOP Conf. Series: Earth and Environmental Science **84** (2017) 012014 doi:10.1088/1755-1315/84/1/012014

New forms of public-private partnership for sustainable development of the fuel and energy sector

E O Pakhomova, E V Goosen and S M Nikitenko

Federal Research Center of Coal and Coal Chemistry, Siberian Branch of the Russian Academy of Sciences, 18 Sovetskiy Ave., Kemerovo, 650000, Russia

E-mail: nsm.nis@mail.ru

Abstract. Public-private partnership (PPP) as a form of interaction between the public and private sectors contributes to the improvement of social and economic situation in the country. PPP is viewed in a broad context and is presented as a system of financial and non-financial relations between government and business. A number of legislative initiatives contributed to the emergence of a new form of PPP that is a special investment contract (SPIC). In the Russian regions there are legislative initiatives that provide differences in terms of contracting. In a number of regions a federal-specific volume of mandatory investment has been reduced to attract investors, while there are regions where this volume is higher than at the federal level. Monitoring of regional legislation showed that the authorities of the constituent entities of the Russian Federation are striving to obtain guarantee obligations to create jobs and receive salaries above the average in the region. Measures to encourage investors have been determined, and the benefits to the government and business from the implementation of a SPIC have been shown.

1. Introduction

On the agenda of the Russian economy there is the task of finding new development instrument s and new management solutions. The formation of a favorable investment climate is now becoming one of the key tasks of the government policy of the Russian Federation, the regions of which have different development potentials and need a differentiated approach to the choice of a development instrument. Such an instrument can be various forms of cooperation between government and business. However, their areas of interaction in key resource-rich sectors of the economy have been limited by legislative changes and need to find new solutions.

2. Methods of research

Methods of analysis, comparison, generalization, and expert evaluation have been used as a methodological basis.

3. Results and descussion

After amending certain legislative acts of the Russian Federation and adopting amendments to the Law on Concession Agreements, PPP projects in the field of subsoil use were excluded from the list of projects as not in accordance with the law. From 1285 PPP projects in 2015 only 5 projects could be attributed to the sphere of subsoil use [1]. Part 1 of Article 3 of the Law on PPP expressly prohibits the conclusion of PPP agreements for subsoil use [2]. The licensing nature of subsoil use limits the ability to create PPP projects in the fuel and energy complex (FEC) and under the Law on Concessions. Another legislative barrier for the development of PPP in the FEC is defining a private partner in PPP projects given in part 3 of Article3 of the PPP Law, according to which state and municipal unitary enterprises, state companies, as well as non-profit organizations and subsidiaries of state companies cannot be private partners in PPP projects. This restriction automatically makes it impossible for most large oil-and-gas and coal companies to participate in PPP projects in which a significant share of the shares belongs to the state. Whereas, some elements of PPP can be used, including those for the development of the fuel and energy complex. The latter include such forms of cooperation between government and business, in which the state's participation is indirect, non-financial. With this approach, the concept of PPP is significantly expanded and becomes a real institution of economic development [3, 4].

PPP is formed as a system of relationships of business entities and authorities, so the idea of public-private partnership is much broader and involves various forms of their interaction [5]. The realization of cooperation can be involved through financial and non-financial mechanisms. Financial mechanisms are represented by the state's financial participation (direct investments, subsidies, etc.), not financial mechanisms are benefits, guarantees, indulgences for business.

Recently, in view of the limited budget possibilities, the most interesting are non-financial mechanisms for implementing forms of partnership between government and business. These include SPIC, rent, concession and production sharing agreement. In Russia, from the non-financial mechanisms, the most interesting is a SPIC, since it is based on equitable terms for all investors (table 1). The rules for concluding a special investment contract with the participation of the Russian Federation are established by the Decree of the Government of the Russian Federation of July 16, 2015 No. 708 [9].

Table 1. The main incentive measures for investors within the framework of SPIC [6].

Directions of incentives / level of benefits	Types of benefits	Objective
Simplified access to government orders / federal	Possibility to be the only supplier of products produced under a SPIC	Creation and promotion of the market for products (especially important for the innovative products implementation
Tax benefits / federal and regional	Non-deterioration of tax conditions at the time of contract conclusion; the possibility of reducing the federal and regional part of the income tax rate to 0%; accelerated depreciation for fixed assets	Creation of stable conditions for the investment environment (especially important for the innovative technologies introduction
Russian product / federal	Accelerated and simplified procedure for obtaining the status of a Russian manufacturer	Production localization and foreign investors attraction
Other measures / sectoral and regional	Other benefits, such as the provision of land owned by state and municipalities without a tender.	Production localization, use of regional reserves.

The implementation of a SPIC is beneficial not only to business; the state as a result of the agreement also receives a number of benefits. Both of them are basically related to the controlled process of active and passive industrial policy (table 2).

Table 2. Benefits for government and business from the implementation of a SPIC [6].

For government	For business
Growth of industrial products, which have no analogues in Russia; Development of industries of priority importance; Implementation of the best available technologies; Creation and development of high-tech products manufacture; Localization of production in the required volume in Russia; Creation of industrial infrastructure.	Privileges on taxes, fees and benefits for customs payments; Privileges on rent payment for use of state and municipal property, including land plots; Other benefits and preferences established by law; Guarantees against adverse changes in the legislation in force.

The SPIC has received its legislative form in the regions. As of May 1, 2017, out of 85 constituent entities of the Russian Federation 68 have a separate normative act defining the rules and procedure for concluding a SPIC on behalf of the entity, which formed regional differences.

The volume of investments in the amount of less than 750 million rubles is presented in 36 regions of the Russian Federation. The lowest volume is in the Republic of Khakassia with 10 million rubles. In 5 subjects of the Russian Federation (Altai Territory, Sevastopol, Bryansk Region, Ivanovo Region, Ryazan Region) it makes up 50 million rubles. The increased volume of investments is set in the Moscow region, where the amount of investment is 1 billion rubles. The largest amount of investment is set at 1 billion 500 million rubles in the Kaliningrad region for 3 years. In seven regions of the Russian Federation (the Altai Territory, the Komi Republic, the Udmurt Republic, the Republic of Chuvashia, Perm Krai, Orenburg Region, Saratov Region), there are the requirements for the number of jobs created. The minimum number of jobs is in Altai Republic and the Orenburg region (50 jobs); the maximum number of workplaces retained until the end of the project is in the Perm Territory and Orel region (250 jobs). Thus, the mechanism for attracting investments becomes an instrument for localizing not only production, but also human capital.

By 2017, 8 special investment contracts had been concluded in Russia and another 10 are under consideration (table 3).

Table 3. List of special investment contracts of federal level (end of 2016) [6].

No.	Name of an industrial investor	Subject of the contract
	Signed and approved SPICs	
1	DMG MORI AG Concern, Ulyanovsk region	Construction of the Ulyanovsk machine-tool plant
2	HMS Livgidromash, JS , Oryol region	Localization of pumps production for oil refining
3	Tomsk Technologies of Mechanical Engineering LLC, Tomsk Region	Mastering the serial production of domestic anti-surge and control valves, electric drives and electrical appliances
4	MAZDA-SOLLERS Manufakchuring Rus LLC, Primorsky Krai	Constriction of a plant producing engines intended for motor vehicles export
5	CLAAS LLC, Krasnodar Kri	Modernization of CLAAS agricultural machinery plant in Krasnodar
6	EuroChem-UKK LLC, Perm Kri	Establishment of the Usolsky potash plant, an industrial production with 2.3 million tons of 95 % KCL from Gremyachinskoye field per year
7	EuroChem-VolgaKaliy,	Mining and enrichment plant for extraction and enrichment

	Volgograd region	of potassium salts with capacity
8	Mercedes-Benz RUS JSC, Moscow region	Creation and development of industrial production of Mercedes-Benz wheeled vehicles
	SPICs under consideration	
1	Transneft-Siberia JSC, Tyumen Region	Creation and development of equipment industrial production for operation, maintenance and repair of main oil pipelines objects
2	Biocad Closed Joint-Stock Company, St. Petersburg	Organization of biological substances and finished dosage forms production
3	GEROPHARM LLC, St. Petersburg	Organization and development of insulin and insulin analogues, organic substances (active pharmaceutical substances and pharmaceuticals) which have no analogues
4	Hyundai Motor Manufacturing Rus LLC, St. Petersburg	Building a multi-model Hyundai factory in St. Petersburg
5	Golitsyn Bus Plant Holding Company JSC, Moscow region	Establishing the production of high-tech agricultural machinery
6	Novo Nordisk Production Support LLC, Kaluga Region	Design, construction, equipping and commissioning of a new workshop for the assembly of pre-filled FlexPen® syringes
7	RED JSC, Chelyabinsk region	Setting up an industrial production and development of high-voltage electric motors
8	CNH Industrial Russia LLC, Republic of Tatarstan	Modernization and production development of New Holland agricultural machinery
9	OKTAFARMA– FARMIMEX LLC, Moscow	Construction of a full cycle plant for the production of biopharmaceuticals and substances from human plasma
10	Hamilton Standard-Science Joint-Stock Company, Moscow	Production of heat exchangers for air conditioning systems

4. Conclusions

The implementation of this instrument in Russia requires a comprehensive and clear study of all the aspects relating to real operation and achievement of the set goals, including [10]:

- developing a mechanism of interest and attracting investors to the industry ready to assume and fulfill social and economic obligations;
- developing a mechanism for detailed identification of investors involved in the development of a certain type of industry;
- developing regional normative legal acts regulating the conditions for the conclusion of SPICs and the standard forms of a SPIC for each individual industry in such a way that ultimately the results achieved can be compared with the stated objectives (possibly through a regulatory impact assessment or an impact assessment).

Nevertheless, the potential of a SPIC for investors is great, namely it is a Russian product, the only supplier, tax incentives; for government is an instrument for attracting investments, localizing the investor on its territory with a 10-year guarantee, jobs and no necessity of budget investment, as well as creating an additional product in the economy with the subsequent development of service industries.

SPIC as a development tool has comparative advantages and can become a full-fledged form of cooperation between government and business in the fuel and energy sector.

Acknowledgements
The article has been written with the financial support of the Russian Science Foundation (RSF) under the agreement No. 16-18-10182.

References

[1] Kontorovich A E, Eder L V and Nemov V Yu 2013 *Neftyanoe khozyaystvo (Oil Industry)* **1** 4–8
[2] Federal Law No. 224-FZ of 13.07.2015 *"On Public-Private Partnership, Municipal-Private Partnership in the Russian Federation and Amendments to Certain Legislative Acts of the Russian Federation"*
[3] Nikitenko S M, Goosen E V and Sablin K S 2016 *IOP Conference Series: Earth and Environmental Science* **45** 012001
[4] Korzhubaev A G, Filimonova I V, Eder L V and Sokolova I 2009 *Neftyanoe Khozyaystvo (Oil Industry)* **3** 14–7
[5] Belitskaya A V 2012 *Legal Regulation of Public-Private Partnership* (Moscow: Statute) p 191
[6] *Official Website of the Industrial Development Fund* http://frprf.ru/
[7] Federal Law from 05.04.2013 N 44-FZ "On the Contract System in the Procurement of Goods, Works, Services for the Provision of State and Municipal Needs"
[8] Federal Law No. 488-FZ of 31.12.2014 "On Industrial Policy in the Russian Federation"
[9] Resolution of the Government of the Russian Federation from 16.07.2015 No. 708 "On Special Investment Contracts for Certain Industries" (together with the "Rules of the Conclusion of Special Investment Contracts")
[10] Kontorovich A E, Korzhubaev A G, Filimonova I V and Eder L V 2008 *Neftyanoe khozyaystvo (Oil Industry)* **5** 24–7

IOP Conf. Series: Earth and Environmental Science **84** (2017) 012015 doi:10.1088/1755-1315/84/1/012015

Intellectual property as an instrument of interaction between government, business, science and society

S M Nikitenko[1], M A Mesyats[1] and O V Rozhkova[2]

[1]Kemerovo Institute (branch) of Plekhanov Russian University of Economics, 39 Kuznetsky Avenue, Kemerovo, 650992, Russia
[2]Dubna State University, 19 Universitetskaya Street, Dubna, 141980, Russia

E-mail: nsm.nis@mail.ru

Abstract. This article is devoted to research the characteristics associated with pledge of intellectual property in foreign and domestic practice. Holding intellectual property objects' pledge transactions accelerates the pace of creating innovative systems in the economy. In present paper the modern scheme for bank loan, financing secured with patented intellectual property is researched. The authors give the brief description of features of pledge security registration for loans in some Europe countries. The Europe Union experience shows that as collateral for monetary loans can be used trademarks, patents on the intellectual property, as well as their registration requests. Russian experience of the pledge operations of the intellectual property is too small. This way of bank lending is at an early stage of development. The main constraint is the difficulty of assessing the value of the pledged intellectual property as intangible assets. However, taking into account world and domestic practice this direction for Russian market is estimated by the authors as promising one. Pledge transactions take place within the framework of the Quadruple-Helix Model concept that involves four participants: "science", "business", "government" and "society". Intellectual property are estimates by the authors as an instrument of interaction between government, business, science and society.

1. Introduction

The main challenge for the world's economies today is the process of building national and regional innovation systems. The methodological basis for this is the Triple Helix Model, describing the directions and forms of cooperation between science, businesses and government in the innovation sphere, and its "evolution" sequel – the Quadruple-Helix Model, which includes the civil society along with the three listed participants.

Since then the conditions have changed. New organizational and economic diffusion (transfer) tools of intellectual activity as the basis for innovative development. Today, technology transfer takes into account the "host regions", their resource security, economic, environmental and social component. The synergistic effect plays an important role due to innovative technologies complementarity.

Practice has shown that harmonization on the basis of pledge mechanisms of intellectual property management processes can significantly speed up the innovation implementation process and serve the equalization of socio-economic development levels of individual subjects and their economy promotion in an innovative way. This is a new phenomenon in Russia. Therefore, the experience of European countries in the intellectual property management, including intellectual property rights as a pledged asset, is so important for Russia.

In recent years an institutional leap towards the formation of institutions of investment in intellectual property (IP) has been taken in Russia. As a result, in 2016 about 20 leading Russian banks concluded 81 pledge agreements over the exclusive rights to trademarks, inventions, utility models and industrial designs. The role of regional funds to support small and medium-sized innovative companies providing on behalf of the government pledges to commercial banks has been growing.

2. Methods of research

In this research general scientific methods of systematization and analysis of the practice of lending secured rights on intellectual property in foreign countries were used, which makes it possible to assess the possibilities of using this experience in Russia.

3. Results and discussion

The main purpose of commercial use of intellectual property objects is to generate profit. The objects of transactions are not the objects themselves, but the property rights to these objects, including patents and patent applications. They are the assets that are used as pledge for loans or other loan agreements [1].

Such assets are especially valuable for businesses engaged in research and results commercializing. In this situation, the cost of a patent portfolio can many times exceed the cost of other assets of a company and act as the most valuable asset. To date, in developed countries investment in IP accounts for more than 15% of GDP and almost half of all investments is made in fixed capital. The share of investment in IP in the largest international corporations is rapidly growing [2]. So, if in the early 80's of the 20th century the share of intangible assets in the structure of the cost of American companies accounted for no more than 40 %, then at the beginning of the 21st century their share exceeded 70 %. At the same time, 30-40 % of non-material property was not identified and reflected in corporate balances. This allows to say that the share of intangible assets can actually be even higher. A similar picture is typical for West European corporations [3]. According to Interbrand consulting company, tangible and intangible assets of the world well-known companies are correlated as follows: British Petroleum – 30:70, IBM – 17:83, Coca-Cola – 4:96. For comparison, in Russia, the volume of investment in intellectual property objects has not exceeded 1 % so far [4].

It is important to note that pledge transactions involving the use of IP's rights along with benefits carry potential high risks for the parties involved. To reduce them, countries are developing special legislation regulating transactions with IPO rights; companies and banks use such protection mechanisms as a preliminary assessment of IP's market value, determining the best way to calculate depreciation charges and depreciation periods, revaluing the cost, using IP as a pledge in combination with other assets of the enterprise, preliminary expert analysis of the economic efficiency of the IP use, etc.

In addition, many countries provide for the possibility of registering transactions with industrial property in the public registry, which guarantees the rights of parties involved in the transaction. The French Intellectual Property Code defines the rules regarding the pledge of the right to use the software. Article L 132-34 of this Code provides for mandatory registration of a pledge in the special register of the National Institute of Industrial Property (INPI) [5, 6, 7, 8, 9]. In Austria and the United States, the exclusive rights are not the subject of the pledge, but the patent itself as movable property is. A pledge transaction must be registered with the Patent Office [10].

In the case of a pledge over formal intellectual property rights the EU countries take into account the following points:

- public registers operate a priority system granting to the first applicant a priority pledge right over Formal IP Rights;
- formalities for creating the pledge over the Formal IP Rights are regulated by the law of the country in which the pledge right needs to take effect;
- the Formal IP Rights remain vested in the company-borrower, which will continue to be responsible for the maintenance of the Formal IP Rights.

- formalities for creating a pledge over the Formal IP Rights are comparable to the ones that are traditionally provided for the immovable assets (real properties) [7].

In accordance with the pledge practice of Great Britain, the pledger will need to consider and estimate carefully the consequences of entering into the pledge agreement and scrutinise its terms. The pledgee will have the right to sell the patents in the event of a default by the pledger. If the pledgee has taken pledge over not just the patents but the other business assets, the pledgee may have other options too, such as the power to appoint an administrator with authority to run the business of the defaulting pledger. The pledgee will wish to ensure that the value of the pledged patents is maintained; in the agreement, it may seek to impose obligations on the pledger to comply with such a guarantee. The pledger would be well advised to seek to reserve the right to allow to lapse, or abandon, patents or patent applications it reasonably considers are no longer of value [9].

In most cases the pledger will want the freedom to continue to exploit the patent portfolio as part of its business. In this regard, it must make sure any pledge document does not its freedom to do this in any material way. It should be emphasized, that there is normally a community of interest between the pledger and pledgee in this context because the pledgee wants the business to generate revenue to repay the debt. The pledger has to negotiate the terms with the pledgee at the outset so as to make clear that any future licensee's interest is not at risk of being defeated by the pledgee exercising its right to sell the patents in the event of default on the loan. In such cases the pledger could offer as a compensation to the pledgee an access to the revenue generated by the licence either by agreeing to pay it directly to the pledgee or by giving the pledgee a pledge over the revenue stream itself [9].

Under the European Patent (EP) regime, EPs may be granted as pledge (see Article 71 of the European Patent Convention) and the relevant pledge needs to be filed with the EPO. National laws apply in relation to the effect and validity of the pledges, including formalities in order to oppose the right to third party at the local level (see Articles 74 of the European Patent Convention) [7].

In Russia, the experience of IP pledge transactions is not big. This version of bank lending in the country began to be used only in 2009 and is at the initial stage of development (table 1). The conclusion of any contracts (including pledge agreements) related to the disposal of an exclusive right to patented objects of intellectual property and trademarks in Russia are the subject to registration in Rospatent [11].

Table 1. Number of IPO pledge agreements registered by Rospatent in 2009-2016.

	2009	2010	2011	2012	2013	2014	2015	2016
Pledge agreements over exclusive rights to the results of intellectual activity	-	8	16	17	20	15	13	9
Pledge agreements over exclusive rights to trademarks	16	62	52	24	60	40	58	72
Number of trademarks referring to which the pledge agreements are concluded	n/d	195	280	82	191	258	362	689
Total number of pledge agreements over exclusive rights to IPOs	16	70	68	41	80	55	71	81

*Source: compiled by the authors on the basis of [11]

The main constraint on the issue of extending the practice of bank lending against IP's pledge is the difficulty in assessing the value of pledged assets as intangible assets [12, 13]. From the authors' point of view, the existing Russian experience with the use of IP's as a pledge for obtaining credit and applying the experience of European countries today can serve as a solid foundation for activating lending in the banking services market; and the pledge over IP rights can be a promising tool for the development of companies actively engaged in the implementation of intellectual activity results.

Thus, the pledge over intellectual property rights is an opportunity to build a close and, most importantly, mutually beneficial interaction between science and business, which are the two participants in the Quadruple-Helix Model of building national and regional innovation systems. The third party is government. The role of government, when using the mechanism of pledging intellectual property rights as the starting point of innovative development of the economy is to ensure, including the legislative provision, the security of pledge transactions. Government that acts as a guarantor in an IP pledge transaction significantly reduces the risks of the pledgee, thereby increasing the attractiveness of such transactions.

Government influences pledge relations through economic and administrative methods which are based on the guarantee mechanism. The basis of legal regulation of pledge relations is the government guarantees of pledgee's rights which are called upon to provide an adequate protection in case of owners' interests violation while carrying out innovative activities. Wherein, government, as a rule, guarantees first of all the stability of rights, which is very important in long-term relations.

In order to define the meaning of "government guarantee mechanism of pledging intellectual property objects" concept, it must be borne in mind that these are a form of social guarantees. These mean material and legal means that ensure the realization of the social and economic rights of society members. Consequently, legal guarantees of investment are legal means ensuring the implementation of the rights of the parties.

If we analyze the state of Russia's notion of "government guarantees as the basis for ensuring a secure IP pledge", the following should be highlighted: IP is a specific object, with a complex estimation of its value and, therefore, very risky.

For full-fledged government involvement as a guarantor, when conducting IP pledge transactions between science and business, it is necessary:

- to expand the pledge law on the basis of the world's leading practices, international agreements on the forms and methods of protecting the rights of participants in IP pledge transactions;
- to define clearly, preferably legally, the ultimate goal of government actions in this area, which is ensuring the security of participants' property in IP pledge transactions and the actual realization of their rights as well.

After implementing the above measures, we will get a clear and transparent model of interaction between science, business and government (authorities) in creating innovative growth points based on the mechanism of IP rights pledging.

At the same time, in Quadruple-Helix Model of building national and regional innovation systems, there is another participant. It is a civil society. Government, science, business and society are not separate subsystems. The same participant can be an element of two subsystems. For example, enterprises that own a significant share of government participation should be classified as both government and business; commercial structures that are owned by public organizations can be attributed as intersection of society and business. The intersection of such spheres as society and government covers the specifics of the functioning of elective authorities, parties, and the electoral system.

If we consider innovation based on the example of an enterprise, it is worthwhile to say that modernization development depends on the involvement of employees in the work process, the use of experience and abilities of employees, as well as motivation and incentives. The more employees of the company are involved in the innovation process, and each of these employees is motivated to innovate, the better the result is [14]. Only with the participation of all the participants within the framework of the Quadruple-Helix Model concept the greatest economic effect can be achieved.

4. Conclusions

Thus, we can say that "society" is inseparable from the "three" which is science, business and government; in case of its isolation it is impossible to consider the interaction mechanisms of the "three" in any sphere, including the sphere of IP pledges. The role of society in developing innovative systems on the basis of IP pledge relations is enormous, because it is connected with all the other participants in the Quadruple helix concept. IP's are created by science for the needs of society; there

IOP Conf. Series: Earth and Environmental Science **84** (2017) 012015 doi:10.1088/1755-1315/84/1/012015

is a relationship between society and science. Business in assessing the cost of IP's will always take into account the attitude of society to the product or service created or provided using IP's; there is a relationship between society and business. Government (authorities) in the development of guarantee mechanisms legislation within the framework of IP pledge transactions will rely on representatives of the society (for example, deputies of various levels); there is a relationship between society and government (authorities).

Summarizing the above, we can conclude the following. Holding IP pledge transactions accelerates the pace of creating innovative systems in the economy. These transactions take place within the framework of the Quadruple-Helix Model concept that involves four participants. They are science which creates, business which invests, government which guarantees the observance of the participants' rights and society which determines the type of IP's and the rules of interaction between science, business and government. So, intellectual property can be used as an instrument of interaction between government, business, science and society.

Acknowledgements
This article was prepared with the financial support of Russian Foundation for Basic Research (No. 16-06-00054, the project "Instrumental and methodical approach to the adaptation of the triple helix model for Russia given the historical retrospective").

References

[1] Nikitenko S M, Goosen E V and Klishin V I 2013 *Innovations* **9(179)** 9–19
[2] Goosen E V 2016 *Fundamental research* **11-2** 362–6
[3] Khotinskaya A V *Intangible Assets as Marketing Resource of the Company* http://elibrary.ru
[4] *Anatomy of Growth Top Growing Brands* 2016 http://interbrand.com/
[5] *Code de la Propriété Intellectuelle* (Version Consolidée au 23 Février 2015) http://www.wipo.int
[6] *Dutch Civil Code, Book 3* http://www.dutchcivillaw.com
[7] *Esposizione L IP Rights and Loan Financing: European Perspective* http://techlaw.org
[8] Zobitz G E et al *Loans & Secured Financing* 2016 (London: Law Business Research Ltd) p 120
[9] Marshall J, Caldwell R and Cain B *Taking Security over Patents* https://united-kingdom.taylorwessing.com
[10] Ruzakova O *Exclusive Rights Pledge* http://superpressa.ru
[11] Report on the Rospatent Activities in 2016 http://www.rupto.ru/about/reports/2016/otchet_2016_ru.pdf
[12] Nikitenko S M *Fundamental Research* **11 (Part 1)** 196–9
[13] Nikitenko S M, Goosen E V and Sablin K S 2016 *IOP Conference Series: Earth and Environmental Science* **45** 012001
[14] Lugovaya E S 2012 *Bulletin of Volgograd University* **2(17)** 103–8

IOP Conf. Series: Earth and Environmental Science **84** (2017) 012016 doi:10.1088/1755-1315/84/1/012016

The problems of identification of resource-type regions

E S Kagan and E V Goosen

Kemerovo State University, 6 Krasnaya Street, Kemerovo, 650000, Russia
Federal Research Center of Coal and Coal Chemistry, Siberian Branch of the Russian Academy of Sciences, 18 Sovetskiy Ave., Kemerovo, 650000, Russia

E-mail: kagknes@mail.ru.

Abstract. The article considers approaches to the allocation of qualitative and quantitative criteria for the allocation of resource-type regions. An enclave character of the development trajectory of the Russian regions is shown. The authors' approach to the allocation and classification of resource regions is proposed, the dynamics of the selected groups over the period of 2005–2014 is shown.

1. Introduction

Much attention has been paid to the problems of development of the resource economies and regions over the recent years. This is because most countries which have significant deposits of natural resources, including Russia, lag behind the countries not possessing such natural wealth in the pace and quality of economic growth, social development, political and economic institutions.

Currently, Russia's economy is on the verge of a systemic crisis, whose key cause is the exhaustion of the resource-based model of development based on the extensive exploitation of large deposits of mineral resources (MR), especially in the energy sector, implementation of large-scale mega-projects and the activities of large export-oriented mining companies. The problem of searching for new effective models of the country's development, based on innovation and accelerated development of manufacturing industries, is paramount.

The task of modernizing the mineral resource development model declared by the Government of the Russian Federation and a change in the spatial specialization are expected to solve this problem. In the new model of development, the resource type regions should be the starting point for the formation of the intensive model of the country's development. All this suggests that the analysis of resource regions is a very urgent task.

The article *aims* at is justifying the qualitative and quantitative approach to the identification of "resource type" regions and the development of their typology.

2. Methods of research

The research methodology consisted in the methods of analysis, comparison, generalization and expert assessment.

3. Results and discussion

Large deposits of mineral resources (MR) in a region are both a source for growing prosperity and the cause of many problems in its development. Resource orientation provides regions with high growth and high income levels, while limiting the development of other industries and enhancing the disintegration processes.

The discussion of the problems and features of resource-type countries' and regions' development was started in R. Auty's research [1]. In his later works with E. Gelb, R. Auty tried to explain the causes and mechanisms of the falling living standards in the countries–exporters of oil in the 1970s-1980s [2]. The fact of slower development of resource-rich countries in the 1990s attracted the attention of John Sachs and A. Warner when performing cross-country comparisons [3]. At the same time, large number of vivid metaphors denoting this phenomenon emerged in the scientific literature: the "resource curse", the "Dutch disease", the "oil curse", "enclave development", etc. [4]. They all reflect a negative influence of the MR factor on the economy of the transitional countries rich in natural resources in three areas: technological, macroeconomic and institutional [4].

The technological and macroeconomic impact is implemented by the following pattern: big revenues from the export of raw materials, with investment mechanisms holding back the development of non-oil sectors of the economy. The labour-intensive, knowledge-intensive, the manufacturing sectors of the economy are particularly affected. The export of resources leads to strengthening of the national currency as a result of increased foreign exchange inflows from exports of raw materials. This, combined with constant growth in wages in capital-intensive extractive industries, leads to a decrease in competitiveness of other sectors, hampers the development of the market, limits the possibilities of intensive growth and the economy development on the innovative basis. The simple dominance of capital-intensive technologies in the resource sector, reduces the need for skilled staff and investments in human capital. Large capital-intensive commodity production has a negative impact on the level of competition and market potential in the country, further worsening the situation. Commodity companies starting acquiring patents not for their implementation, but for limiting other companies' access to new technologies. The country starts investing fewer resources in the development of education, science and new technologies.

High revenues from commodity exports reduce the quality of institutions. The desire to limit competition by means of the state leads to the formation of rent-seeking behavior that contributes to the selection and preservation of the inefficient economic and political resources for maximum profit and its distribution among the representatives of the political elite and big businesses. As noted by M. Ross, the vast majority of resource-rich countries have been unable to avoid reducing the market potential and deterioration of the quality of economic institutions, they have underdeveloped democratic institutions and civil society [5]. In turn, inefficient institutional environment impedes economic growth, reducing its quality indicators and, accordingly, affects the quality of life of the population, leading to increased social stratification, deterioration of the environment, increasing inequalities within and between countries and regions [6].

In the late 1990 – early 2000s, along with the works describing the negative impact of resource wealth on national economies, new works emerged attempting at finding a way out of resource-dependent development. The interest to this problem was fueled by the active search for a new model of development of Russia. V. Polterovich, V. Popov, A. Tonis and other researchers attempted in their works to develop institutional mechanisms creating the conditions to overcome the resource curse and the conversion of revenues from exports of natural resources into a source of growth and development of the Russian economy. The authors expressed disagreement with the fact that there is an inevitable direct relationship between resource wealth and underdevelopment of countries and noted that the vector of the resource wealth influence on the economy is determined by the quality of institutions [7]. The authors emphasized that the "resource curse" is the manifestation of the negative influence of the economic structure on the economic growth (rather than the development) [8].

In general we agree with this approach, but we believe that in addition to the institutional factors, the of countries' and regions' ability to create the industries alternative to the extractive sector based on revenues from hydrocarbon exports is of great importance for overcoming the resource curse.

In addition, we believe that with regard to Russia, the "resource curse" should be studied at the regional level. This is due to the fact that it is a powerful factor in the preservation and deepening of

IOP Conf. Series: Earth and Environmental Science **84** (2017) 012016 doi:10.1088/1755-1315/84/1/012016

regional disparities in the levels of socio-economic development. Commodity specialization is the main factor determining the peculiarities of the system of settlement and spatial organization of the Russian Federation. Commodity regions attract investment and skilled and mobile workforce, becoming the leading regions as manifested by cumulative socio-economic indicators. At the same time, the high cost of maintaining the infrastructure and difficult climatic conditions become an obstacle for their sustainable development. Thus, while in 1998 the per capita GRP production in the ten most economically developed regions of Russia exceeded the average level by 2.5 times, in 2000, it was more than 3.2 times. Among the developed regions, the regions of raw material specialization dominated [9].

The concept of "resource region" has not yet received a clear definition, neither in scientific literature nor in practice. In foreign studies, the resource regions are studied within the scope of national economies. Thus, R. Auty defined resource economies (mineral economies) as developing countries where at least 8 % of the GDP and 40 % of exports are produced in the industries related to mining. [1]. In Russian researches, resource economies are analyzed both within national economies [7], and from the standpoint of a particular region [8]. Most often, resource regions are understood as the regions within the borders of a subject of the Russian Federation, which have significant deposits of natural resources and actively explore them [10].

Despite the fact that the features of resource regions development largely determine the trajectory of the country as a whole, in Russia there is no universally accepted approach to the identification and classification of the "resource regions". For example, according to the classification of A.O. Polynia [11], one of the key factors in identifying resource-type regions is the index of "the level of the region's MR potential", defined as the ratio of the regional and national average values of the mineral resource base valuation per capita (at market prices on the valuation date). In M. Grigoriev, Yu. Urozhaeva, and D. Ivanov's synthetic classification of regions, all regions of Russia are divided into 9 types forming 4 groups, with commodity regions identified within each group [12]. N.N. Mikheeva identifies 21 commodity regions in Russia, focusing on the share of mining in GRP [13]. The same approach was followed by V.P. Orlov who defines the resource region as a region "where more than 5 % of shipped products are mineral resources" [14]. I.N. Ilyina suggests using the share of gross value added from mining in the GRP of more than 30% as the key criterion for identifying resource regions of Russia [8]. N.N. Mikheeva and S.V. Belousova suggest using the same rate, but they justify the index level with a cutoff at 10.5 % and 10.8 % [13, 15]. L.A. Tolstolesova defines a resource region as the subject of the Russian Federation (or several subjects) where mineral complex production make more than 50 % in the structure of industrial production due to its geographical location and the availability of significant natural resources [16]. Some authors attempt giving a quality definition of resource regions [17].

In this article, the resource regions ("resource type" regions) are understood as regions specializing in the extraction and processing of mineral complex products due to their geographical location and the availability of substantial MR potential, where the largest export-oriented vertically-integrated commodity companies, determining the direction and nature of the regional economic development.

To identify the Russian resource regions and suggest their classification depending on the degree of resources dependence, the authors chose the time period of 2005-2014. The year 2005 was taken as a reference point, since it was then that the economy recovery after the recession of the 1990s was over. The periods of rapid growth of the Russian economy in 2006-2007 and the period of sharp recession during the crisis of 2008-2009 were excluded from the analysis. The latest statistical data are available for the year 2014.

To determine the lower boundary of the resource dependence of the Russian regions, the authors took V.P. Orlov's approach, determining the "resource type" of the Russian regions on the basis of the share of the value added in the section "Resource extraction" in the GRP. [14]. Non-resource "border" regions were included into the group of the analyzed regions to account for the dynamic changes in the level of resource dependence of Russia as a whole (the share of extractive industries in the GRP fluctuated in the range from 8.9 % to 11.4 % during this period) and its regions (some regions crossed the border several times, moving into to group of resource regions, and then coming out of the group again). The "border" regions were defined as non-resource regions that crossed the border of 5 % at least once in the analyzed period. As a result, the number of selected regions amounted 36 regions.

To ensure the comparability of indicators based on the share of extractive industries in the GRP the K-rate was calculated. This rate represents the ratio of the share of extractive industries in the GRP (K_{reg}) to the share of extractive industries in the Russian Federation's GDP (K_{Rus}):

$$K = \frac{K_{reg}}{K_{Rus}}$$

The results of the K-rate calculations for the selected 36 regions in 2005-2014 are provided in table 1.

Table 1. Dynamics of resource dependence of the regions (2005-2014).

	Region / year	2005	2010	2011	2012	2013	2014
1	Nenets Autonomous Okrug	5.80	7.56	6.60	6.73	7.03	7.01
2	Khanty-Mansiysk Autonomous Okrug - UGRA	5.85	6.06	5.98	6.06	6.07	6.31
3	Sakhalin Region	1.73	5.70	5.41	5.49	5.66	6.20
4	Tyumen Region	4.68	4.80	4.63	4.87	4.86	5.11
5	Yamalo-Nenets Autonomous Okrug	4.80	4.61	4.29	4.64	4.87	4.74
6	The Republic Of Sakha (Yakutia)	3.09	3.86	3.88	3.82	3.98	4.20
7	Chukotsky Autonomous Okrug	0.59	3.67	3.71	3.38	2.90	4.05
8	Orenburg Region	2.89	3.45	3.16	3.30	3.80	3.40
9	Republic Of Komi	2.68	3.22	2.98	2.88	3.00	3.17
10	Tomsk Region	2.77	2.30	2.69	2.79	2.69	2.69
11	Udmurt Republic	2.09	2.25	2.30	2.29	2.33	2.28
12	Kemerovo Region	2.12	3.02	3.09	2.39	2.06	2.04
13	Astrakhan Region	0.21	0.34	1.56	1.74	1.97	2.03
14	Irkutsk Region	0.30	0.72	1.16	1.40	1.56	1.88
15	Republic Of Tatarstan	2.41	2.08	1.98	1.95	1.90	1.87
16	Magadan Region	2.12	1.98	2.24	1.63	1.61	1.62
17	Krasnoyarsk Krai	0.30	1.74	1.48	1.38	1.59	1.59
18	Perm Krai	1.23	1.30	1.39	1.62	1.58	1.46
19	Samara Region	0.84	1.14	1.25	1.31	1.24	1.34
20	Belgorod Region	1.70	1.62	1.87	1.43	1.40	1.17
21	Murmansk Region	0.84	1.46	1.66	1.44	1.67	1.15
22	The Republic Of Karelia	1.51	1.23	1.27	1.21	1.15	1.13
23	Amur Region	0.41	0.98	1.40	1.33	1.07	1.06
24	The Republic Of Khakassia	0.57	1.48	1.40	1.06	1.10	1.03
25	Kursk Region	1.41	1.16	1.33	1.13	1.09	0.88
26	Transbaikal Region	0.59	1.23	0.71	0.81	0.93	0.73
27	Tyva Republic	0.40	0.66	0.55	0.54	0.43	0.61
28	Volgograd Region	0.53	0.54	0.55	0.53	0.55	0.43
29	Khabarovsk Krai	0.37	0.43	0.49	0.58	0.53	0.42
31	Kaliningrad Region	1.15	0.57	0.51	0.43	0.40	0.33
32	The Republic Of Buryatia	0.35	0.47	0.54	0.45	0.38	0.33

33	Republic Of Bashkortostan	0.99	0.85	0.43	0.26	0.26	0.30
34	The Republic Of Ingushetia	0.80	0.18	0.17	0.17	0.19	0.16
35	Chechen Republic	0.79	0.26	0.21	0.17	0.18	0.16
36	Republic Of Kalmykia	0.51	0.33	0.26	0.27	0.27	0.13

*Note: The intensity of shading reflects the degree of the region's resource dependence.
- monoregions with maximum resource dependence (monoregions);
- resource regions with high resource dependence (high dependence regions)
- regions with medium resource dependence (medium dependence regions)
- resource regions with low resource dependence (low dependence regions)
- non-resource regions

Source: [9], authors' calculations

The table shows the presence of 5 clearly defined groups of regions that have different level of MR dependence have been respectively defined by the authors as monoregions, high dependence regions, medium dependence regions, low dependence regions, and non-resource regions.

The distribution of regions into groups and with the account for years is shown in tables 1 and 2. In table 1, it is reflected in the intensity of shading, while table 2 shows the number of regions in each group in each year.

Table 2. Classification of the regions depending on the share of the extraction industries in the GRP and the K-rate.

Type of a region	K	2005	2010	2011	2012	2013	2014
Monoregions: (> 30%)	>3	5	11	9	8	8	9
High dependence regions (20-30%)	[2-3)	7	3	4	4	4	4
Medium dependence regions (10-20%)	[1-2)	7	10	12	13	13	11
Low dependence regions (5 до 10%)	[0.5-1)	10	6	5	4	3	7
Resource regions, total	>0.5	29	30	30	29	28	31
Non-resource regions, with the share of extracting industries within 1 % – 5 %	[0.1-0.5)	7	6	6	7	8	5

*Source: [9], authors' calculations

It is significant that the number of regions that were included in the resource regions group was quite stable, their number in the considered period was around 30. In general, in 2005 – 2014, there was a slight increase in the number of resource regions – from 29 to 31.

One can also note the increase in the share of monoregions from 5 in 2005 to 9 in 2014; the increase in the number of monoregions was accompanied by the increase in their resource dependence, the upper value of the K-rate increased by almost 2 times. (see table 2). The increase in the number of monoregions occurred due to the increase in the degree of the high dependence regions' dependence (Orenburg Region and the Republic of Komi) and the conversion of medium dependence regions (Sakhalin Region) and previously non-resource regions (Chukotsky Autonomous Okrug) into monoregions. This was due to the rising price of hydrocarbons at the world markets and the overall growth of the resource dependence of the Russian economy.

The high dependence group was reduced due to processes of polarization. Two regions increased their dependence and became monoregions, two more regions contrarily reduced their resource dependence and went to the less resource-dependent groups. One monoregion (Kemerovo Region) moved to the high dependence group (see tables 2 and 3)

The biggest changes occurred in the largest group of medium dependence regions. Over the period of 2005 – 2104, the total number of the regions increased from 7 to 11. It is particularly noticeable in the change of intensity of shading of the regions of this group in table 3. In addition, the regions constituting this group were constantly changing. Only 2 regions constantly belonged to the group of medium dependence regions: Perm Krai and the Republic of Karelia. The rest moved from high dependence regions (the Republic of Tatarstan and Magadan Region), low dependence regions (Samara and Murmansk Regions) and non-resource regions (Tuva Republic and Amur Region).

The impact of the dominant MR on the status of the regions is also significant. Table 3 shows that the most resource-rich regions, possessing large deposits of hydrocarbons (oil and gas) tended to increase their resource dependence, with the exception of the Republic of Tatarstan, Tomsk Region and the Republic of Bashkartostan. This reduction of resource dependence reflects a positive trend in the structure of the GRP of all the regions, the gradual growth of the share of processing industries and high-tech production. However, these positive trends do not yet have a significant impact on the other resource regions.

Table 3. Classification of Russian resource regions and the dynamics of group transition.

No.	Region	Dominant MR	Resource dependence dynamics	Transition from another group compared to 2005.
		Resource monoregions K>3		
1	Nenets Autonomous Okrug	oil and gas	↑	-
2	Khanty-Mansiysk Autonomous Okrug - UGRA	oil and gas	↗	-
3	Sakhalin Region	oil and gas	↑↑	from medium dependence group
4	Tyumen Region	oil and gas	↗	-
5	Yamalo-Nenets Autonomous Okrug	oil and gas	↘	-
6	The Republic Of Sakha (Yakutia)	oil and gas	↗	-
7	Chukotsky Autonomous Okrug	oil and gas	↑↑	from low dependence group
8	Orenburg Region	oil and gas	↑	from high dependence group
9	Republic Of Komi	oil and gas	↑	from high dependence group
		High dependence regions K= 2-2.99		
10	Tomsk Region	oil and gas	↘	-
11	Udmurt Республика	oil and gas	↗	-
12	Kemerovo Region	coal	↓	from monoregions
13	Astrakhan Region	oil and gas	↑↑	from non-resource regions
		Medium dependence regions K =1-1.99		
14	Irkutsk Region	oil and gas	↑↑	from non-resource regions
15	Republic Of Tatarstan	oil and gas	↓	from high dependence group
16	Magadan Region	oil and gas	↓	from high dependence group
17	Krasnoyarsk Krai	oil and gas	↑↑	from non-resource regions
18	Perm Krai	oil and gas	↗	-
19	Samara Region	oil and gas	↑	from low dependence group
20	Belgorod Region	oil and gas	↓	-
21	Murmansk Region	other mineral resources	↑	from low dependence group
22	The Republic Of Karelia	other mineral resources	↘	-
23	Amur Region	coal and other mineral resources	↑↑	from non-resource regions

24	The Republic Of Khakassia	oil and gas	↑	from low dependence group
		Low dependence regions К =0,5-0,99		
25	Kursk Region	other mineral resources	↓	from medium dependence group
26	Transbaikal Region	coal	↗	-
27	Tyva Republic	coal	↑	from non-resource regions
		Non-resource regions К =0,1-0,49		
28	Volgograd Region	oil and gas	↓	from low dependence group
29	Khabarovsk Krai	coal and other mineral resources	↗	-
31	Kaliningrad Region	other mineral resources	↓↓	from medium dependence group
32	The Republic Of Buryatia	coal	↘	-
33	Republic Of Bashkortostan	oil and gas	↓	from low dependence group
34	The Republic Of Ingushetia	oil and gas	↓	from low dependence group
35	Chechen Republic	oil and gas	↓	from low dependence group
36	Republic Of Kalmykia	oil and gas	↓	from low dependence group

*Source: [9], authors' calculations

The data of table 3 also indicate that the regions rich in coal and other mineral resources (iron ore, nonferrous metal ores, etc.) reduced the level of their resource dependence during the analyzed period. However, these changes did not reflect the positive changes in the sectoral structure of the GRP (growth of the manufacturing industries and knowledge-intensive production), but the decrease in the overall decline of the GRP under the impact of lower raw materials prices in the world markets. Most likely, the real level of resource dependence in these regions remains at a high level. The exception is the Republic of Tyva. It has seen an increase in resource dependence, reflecting the development of new coal deposits.

4. Conclusion

The analysis showed that despite the fact that Russia is a resource economy, only 27 of the 85 Russian regions can be identified as resource-dependent, they have a decisive impact on the character and trajectory of the Russian economy's development. An important indicator of the region's resource dependence may be the share of the mining sector in the GRP structure.

There is a significant differentiation in the level of resource dependence of Russian regions. This allows suggesting a classification of resource regions and to distinguishing among four sustainable groups: monoregions, high dependence regions, medium dependence regions, and low dependence regions. The boundary of the resource dependence of the Russian regions is movable, thus it makes sense to include "border" regions (shifting from resource to non-resource regions). It is important to understand the reason for their "border" position.

Resource regions is not uniform, which is manifested in the major differences in the degree of dependence of the regional economy on the industries engaged in the extraction of resources, trends and mechanisms for the development and implementation of the regions' resource dependence.

Acknowledgements
The authors are grateful for the financial support to the Russian Science Foundation (Agreement No. 16-18-10182), as well as to the Russian Foundation for Basic Research (Agreements No. 15-06-090994, 16-36-00369, 17-06-00537).

References

[1] Auty R 1993 *Sustaining Development in Mineral Economies: The Resource Curse Thesis.* (L.: Routledge) p 283
[2] Auty R M and Gelb A H 2001 *Political Economy of Resource Abundant States* http://documents.worldbank.org/curated/en/926361468781759170/pdf/28750.pdf
[3] Sachs J D and Warner A M 1995 *Natural Resource Abundance and Economic Growth. NBER Working Paper Series, Working Paper No. 5398* (Cambridge: National Bureau of Economic Research) http://www.nber.org/papers/w5398.pdf
[4] Gylfason T 2001 *European Economic Review* **45**
[5] Ross M L 2007 *Escaping the Resource Curse* (N.Y.: Columbia University Press) p 237–55
[6] Guriev S, Plekhanov A and Sonin K 2010 *Issues of Economics* **3** 4–23
[7] Polterovich V, Popov V and Tonis A 2007 *Issues of Economics* **6** 4–27
[8] Ilyina I N 2013 *Issues of State and Municipal Management* **2** 91–102
[9] Official website of the Federal State Statistics Service http://www.gks.ru
[10] Decree of the President of the Russian Federation No. 13 of January 16, 2017 "*On the Approval of the Fundamentals of the State Policy of Regional Development of the Russian Federation for the Period until 2025*"
[11] Polynev A O 2010 *Competitive Opportunities of Regions. Methodology of Research and Ways* to Improve it (Moscow: KRASAND) 2010 p 13–5
[12] Golyashev A V and Grigoriev L M 2014 *Types of Russian Regions: Sustainability and Shifts in 2003-2013* (Moscow: Analytical Center under the President of the Russian Federation) p 47
[13] Mikheeva N N 2009 *Region: Economics and Sociology* **2** 23–42
[14] Orlov V P 2007 *Mineral resources of Russia. Economics and Management* **2** 2–3
[15] Belousova S V 2015 *ECO* **6** 40–8
[16] Tolstolesova L A 2010 *Problems of Modern Economy* **4(36)** 189–93
[17] Levin S N, Kagan E S and Sablin K S 2015 *J. of Institutional Studies* **7(3)** 92–101

IOP Conf. Series: Earth and Environmental Science **84** (2017) 012017 doi:10.1088/1755-1315/84/1/012017

Development of a wear-resistant flux cored wire of Fe-C-Si-Mn-Cr-Ni-Mo-V system for deposit welding of mining equipment parts

I V Osetkovsky, N A Kozyrev, R E Kryukov, A A Usoltsev and A I Gusev

Siberian State Industrial University, 42 Kirova Street, Novokuznetsk, 654007, Russia

E-mail: kozyrev_na@mtsp.sibsiu.ru

Abstract. The effect of introduction of cobalt in the charge of the flux cored wire of Fe-C-Si-Mn-Cr-Ni-Mo-V system operating under abrasive and abrasive-shock loads is studied. In the laboratory conditions samples of flux cored wires were made, deposition was performed, the effect of cobalt on the hardness and the degree of wear was evaluated, metallographic studies were carried out. The influence of cobalt introduced into the charge of the flux cored wire of Fe-C-Si-Mn-Cr-Ni-Mo-V system on the structure, nature of nonmetallic inclusions, hardness and wear resistance of the weld metal was studied. In the laboratory conditions samples flux cored wire were made using appropriate powdered materials. As a carbon-fluorine-containing material dust from gas cleaning units of aluminum production was used. In the course of the study the chemical composition of the weld metal was determined, metallographic analysis was performed, mechanical properties were determined. As a result of the metallographic analysis the size of the former austenite grain, martensite dispersion in the structure of the weld metal, the level of contamination with its nonmetallic inclusions were established.

1. Introduction

Repair of mining equipment requires significant costs for its replacement or restoration [1]. One of the most effective technologies for hardening and restoring parts is arc welding by flux cored wire [2]. Therefore, the development of materials and the use of innovative recovery technologies that significantly increase the wear resistance of products is today an urgent task. For this purpose, the development and manufacture of special surfacing flux cored wires is being carried out [3-16]. Due to optimally selected chemical composition of flux cored wires, the welded coatings have high hardness, as well as abrasive and shock abrasive wear resistance. Welding wires of Fe-C-Si-Mn-Cr-Ni-Mo systems – type A and B according to the classification of MIS are widely used for surfacing abrasive-wear parts [17]. At present such flux cored wires of various foreign productions are widely applied in our country.

2. Methods of research

The influence of cobalt introduced into the burden of the flux-cored wire of the Fe-C-Si-Mn-Cr-Ni-Mo-V system in the manufacture of wire test samples, the degree of wear of the samples and the hardness of the deposited layer. The production of a flux-cored wire and its surfacing for the preparation of samples was carried out according to the technology given in [18, 19].

The wire was manufactured on a laboratory machine. Diameter of the wire is 5 mm, the cover is made from a strip St3. As the filler the appropriate powdered materials were used: iron powder PZhV1 according to GOST (Russian State Standards) 9849-86, ferrosilicon powder FS 75 according to GOST1415-93, high-carbon ferrochromium powder FKh900A according to GOST 4757-91, carbon ferromanganese powder 78 FMn(A) according to GOST 4755-91, nickel powder PNA-1L5 according to GOST 9722-97, ferromolybdenum powder FMo60 according to GOST 4759-91, ferrovanadium powder FV50U 0.6 according to GOST 27130-94, cobalt powder PK-1U according to GOST 9721-79, and as a carbon-containing component dust from gas-cleaning units at aluminum production was used with the following component composition, wt%: Al_2O_3 = 21-46.23; F = 18-27; Na_2O = 8-15; K_2O = 0.4-6; CaO = 0.7-2.3; Si_2O = 0.5-2.48; Fe_2O_3 = 2.1-3.27; C_{tot} = 12.5-30.2; MnO = 0.07-0.9; MgO = 0.06-0.9; S = 0.09-0.19; P = 0.1-0.18.

The hardness was measured using a MET-UD hardness tester. Five hardness measurements were made on the surface of each sample. Table 1 shows the averaged values for the hardness of the weld metal averaged over five measurements. Wear-resistance tests were carried out on the machine 2070 SMT-1 as follows. Samples were weighed on scales, with the possibility of measuring masses up to 10^{-4} g. After recording the weight of samples, a wear-resistance test was carried out by rotating the disc welded by tungsten alloys on a flat surface of the weld metal for 6 hours with the following conditions: load – 30 mA, rotation speed – 20 rpm. After the wear-resistance tests, the samples were weighed and the difference between the initial and final weight was determined, and also the number of disk revolutions of the installation 2070 CMT-1 was recorded. Wear-resistance was determined on the basis of a decrease in the sample weight per one revolution of the disk. The results of the tests, hardness measurements and chemical analysis of samples are given in table 1.

Table 1. Chemical composition, wear and hardness of deposited metal.

Sample number	Mass fraction of elements, %								Hardness of samples HRC	Wear of samples g/r*10^{-4}
	C	Si	Mn	Cr	Mo	Ni	V	Co		
1	0.24	0.19	0.93	1.83	0.69	0.14	0.60	-	41	0.37
2	0.25	0.27	0.96	1.65	0.68	0.33	0.60	-	40	0.358
3	0.25	0.28	0.93	1.67	0.57	0.54	0.58	-	40	0.359
4	0.29	0.15	0.92	1.65	0.60	0.65	0.59	-	41	0.449
5	0.23	0.12	0.85	1.45	0.53	0.51	0.68	0.08	25	1.65
6	0.21	0.23	0.89	1.45	0.54	0.55	0.54	0.03	21	1.15
7	0.17	0.18	0.85	1.40	0.52	0.54	0.63	0.05	21	1.11
8	0.17	0.28	0.91	1.32	0.45	0.46	0.59	0.06	21	1.12

The metallographic analysis of the samples was carried out using OLYMPUS GX-51 optical microscope in the bright field with the magnification range × 100-1000. An alcoholic solution of nitric acid was used as a reagent for etching the samples surface. Investigation of longitudinal samples of the welded layer for the presence of nonmetallic inclusions was made according to GOST 1778-70 with magnification × 100. The value of the former austenite grain was determined according to GOST 5639-82 with magnification × 100. The size of the martensite needles was determined according to GOST 8233-56 with magnification × 1000.

3. Results and discussion

Metallographic analysis showed that the microstructure of the deposited layer with a carbon content of 0.24-0.29% (samples No. 1-4) is uniform, thin branches of dendrites are observed. The microstructure represents a fine-needle and medium-needle martensite (point No. 3-6) in the former austenite grains, along the boundaries of which thin δ-ferrite interlayers are located, and a small amount of residual austenite in the form of separate islets. The size of the martensite needles in the structure of the examined samples is in the range of 2-10 μm (table 2, figure 1).

In the microstructure of sample No. 1 there is medium-needle martensite (score No. 5, 6) (figure 1a, b). The value of primary grains of austenite on the grain scale corresponds to No. 5 and No. 6.

Table 2. Characteristics of nonmetallic inclusions and samples structure.

Sample number	Contamination with non-metallic inclusions, points		Austenite grain size, point	Size of martensite needles, μm
	Silicates nondeforming (brittle)	Spot oxides		
1	2b, 2a, 3a	1a	5, 6	7-10
2	1b, 2b, 3a	1a	5, 6	4-8
3	2b, 3a	1 a	5, 6	5-8
4	2b, 3a (1b)	1 a	6, 5	2-5
5	1b, 2b, 3a	1 a	6, 5	2-5
6	1b, 2b, 2a	1 a, 2a	6	2-4
7	1b, 2b, 3a	1 a	6	2-5
8	1b, 2b, 3a	1 a	6	2-4

When the nickel content is increased to 0.33-0.54% the austenite grain size remains unchanged. In the microstructure of samples No. 2 and 3 medium-needle martensite (score No. 5) is also present, however, in some of its zones fine-grained martensite is observed (point No. 3) (figure 1 c-f, Table 2).

An increase in the nickel content to 0.65% (sample No. 4) significantly crushes the martensite needles, and also reduces the size of the former austenite grain. In the microstructure of sample No. 4, fine-grained martensite (point No. 3) is observed formed within the boundaries of the former austenite grain, the value of which corresponds to No. 6 and No. 5 on the grain size scale (figure 1 g, h, table 2).

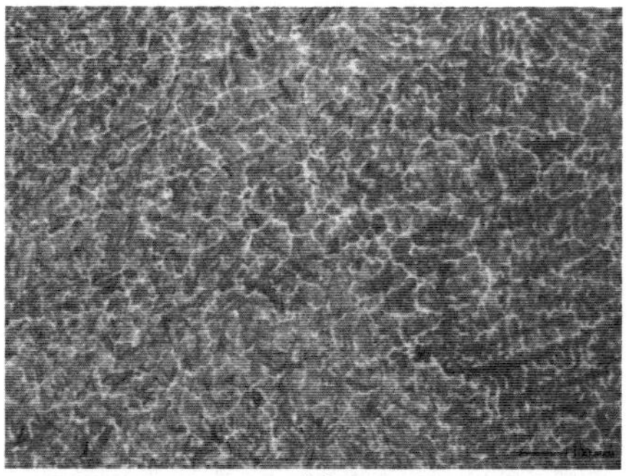
(a)

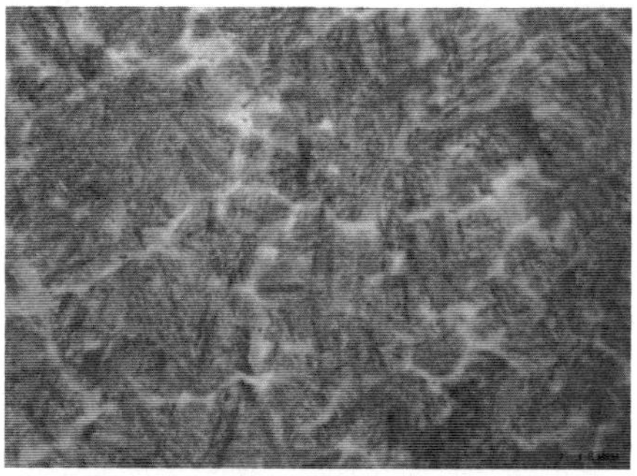
(b)

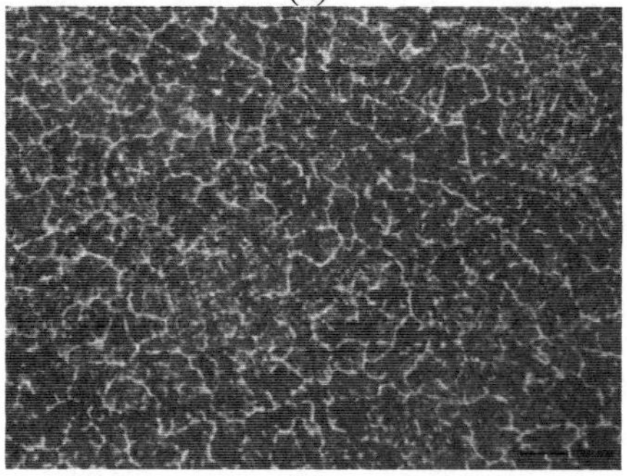
(c)

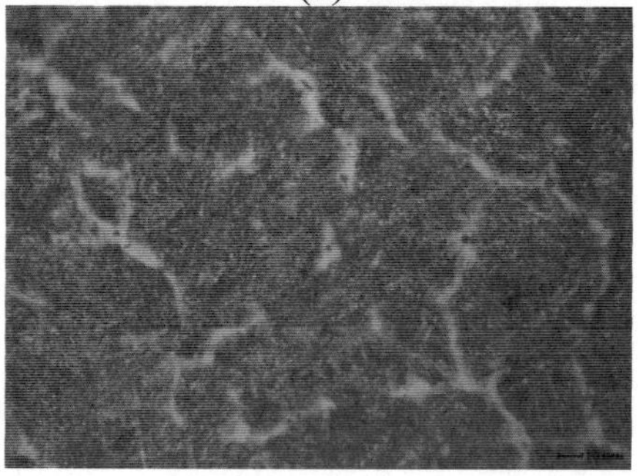
(d)

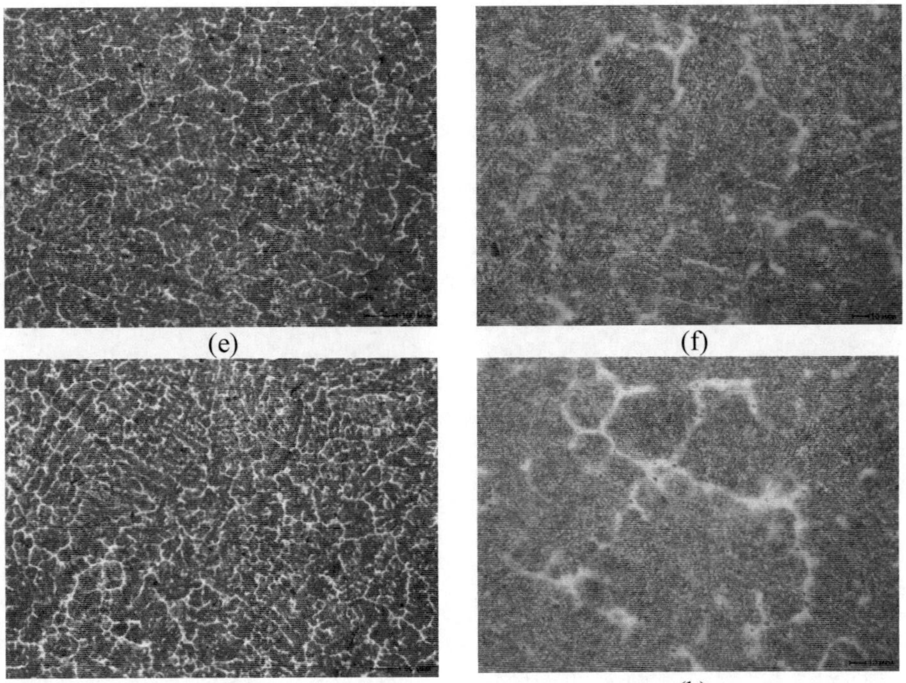

(e) (f)

(g) (h)

Figure 1. Deposited layer microstructure of flux cored wire of Fe-C-Si-Mn-Cr-Ni-Mo-V system with the improved composition (a, c, e, g × 100), (b, d, f, h × 500). a, b – sample No. 1; c, d – sample 2; e, f – sample No. 3; g, h – sample No. 4.

The introduction of cobalt into the deposited flux cored wire with simultaneous reduction of carbon content up to 0.17-0.23% (samples No. 5-8) provides a uniform structure with fine-needle martensite (score No. 3) in the former austenite grains, the residual austenite present in a small amount in the form of separate islands, and δ-ferrite in the form of thin interlayers along the boundaries of the primary austenite grains (figure 2).

The size of martensite needles in the structure of the samples under study is in the range of 2-5 μm (table 2). The value of the former austenite grain corresponds to No. 6.

Thus, it was established that an increase in nickel content up to 0.65% in the weld steel as well as introduction of cobalt with simultaneous reduction of carbon content up to 0.17-0.23% helps to crush refine and reduce the size of the former austenite grain.

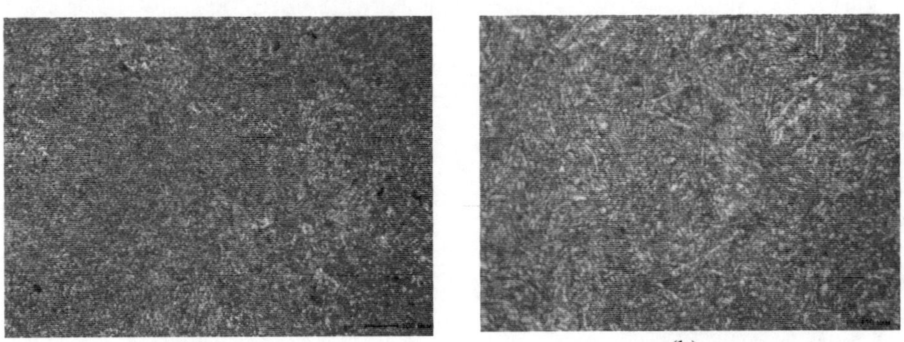

(a) (b)

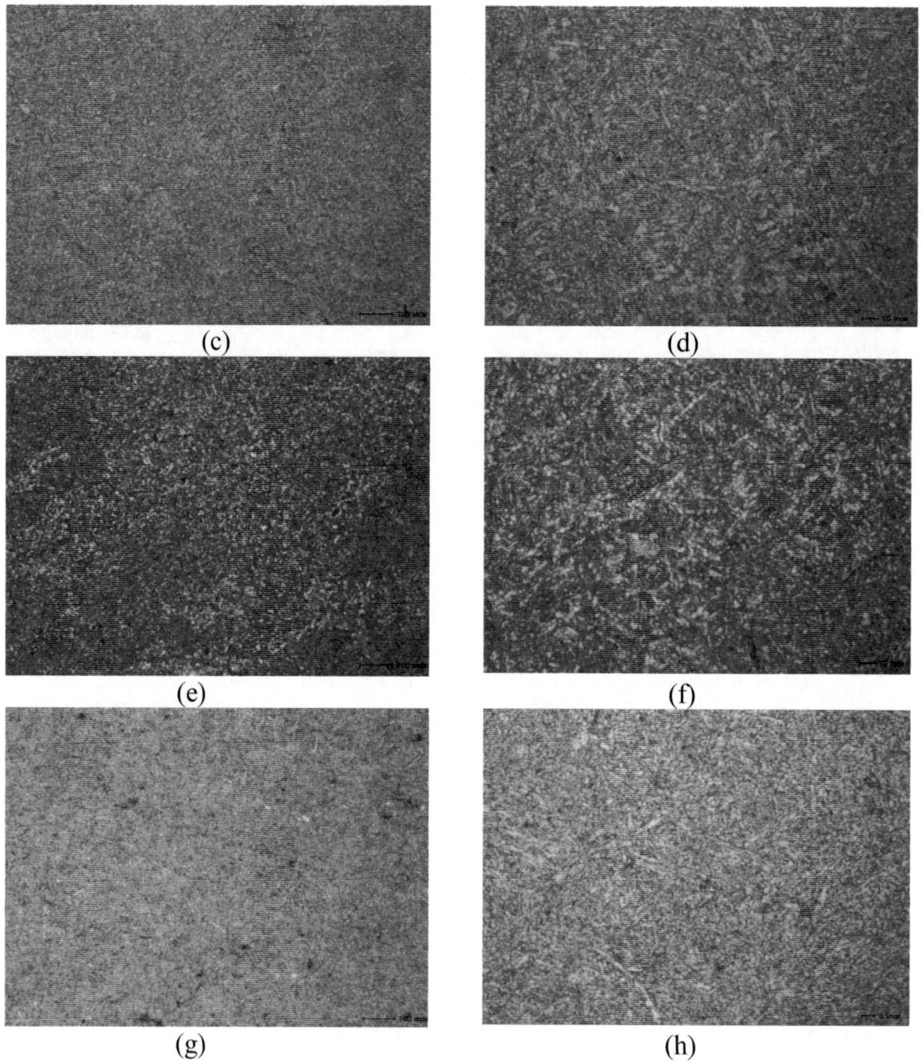

Figure 2. Deposited layer microstructure of flux cored wire of Fe-C-Si-Mn-Cr-Ni-Mo-V system with cobalt (a, c, e, g × 100), (b, d, f, h × 500). a, b – sample No. 5; c, d – sample No. 6; e, f – sample No. 7; g, h – sample No. 8.

As a result of studying the nature of nonmetallic inclusions of the welded layer by the flux-cored wire of Fe-C-Si-Mn-Cr-Ni-Mo-V system the contamination with oxide nonmetallic inclusions was noted (table 2).

It was established that the change in the degree of alloying of the flux-cored wire of Fe-C-Si-Mn-Cr-Ni-Mo-V system has practically no effect on the contamination level with nonmetallic inclusions of the layer deposited by it.

The effect of chemical composition of the flux cored wire of Fe-C-Si-Mn-Cr-Ni-Mo-V system on the degree of wear and the hardness of the welded layer was estimated using multivariate correlation analysis, which allows the patterns of the change in the resultant index depending on the behavior of various factors described in [20, 21, 22] to be studied.

The factors that influence the examined indicator were determined for the analysis, and the most significant of them were selected (table 1). After that, the initial information was checked for reliability, uniformity, compliance with the law of normal distribution. Then a model of the factor system was constructed. Since the above systems have independent factor characteristics, the deterministic factor analysis is used.

The type of connection was determined using pair and partial correlation coefficients. It is revealed that the relationship has a rectilinear character, which is also confirmed by the determinism coefficient, which equals to 1 for the models of reduced factor systems.

The calculation of main connection indicators of the correlation analysis was carried out in stages. First, one factor was taken into account which has the most significant effect on the resultant index, then the second, third, etc. At each stage, the equation of connection and indicators were calculated, with the help of which its reliability is estimated.

Based on the results of the calculations we obtained dependencies, the adequacy of which was checked for actual values by the average error of approximation:

$$\bar{\varepsilon} = \frac{1}{m}\sum_{i=1}^{m}\left|\frac{Y_i - \tilde{Y}_i}{Y_i}\right| \cdot 100, \qquad (1)$$

where m – the number of observations; $\tilde{Y}_i$ is the calculated value of the resulting indicator; Y_i – the actual value of the resulting indicator.

As a result of the analysis, the following dependencies were obtained:

- hardness of the deposited metal when Co is added to the system Fe-C-Si-Mn-Cr-Ni-Mo-V:y=156.66*C+10.92*Si-6.94*Mn-7.51*Cr-75.67*Mo-28.10*Ni+129.45 *V-364.27*Co (approximation error 0.01%);
- wear resistance of the deposited metal when Co is added to the system Fe-C-Si-Mn-Cr-Ni-Mo-V:y=-0,000101*C+0,000052*Si-0,000126*Mn+0,000117* Cr+0,000413*Mo+0,000130*Ni-0,000583*V+0,002874*Co (approximation error 1.59 %);

The calculated values of the mean error of approximation show that the obtained dependences are adequate and they can be used to determine the resulting indicators.

4. Conclusions

1. It was established that an increase in nickel content up to 0.65% in the composition of deposited steel, as well as the introduction of cobalt, simultaneous with the reduction of carbon content up to 0.17-0.23% provides refinement of martensite needles and decrease the size of the former austenite grain.

2. Change in the chemical composition of the flux cored wire of the Fe-C-Si-Mn-Cr-Ni-Mo-V system practically does not affect the contamination level with nonmetallic inclusions of the layer deposited by it.

3. Based on the results of the multifactor correlation analysis the hardness of the deposited layer and its wear resistance were determined from the mass fraction of the elements in the composition of flux cored wires of Fe-C-Si-Mn-Cr-Mo-Ni-V-Co system. The obtained dependences can be used to predict the hardness of the deposited layer and its wear resistance when the chemical composition of the weld metal changes.

References

[1] Patricio F et 2014 *J. Manufacturing Proc.* **16** 4 –25
[2] Malinov V L 2011 *Bulletin of PSTU. Series: Eng.* **2** 107–17
[3] Dashuang Liu et al 2013 *Materials and Manuf. Proc.* **28** 643–9

[4] Atamert S and Stekly J 1993 *Surface Eng.* **9** 231–40
[5] Liu D S et al 2012 *Surface Eng.* **12** 338–44
[6] Liu D S 2013 *Surface Eng.* **29** 627–32
[7] Ramachandran C S et al 2009 *Surface Engineering* **25** 458–69
[8] Foley R T et al 1963 *ASLE Trans.* **6** 29–39
[9] Rajiv E P et al 1996 *Surface Engineering* **12** 163–6
[10] Giusti R et al 2016 *Surface Engineering* **32** 677–85
[11] Exner H et al 1970 *Powder Metallurgy* **13** 13–31
[12] Li T et al 2007 *Philosophical Magazine* **87** 5657–71
[13] Yaedu A E et al 2005 *Materials Science and Technology* **21** 459–66
[14] Cârâc G et al 2004 *Surface Engineering* **20** 353–9
[15] Ma C et al 2015 *Transactions of the IMF* **93** 8–17
[16] Anand E E and Natarajan S 2014 *Surface Engineering* **30** 716–21
[17] Paton B E 1974 *Technology of Electric Welding of Metals and Alloys by Fusion* (Moscow: Metallurgy) p 768
[18] Gusev A I and Osetkovskiy I V 2016 *Proc. VII All-Russian Sci. Conf. on Progressive Technologies and Economics in Mechanical Engineering* (Tomsk: TPU) pp 91–4
[19] Osetkovsky I V and Gusev A I 2016 *Proc. VII All-Russian Sci. Conf. on Progressive Technologies and Economics in Mechanical Engineering* (Tomsk: TPU) pp 143–6
[20] Wentzel E S 1999 *Theory of Probability* (Moscow: Higher School) p 576
[21] Wentzel E S 2003 *Theory of Probability and its Engineering Applications* (Moscow: Academiya) p 464
[22] Gmurman V E 2003 *Theory of Probability and Mathematical Statistics* (Moscow: Higher School) p 479
[23] GOST 1778-70. *Steel. Metallographic Methods for Determining Nonmetallic Inclusions*
[24] GOST 5639-82. *Steels and Alloys. Methods for Determination of Grain Size*

IOP Conf. Series: Earth and Environmental Science **84** (2017) 012018 doi:10.1088/1755-1315/84/1/012018

Study of the properties of flux cored wire of Fe-C-Si-Mn-Cr-Mo-Ni-V-Co system for the strengthening of nodes and parts of equipment used in the mineral mining

A I Gusev, N A Kozyrev, A A Usoltsev, R E Kryukov and I V Osetkovsky

Siberian State Industrial University, 42 Kirova Street, Novokuznetsk, 654007, Russia

Email: kozyrev_na@mtsp.sibsiu.ru

Abstract. The effect of the introduction of vanadium and cobalt into the charge of the powder surfacing wire of Fe-C-Si-Mn-Cr-Mo-Ni system is studied. In the laboratory conditions, the samples of flux cored wires were produced. The surfacing made by the prepared wire was produced under the flux AN-26C, on the plates of steel St3 in 6 layers with the help of ASAW-1250 welding tractor. Reduction of carbon content in the deposited layer to 0.19-0.2% with simultaneous change in the content of chromium, nickel, molybdenum and other elements present in it contributes to the enlargement of the martensite needles and the increase in the size of the former austenite grain. The obtained dependences of hardness of the deposited layer and its wear resistance on the mass fraction of elements, included in the composition of powder wires of the proposed system, can be used to predict the hardness of the welded layer and its wear resistance under different operating conditions for mining equipment and coal mining equipment.

1. Introduction

The mechanisms of machines of mining equipment, which experience abrasive and impact wear during operation, fail prematurely. Wear of their working surfaces necessitates restoration. Therefore, the development of materials that significantly increase the wear resistance of such parts and the use of technology for their recovery is an important task. The most promising is the use of surfacing by flux cored wire on the wear surfaces of parts. For these purposes in our country and abroad the development and manufacture of special surfacing powder wires is being carried out [1 - 14]. Thanks to the optimal selected method of alloying the welded coatings have high values of hardness, abrasive and shock abrasive wear resistance.

Welding wires of Fe-C-Si-Mn-Cr-Ni-Mo systems of type A and B according to the classification of MIS are widely used for surfacing abrasive-wearing products. At present, the powder wires of this system produced by DRATEC (Germany) DT-SG 600 F and powdered wires by ESAB OK Tubrodur 15.52, OK Tubrodur 58 O/G M are widely used in our country. By the results of researches it is necessary to define a field of application of wires of different types in details of mining machines.

2. Methods of research

The chemical composition of the weld samples under investigation was determined according to GOST 10543-98 (Russian National Standard) by the X-ray fluorescence method on XRF-1800 spectrometer and by the atomic-emission method using DFS-71 spectrometer. The hardness of the samples studied was measured with MET-DU hardness tester. The wear resistance tests were carried

IOP Conf. Series: Earth and Environmental Science **84** (2017) 012018 doi:10.1088/1755-1315/84/1/012018

out on the machine 2070 SMT-1. The tests were carried out in the following regimes: load 30 mA, frequency 20 rpm. The metallographic analysis was carried out using OLYMPUS GX-51 optical microscope in a light field in the magnification range of 100-1000 after etching in alcoholic nitric acid solution. The grain size was determined in accordance with GOST 5639-82 at magnification ×100. The size of the martensite needles was determined in accordance with GOST 8233-56 at magnification ×1000. Investigation of longitudinal samples of the surfaced layer for the presence of nonmetallic inclusions was made in accordance with GOST 1778-70 at magnification × 100.

3. Results and discussion

This work continues the research devoted to the development of new compositions of powder wires used for surfacing products working under abrasive wear in the mining industry [16 - 18], in particular, the study of the effect of the use of vanadium and cobalt in the production of wire samples of Fe-C-Si-Mn-Cr-Mo-Ni system on the degree of wear and hardness of the deposited layer.

The wire was manufactured on a laboratory machine. Diameter of wire 5 mm, the cover is made from a strip St3. As a filler the appropriate powdered materials were used: PZhV1 iron powder according to GOST 9849-86, ferrosilicon powder FS 75 according to GOST1415-93, high-carbon ferrochromium powder FH900A according to GOST 4757-91, powder carbon ferromanganese 78 FMn (A) according to GOST 4755-91, PNK-1L5 nickel powder according to GOST 9722-97, ferromolybdenum powder FMo60 according to GOST 4759-91, ferrovanadium powder FV50U 0.6 according to GOST 27130-94, cobalt powder PC-1U according to GOST 9721-79. As carbon-containing component dust from gas cleaning of aluminum production was used with the following component composition, wt%: Al_2O_3 = 21-46,23; F = 18-27; Na_2O = 8-15; K_2O = 0.4-6; CaO = 0.7-2.3; Si_2O = 0.5-2.48; Fe_2O_3 = 2.1-3.27; C_{tot} = 12.5-30.2; MnO = 0.07-0.9; MgO = 0.06-0.9; S = 0.09-0.19; P = 0.1-0.18.

Surfacing by the produced wire was made under the flux of AN-26C, on plates of steel grade St3 in 6 layers (to exclude the intermixing of the welded metal with the substrate), using ASAW-1250 welding tractor in the deposition regime: I = 450 A, U = 30 B , V = 10 cm/min. The plates were then cut into the appropriate test samples. The chemical composition of the deposited layers by powdered wires is given in Table 1.

Metallographic studies showed that the microstructure of the deposited layer by the flux-cored wire of Fe-C-Si-Mn-Cr-Mo-Ni-V-Co system is uniform, thin branches of dendrites are observed. The microstructure consists of martensite formed within the boundaries of the former austenite grain, austenite residual, present in a small amount in the form of separate islets, and thin δ-ferrite layers located along the boundaries of the former austenite grains.

It was established that in the microstructure of samples with a carbon content of 0.22-0.55% (samples No. 1-6), fine-needle martensite with needle size up to 6 μm is observed (point No. 4). The value of the former austenite grains on the grain scale is within the limits of No. 6-7 (figure 1, 2, table 2).

Table 1. Chemical composition, hydrogen content, wear and hardness of the deposited layers.

Sample number	Element mass fraction, %														[H], cm³/100g	Samples hardness, HRC	Samples wear g/r
	C	Si	Mn	Cr	Ni	Mo	B	V	Co	Al	Cu	Ti	S	P			
1	0.22	0.35	0.65	2.78	0.09	0.25	0	0.02	0.04	0.01	0.06	0.01	0.036	0.020	2.0	36	0.000040
2	0.43	0.37	0.84	7.04	0.42	0.49	0.01	0.03	0.06	0.02	0.08	0.01	0.038	0.020	2.0	56	0.000020
3	0.5	0.68	0.75	5.57	0.44	0.55	0.01	0.04	0.1	0.03	0.07	0.003	0.037	0.025	2.0	50	0.000015
4	0.55	0.81	0.7	5.59	0.6	0.58	0.01	0.03	0.11	0.07	0.07	0.005	0.044	0.023	2.4	53	0.000005
5	0.46	0.68	0.75	5.04	0.72	0.5	0.01	0.03	0.08	0.05	0.1	0.001	0.042	0.019	2.1	51	0.000071
6	0.44	0.74	0.73	5.59	0.86	0.53	0.01	0.03	0.09	0.06	0.07	0.02	0.038	0.020	2.3	52	0.000017
7	0.19	0.77	0.61	4.17	0.34	0.38	0.01	0.02	0.05	0.01	0.07	0.02	0.054	0.024	2.4	44.5	0.000071
8	0.19	0.63	0.65	4.06	0.3	0.38	0.01	0.03	0.06	0.01	0.08	0.03	0.056	0.019	1.7	43	0.000039
9	0.2	0.59	0.61	4.12	0.3	0.38	0.01	0.02	0.12	0.02	0.06	0.04	0.049	0.019	1.9	46	0.000044
10	0.2	0.64	0.6	4.03	0.3	0.39	0.01	0.03	0.2	0.01	0.08	0.03	0.058	0.021	2.0	30	0.000073

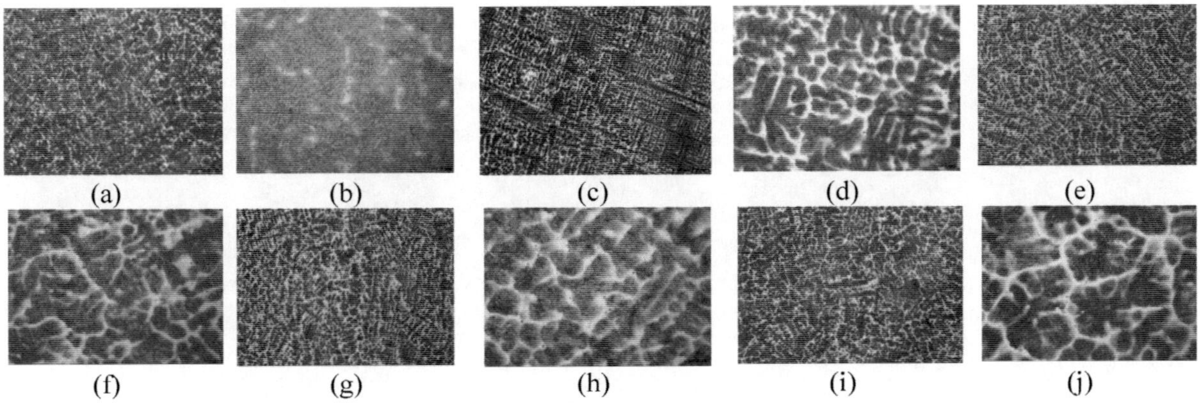

a, b – No. 1; c, d – No. 2; e, f – No. 3; g, h – No. 4; h, j – No. 5

Figure 1. Microstructure of the examined samples, (a, c, e, g, i × 100), (b, d, f, h, j × 500).

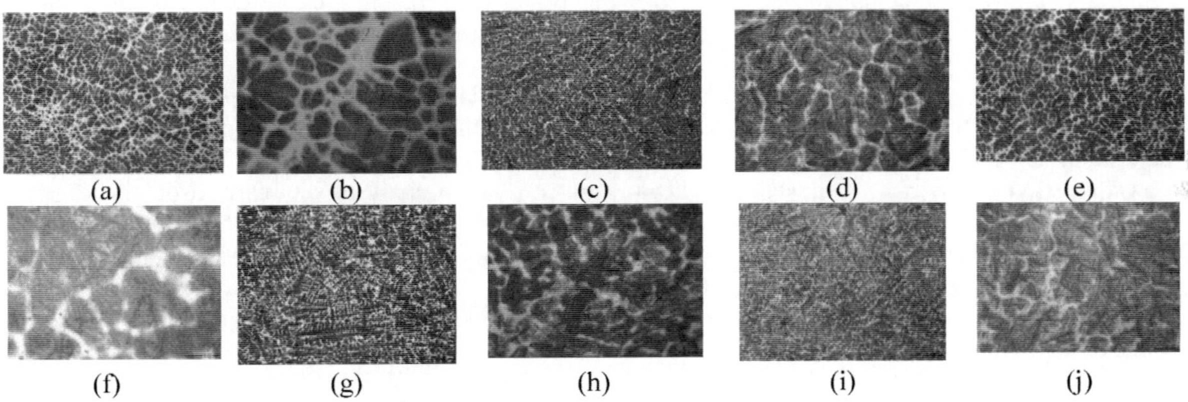

a, b – No. 6; c, d – No. 7; e, f – No. 8; g, h – No. 9; i, j – No. 10

Figure 2. Microstructure of the examined samples, (a, c, e, g, i × 100), (b, d, f, h, j × 500).

As a result of a decrease in the carbon content in the deposited layer to 0.19-0.2% with a simultaneous change in the content of chromium, nickel, molybdenum and other elements present in its composition, the martensite needles are enlarged and the size of the former austenite grains is increased (figure 2).

It is shown that in the microstructure of samples No. 7-10 there is a medium needle (point No. 5) and a large needle (point No. 7) martensite with a needle size of 8-13 μm. The value of the primary austenite grain corresponds to No. 6 (figure 2, table 2).

As a result of the assessment of contamination of the welded layer with nonmetallic inclusions, the presence of oxide nonmetallic inclusions, in particular non-deforming silicates and point oxides (Table 2), was identified.

The influence of chemical composition of powdered wires belonging to Fe-C-Si-Mn-Cr-Mo-Ni-V-Co system on the degree of wear and the hardness of the deposited layer was estimated by means of multivariate correlation analysis, which makes it possible to study the regularities of the change in the resulting index depending on the behavior of various factors by the methods described in [19, 20, 21].

IOP Conf. Series: Earth and Environmental Science **84** (2017) 012018 doi:10.1088/1755-1315/84/1/012018

Table 2. Characteristics of nonmetallic inclusions and structure of the examined samples.

Sample number	Contamination with non-metallic inclusions, points		Value of austenite grain, point	Size of martensite needles, μm
	Silicates non-deforming (brittle)	Spot oxides		
1	2b, 1b, 2a	1a	6, 7	3-6
2	2b, 1b, 3a	1a	7	2-6
3	1b, 2b, 3a	1 a	7	2-6
4	2b, 1б, 2a, 3a	1 a	6, 7	4-6
5	1b, 2b, 3a	1 a	6, 7	3-6
6	1b, 2b, 3a	1 a	7, 6	2-6
7	1b, 2a	1 a	6	8-13
8	1b, 2b, 2a, 3a	1 a, 2a	6	8-10
9	1b, 2b, 2a, 3a	1 a, 2a	6	8-12
10	1b, 2b, 3a	1 a, 2a	6	8-11

For the analysis the factors that influence the investigated indicator were identified, and the most significant of them were selected (table 1). After that, the initial information was checked for reliability, uniformity, compliance with the law of normal distribution. Next, a model of the factor system was constructed. Since the above systems have independent factors the deterministic factor analysis is used.

The type of connection was determined using pair and partial correlation coefficients. It is revealed that the relationship has a straightforward character, which is confirmed by the coefficient of determinacy, which for models of reduced factor systems is equal to 1.

Calculation of the main connection indicators of correlation analysis was carried out in stages. First, one factor was taken into account, which has the most significant effect on the resultant index, then the second, third, etc. At each stage the connection equation and indicators were calculated, with the help of which its reliability is estimated.

Based on the results of the calculations we obtained dependencies, the adequacy of which to actual values was checked by the average error of approximation:

$$\bar{\varepsilon} = \frac{1}{m}\sum_{i=1}^{m}\left|\frac{Y_i - \tilde{Y}_i}{Y_i}\right| \cdot 100 , \quad (1)$$

where m is the number of observations; $\tilde{Y}_i$ is the calculated value of the resulting indicator; Y_i – the actual value of the resulting indicator.

Dependences of the deposited layer hardness and its wear resistance on the mass fraction of the elements that form composition of the powder wires of Fe-C-Si-Mn-Cr-Mo-Ni-V-Co system obtained as a result of the analysis:

- hardness of the deposited layer (without taking into account the hydrogen content): $y=535.343 + 168.120 \cdot C - 276.437 \cdot Si - 890.442 \cdot Mn + 6.037 \cdot Cr + 108.957 \cdot Ni + 445.851 \cdot Mo - 433.688 \cdot Co - 1014.594 \cdot Al + 477.567 \cdot Cu$ (approximation error 0.01%);
- hardness of the deposited layer (taking into account the hydrogen content): $y=214.819 - 163.253 \cdot C - 307.499 \cdot Si - 353.293 \cdot Mn - 21.239 \cdot Cr + 33.871 \cdot Ni + 858.517 \cdot Mo - 280.917 \cdot Co + 321.784 \cdot Cu + 13.830 \cdot H$ (approximation error 0.01%);
- wear resistance of samples (without taking into account the hydrogen content): $y=0.002005 + 0.001110 \cdot C - 0.000804 \cdot Si - 0.003522 \cdot Mn + 0.000040 \cdot Cr + 0.000500 \cdot Ni + 0.000460 \cdot Mo - 0.001061 \cdot Co - 0.005525 \cdot Al + 0.003132 \cdot Cu$ (approximation error 0.32%);
- wear resistance of samples (taking into account the hydrogen content): $y=0.000260 - 0.000695 \cdot C - 0.000973 \cdot Si - 0.000597 \cdot Mn - 0.000108 \cdot Cr + 0.000091 \cdot Ni + 0.002707 \cdot Mo - 0.000229 \cdot Co + 0.002283 \cdot Cu + 0.000075 \cdot H$ (approximation error 10.74%).

The calculated values of the mean error of approximation show that the obtained dependences are adequate and they can be used to determine the resulting indicators.

As a result of the multifactor correlation analysis, the dependencies of deposited layer hardness and its wear resistance on the mass fraction of the elements in the composition of the powder wires of Fe-C-Si-Mn-Cr-Mo-Ni-V-Co system were determined. The obtained dependences can be used to predict the hardness of the deposited layer and its wear resistance when the chemical composition of the weld metal changes.

4. Conclusions

1. It was established that the decrease in the carbon content in the deposited layer up to 0.19-0.2% with a simultaneous change in the content of chromium, nickel, molybdenum and other elements present in it contributes to the enlargement of martensite needles and an increase in the size of the former austenite grain.

2. Based on the results of multifactor correlation analysis, the dependencies of the deposited layer hardness and its wear resistance on the mass fraction of the elements included in the powder wires of the Fe-C-Si-Mn-Cr-Mo-Ni-V-Co system were determined. The obtained dependences can be used to predict the hardness of the deposited layer and its wear resistance under different modes of operation of parts in mining and coal-mining equipment.

References

[1] Kirchgaßner M, Badisch E and Franek F 2008 *Wear J.* **265** 772–9
[2] Azzoni M 2009 *Weld Int.* **23** 706–16
[3] Klimpel A et al 2005 *Materials Processing Tech.* **164-165** 1056–61
[4] Wang Q and Li X 2010 *Welding* **89** 133–9
[5] Metlitskii V A 2008 *Welding Int.* **22** 796–800
[6] Kejžar R and Grum J 2005 *Welding Int.* **20** 961–76
[7] Li R et al 2014 *Surface Eng.* **30** 784–90
[8] Ma H R et al 2016 *Surface Eng.* **46** 1–7
[9] Filippov M A et al 2015 *Welding Int.* **29** 819–22
[10] Liu D S et al 2013 *Materials Science and Tech.* **30** 316–22
[11] Lim S C et al 1997 *Surface Eng.* **13** 247–50
[12] Zhuk Yu 1999 *Materials Tech.* 1999 **14** 126–9
[13] Hardell J et al 2014 *Tribology - Materials, Surfaces & Interfaces* **8** 90–7
[14] Deng X T et al 2016 *Materials Sci. and Tech.* **32** 320–7
[15] Paton B E 1974 *Technology of Electrical Welding of Metals and Alloys by Fusion* (Moscow: Metallurgy) p 768
[16] Gusev A I et al 2016 *Proc. VIIth All-Russian Sci. Conf. on Progressive Technologies and Economics in Machine Building* (Tomsk: TPU) pp 91–4
[17] Gusev A I et al 2016 *IOP Conf. Series: Materials Science and Engineering* **150** 012033
[18] Gusev A I et al 2016 *Bulletin of the Mining and Metallurgical Section of the Russian Academy of Natural Sciences. Department of Metallurgy* **36** 174–181
[19] Venttsel E S 1999 *Theory of Probabilities* (Moscow: Higher School) p 576
[20] Wentzel E S and Ovcharov L A 2003 *Theory of Probability and its Engineering Applications* (Moscow: Akademiya) p 464
[21] Gmurman V E 2003 *Theory of Probabilities and Mathematical Statistics* (Moscow: Higher School) p 479

IOP Conf. Series: Earth and Environmental Science **84** (2017) 012019 doi:10.1088/1755-1315/84/1/012019

New welding fluxes based on silicomanganese slag for deposition and welding of canopies and crib bed of mine support

R E Kryukov, N A Kozyrev, A A Usoltsev and O E Kozyreva

Siberian State Industrial University, 42 Kirova Street, Novokuznetsk, 654007, Russia

E-mail: kozyrev_na@mtsp.sibsiu.ru

Abstract. The paper considers the possibility of efficient use of silicomanganese slag for the production of welding fluxes. The results of studying the use of metallurgical wastes as components of welding fluxes are given. Analysis of the results of mechanical properties of the samples made it possible to determine the optimum content of the pulverized fraction less than 0.45 mm in the flux. The composition and technology of manufacturing a new welding flux using slag of silicomanganese production was developed. The effect of fractional composition on the welding-technological properties of fluxes was studied. The optimal content of liquid glass in the flux, which allows a favorable complex of mechanical properties to be obtained, is 20-30%. To reduce the level of contamination of the weld metal with non-metallic oxide inclusions and to increase the mechanical properties of the welded joint, it is proposed to introduce a carbon-fluorine-containing additive FD-UFS into fluxes based on the slag.

1. Introduction

Today much attention is paid to creation, research and development of new welding fluxes both in the Russian Federation and abroad [1 - 18]. The use of silicomanganese slag for the manufacture of welding fluxes is proposed [19, 20], the technology is protected by patents [21, 22].

2. Methods of research

The chemical composition of the examined surfaced samples was determined according to GOST 10543-98 (Russian National Standard) by the X-ray fluorescence method on XRF-1800 spectrometer and by the atomic-emission method using DFS-71 spectrometer. The metallographic analysis was carried out using OLYMPUS GX-51 optical microscope in a bright field in the magnification range ×100-1000 after etching in alcoholic nitric acid solution. The grain size was determined in accordance with GOST 5639-82 at magnification ×100. Investigation of longitudinal samples of the surfaced layer for the presence of nonmetallic inclusions was made in accordance with GOST 1778-70 at magnification of × 100.

3. Results and discussion

For flux preparation slag of silicomanganese production was used (its chemical composition is given in table 1), while in the first series of experiments the possibility of using a different ratio of slag fractions was investigated (table 2). Butt-welding under fluxes was performed without a double bevel on the samples with dimensions 500×75 mm, thickness 16 mm, from sheet steel grade 09G2S. The process was carried out using S-08GA wire and ASAW-1250 welding tractor in the following modes: $I = 700$ A; $U = 30$ V; $V = 35$ m/h.

From the welded plates the samples were cut and X-ray spectral analysis of the composition of weld metal and metallographic examinations of weld metal were performed. The chemical composition of welding fluxes is given in table 3. The chemical composition of the slag crust is given in table 4, the chemical composition of weld metal is given in table 5.

Table 1. Chemical composition of slag of silicomanganese production.

Content, %										
Al_2O_3	CaO	SiO_2	FeO	MgO	MnO	F	Na_2O	K_2O	S	P
6.91-9.62	22.85-31.70	46.46-48.16	0.27-0.81	6.48-7.92	8.01-8.43	0.28-0.76	0.26-0.36	below 0.62	0.15-0.17	0.01

Table 2. Fractional and component compositions of the investigated fluxes.

Sample	Ratio, %, fraction, mm
1	100 % fractions 0.45 – 2.5
2	95 % fractions 0.45 – 2.5 + 5 % fractions <0.45
3	90 % fractions 0.45 – 2.5 + 10 % fractions < 0.45
4	85 % fractions 0.45 – 2.5 + 15 % fractions < 0.45
5	80 % fractions 0.45 – 2.5 + 20 % fractions < 0.45
6	70 % fractions 0.45 – 2.5 + 30 % fractions < 0.45
7	60 % fractions 0.45 – 2.5 + 40 % fractions < 0.45
8	60 % silicomanganese slag + 40 % liquid glass
9	70 % silicomanganese slag + 30 % liquid glass
10	80 % silicomanganese slag + 20 % liquid glass
11	85 % silicomanganese slag + 15 % liquid glass

Table 3. Chemical composition of welding fluxes.

Sample	Content, %									
	Al_2O_3	CaO	SiO_2	FeO	MgO	MnO	F	Na_2O	S	P
8	5.29	25.84	51.75	0.55	5.02	7.39	0.36	4.66	0.12	0.01
9	5.48	26.68	51.73	0.57	5.16	7.59	0.39	4.19	0.13	0.01
10	5.88	25.53	52.53	0.56	5.07	7.75	0.31	4.07	0.13	0.01
11	6.55	26.81	51.14	0.56	5.78	8.10	0.35	2.62	0.14	0.01

Table 4. Chemical composition of slag crusts.

Sample	Content, %										
	MnO	SiO_2	CaO	MgO	Al_2O_3	FeO	Na_2O	K_2O	F	S	P
1	7.90	46.04	23.38	6.77	10.08	2.07	0.37	0.65	0.73	0.13	0.01
2	7.87	45.58	31.82	6.62	6.77	1.35	0.26	-	0.32	0.11	0.01
3	7.83	44.54	23.84	6.43	9.64	3.59	0.37	0.65	0.69	0.12	0.008
4	8.09	45.91	31.15	6.60	6.79	1.39	0.27	-	0.29	0.11	0.01
5	7.93	45.67	23.84	6.54	9.87	2.86	0.37	0.65	0.72	0.12	0.008
6	8.16	45.74	29.39	6.22	6.93	1.99	0.26	-	0.36	0.12	0.01
7	8.23	45.52	29.12	6.29	6.65	1.88	0.28	-	0.26	0.12	0.01
8	8.19	48.79	24.42	4.82	5.14	2.45	3.64	-	0.35	0.09	0.01
9	8.29	49.92	26.12	5.37	5.60	2.64	3.25	-	0.37	0.10	0.01
10	8.16	48.25	26.32	5.22	6.02	2.17	2.12	-	0.33	0.12	0.01
11	8.18	48.09	27.24	5.67	6.36	1.97	1.64	-	0.34	0.12	0.01

The results of the analysis for the presence of nonmetallic inclusions in the weld zone, carried out in accordance with GOST 1778-70, are shown in figure 1 and in table 5.

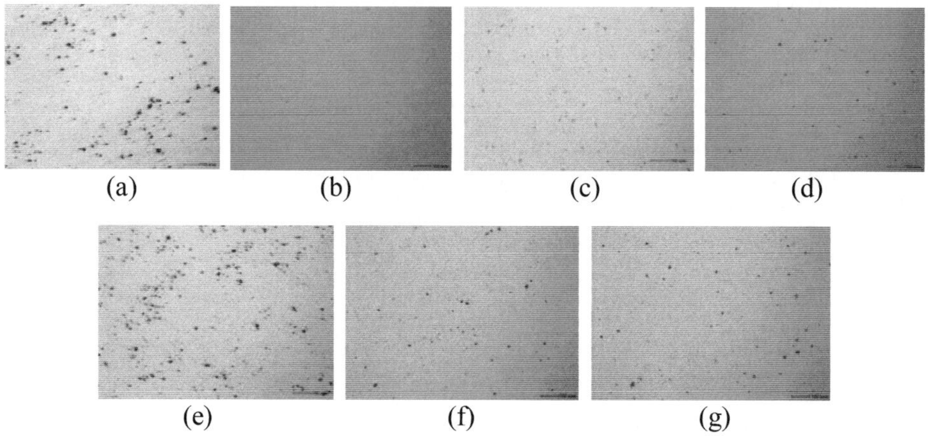

Figure 1. Nonmetallic inclusions in the weld zone of samples:
a – 1; b – 2; c – 3; d – 4; e – 5; f – 6; g – 7

Table 5. Chemical composition of weld seams metal.

Sample	Content, %										
	C	Si	Mn	Cr	Ni	Cu	V	Nb	Al	S	P
1	0.09	0.71	0.51	0.03	0.10	0.11	0.001	0.014	0.023	0.018	0.012
2	0.08	0.54	1.33	0.04	0.05	0.08	0.003	0.014	0.015	0.008	0.008
3	0.09	0.61	1.49	0.04	0.11	0.11	0.01	0.013	0.018	0.016	0.010
4	0.07	0.45	1.24	0.02	0.05	0.07	0.002	0.014	0.014	0.006	0.007
5	0.08	0.66	1.42	0.03	0.10	0.11	0.002	0.015	0.023	0.018	0.012
6	0.08	0.61	1.42	0.02	0.06	0.08	0.003	0.014	0.029	0.010	0.011
7	0.08	0.59	1.39	0.02	0.02	0.05	0.004	0.018	0.091	0.014	0.009
8	0.05	0.52	1.25	0.02	0.04	0.05	0.003	0.017	0.020	0.005	0.007
9	0.03	0.51	1.23	0.02	0.04	0.06	0.002	0.017	0.017	0.007	0.008
10	0.06	0.53	1.31	0.02	0.04	0.06	0.004	0.016	0.018	0.012	0.009
11	0.09	0.52	1.31	0.02	0.04	0.06	0.003	0.015	0.013	0.010	0.008

The microstructures of weld seams metal are shown in figure 2.

Samples were cut from the welded plates and mechanical properties were determined. After studying their mechanical properties the results were obtained indicating an increase in the values of impact hardness (figure 3).

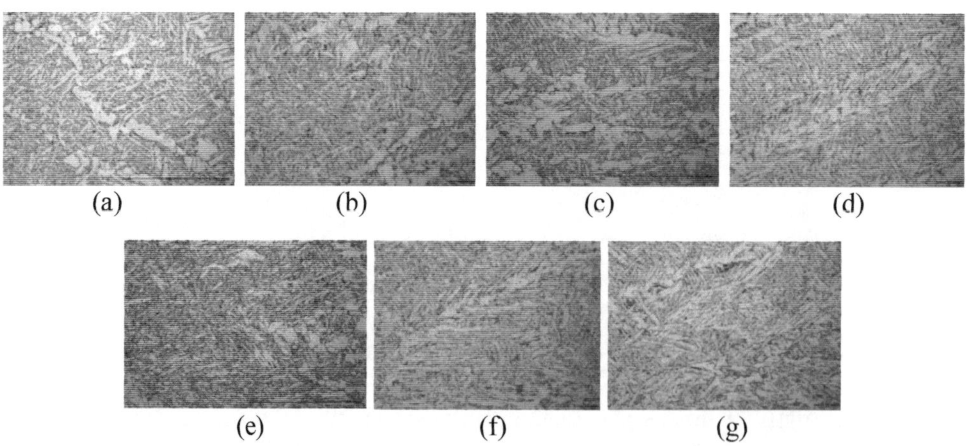

Figure 2. Microstructures of welded joints in the samples: a – 1; b – 2; c – 3; d – 4; e – 5; f – 6; g – 7

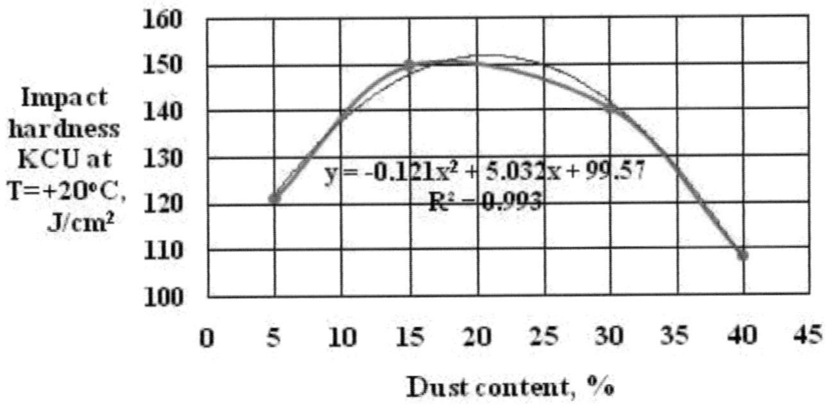

Figure 3. Influence of dust fraction content in the flux on the impact hardness.

Analysis of the mechanical properties of samples cut from the welded plates made it possible to establish that the optimum content of dust fraction less that 0.45 mm in the flux is 20-30%. With such content of dust fraction less than 0.45 mm in the flux, a favorable complex of mechanical properties of the samples cut from the welded plates is achieved.

In the structure of the weld metal of all samples ferrite is present in the form of non-equiaxed grains stretched in the direction of heat removal. The transition from a uniform ferrite-pearlite structure to the perlite and ferrite structure of vidmanstetten orientation can be seen. At the same time, the samples did not show a significant change in grain size on the grain scale (tables 6 and 7).

In the second series, the possibility of using a ceramic flux made of silicomanganese slag dust with a fraction up to 0.45 mm bound by liquid glass was studied. The manufacturing process consisted of mixing silicomanganese slag with liquid glass in various ratios (table 2), drying, crushing, sieving and obtaining a fraction of 0.45-2.5 mm.

Analysis of the results of mechanical properties of samples cut from welded plates made it possible to determine the optimal content of liquid glass in the flux (up to 20-30%) in order to achieve a favorable complex of mechanical properties of samples cut from welded plates (figure 4, 5).

However, the investigated fluxes are oxidizing and are based on the principles of silica-manganese-oxidation-reduction processes, in connection with which the products of these reactions are oxide compounds of silicon and manganese, thereby increasing the level of contamination of the weld metal with nonmetallic inclusions and, as a consequence, mechanical properties deteriorate, especially at low negative temperatures. To reduce the contamination of weld metal and improve mechanical properties, we considered the possibility of introducing the developed FD-UFS additive into a new flux.

Table 6. Non-metallic inclusions in the weld zone.

Sample	Nonmetallic inclusions, point		
	non-deforming silicates	non-deforming silicates	non-deforming silicates
1	4b; 3b; 4a	3b	1a
2	2b; 1b; 3a; 4a	-	1a; 2a
3	4b; 2b	-	1a; 2a
4	2b; 4b	-	1a; 2a
5	4b; 5b; 3b	-	1a; 2a
6	2b; 1b; 2a; 2.5a	-	1a; 2a
7	2b; 2a; 2,5a	-	1a; 2a
8	2b; 1b; 2a; 2.5a	-	1a
9	2b; 1b; 2a; 2.5a	-	1a
10	2b;1b; 2a; 2.5a	-	1a; 2a
11	2b; 2.5a	-	1a, 2a

IOP Conf. Series: Earth and Environmental Science **84** (2017) 012019 doi:10.1088/1755-1315/84/1/012019

Table 7. The value of weld seam grains in accordance with GOST 5639-82.

Sample	Grain size according to grain scale
1	No. 4, No. 5
2	No. 5, No. 4
3	No. 4, No. 5, No. 6
4	No. 4
5	No. 5, No. 4
6	No. 4
7	No. 4
8	No. 5, No. 4
9	No. 4, No. 5
10	No. 4
11	No. 4, No. 5

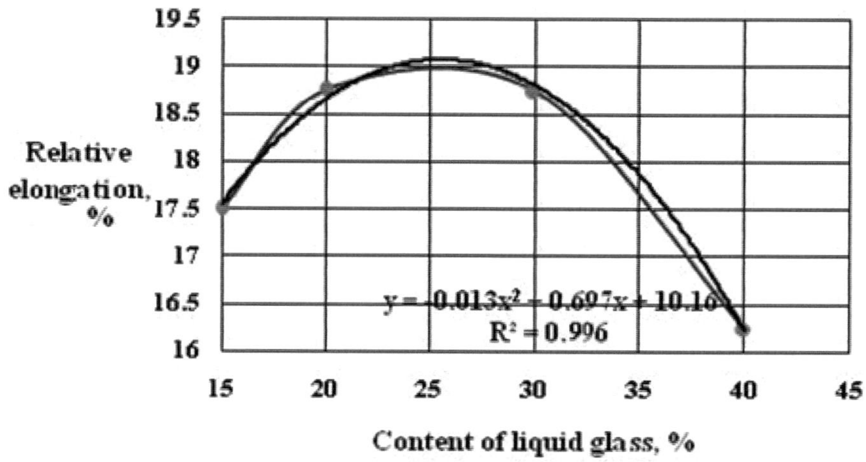

Figure 4. Influence of the content of liquid glass in the flux on the relative elongation.

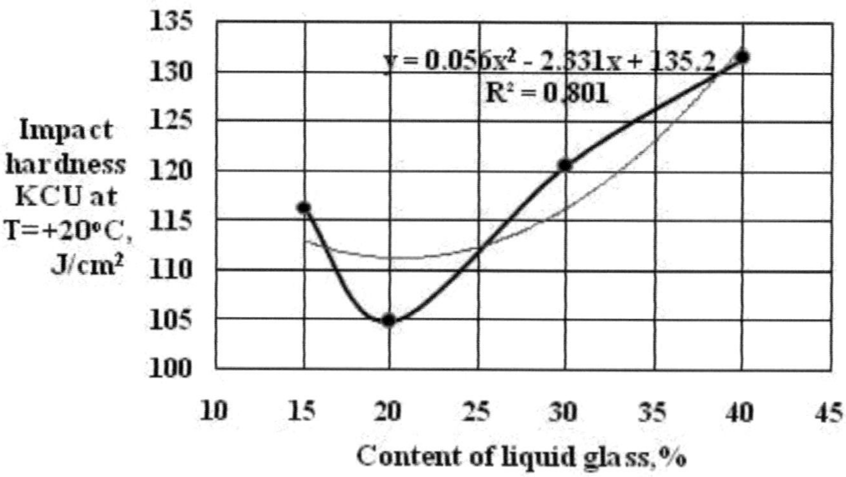

Figure 5. Influence of the content of liquid glass in the flux on the impact hardness.

In the experiments a flux additive to the new flux was used mixed in a ratio of 2, 4, 6, 8 %, respectively.

The chemical composition of the investigated crusts is given in table. 8, the slag crust is given in table 9, the chemical composition of weld metal is given in table 10.

Thus, the highest level of contamination with nonmetallic inclusions is observed in the weld metal, made with a flux without an additive. The addition of FD-UFS additive reduces the level of contamination with non-metallic inclusions, reducing their size and quantity. In the investigated intervals the use of an additive in the amount of 8 % is more effective for reducing the level of contamination with nonmetallic inclusions.

Table 8. Chemical composition of the investigated flux mixtures, %.

Content of FD-UFS in the flux, %	FeO	MnO	Ca	SiO_2	Al_2O_3	MgO	Na_2O	K_2O	S	P	ZnO	F
2	0.40	8.01	15.80	50.08	11.55	7.39	0.77	0.63	0.22	0.008	0.002	1.30
4	0.91	7.90	17.72	46.63	10.32	6.63	1.10	0.68	0.24	0.01	-	1.95
6	0.81	7.68	16.79	43.64	11.27	5.71	2.25	0.65	0.34	0.01	0.003	4.04
8	0.46	7.46	16.00	43.64	11.86	5.56	2.30	0.60	0.33	0.01	0.002	3.96

Table 9. Chemical composition of slag crusts, %.

Content of FD-UFS in the flux, %	FeO	MnO	Ca	SiO_2	Al_2O_3	MgO	Na_2O	K_2O	S	P	ZnO	F
2	2.21	7.25	15.55	38.09	9.39	8.63	0.49	0.57	0.12	0.006	0.002	0.94
4	2.28	7.39	16.90	42.00	9.76	5.77	0.76	0.62	0.15	0.008	0.002	1.12
6	2.24	7.20	16.06	39.94	11.15	7.14	1.09	0.60	0.17	0.008	0.002	1.53
8	2.36	7.14	14.70	42.87	12.40	5.57	1.34	0.57	0.20	0.008	0.002	1.88

Table 10. Chemical composition of weld seams.

Content of FD-UFS in the flux, %	Mass fraction of elements, %									
	C	Si	Mn	Cr	Ni	Cu	Nb	Al	S	P
2	0.09	0.62	1.40	0.02	0.06	0.09	0.014	0.023	0.020	0.008
4	0.10	0.60	1.34	0.02	0.07	0.08	0.010	0.013	0.023	0.009
6	0.12	0.66	1.43	0.02	0.06	0.10	0.011	0.012	0.027	0.008
8	0.13	0.65	1.36	0.03	0.06	0.09	0.013	0.013	0.024	0.008

The results of the analysis for the presence of nonmetallic inclusions in the weld zone, carried out in accordance with GOST 1778 - 70, are shown in figure 6 and in table 11.

The microstructure of the weld seam of samples is shown in figure 7. It is established that the introduction of an additive in the amount up to 8 % does not affect the size and morphology of the structural components.

(a) (b) (c) (d)

Figure 6. Nonmetallic inclusions in the weld zone of samples with additive %: a – 2; b – 4; c – 6; d – 8.

Table 11. Nonmetallic inclusions in the weld zone.

Content of FD-UFS in the flux, %	Nonmetallic inclusions, point		
	non-deforming silicates	brittle silicates	spote silicates
2	2b, 4b, 5b	-	1a, 2a
4	2b, 4b	-	1a, 2a
6	2b, 4b, 1b	-	1a, 2a
8	2b	-	1a, 2a

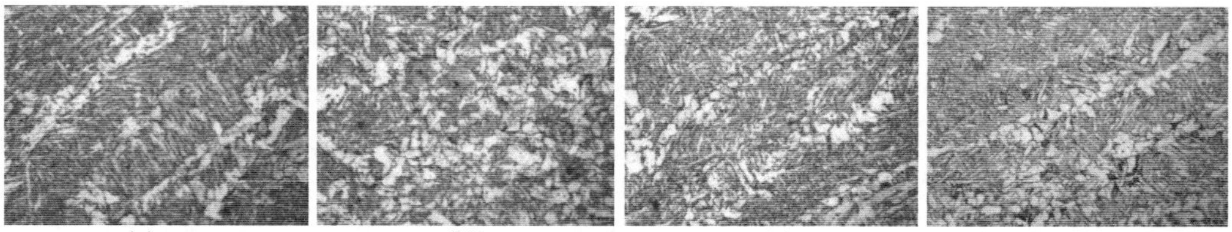

(a) (b) (c) (d)

Figure 7. Microstructure of the weld seams of samples with an additive, %: a – 2; b – 4; c – 6; d – 8.

The study of the values of mechanical properties showed that they improve with the increase in the amount of FD-UFS additive (figure 8). The conducted researches formed the basis of patents of the Russian Federation [21, 22].

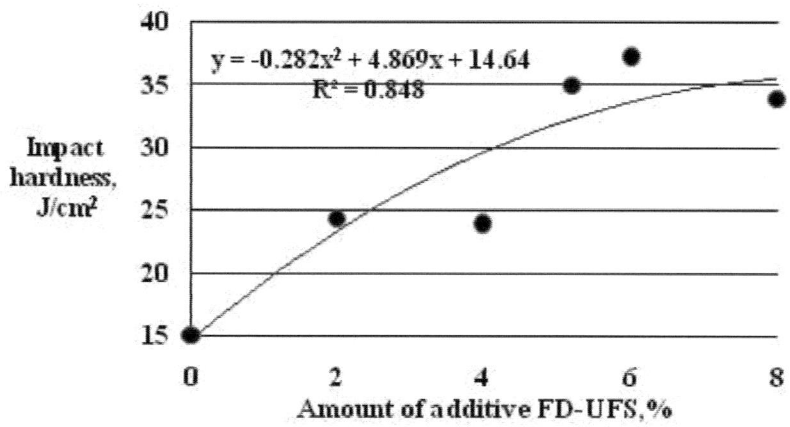

Figure 8. Influence of the content of FD-UFS additive in flux on the impact hardness (KCV at -20 °C).

4. Conclusions

1. The principal possibility of using slag of silicomanganese production for manufacturing welding fluxes is shown.

2. It is possible to use up to 30 % of fine fraction (less than 0.45 mm) in fluxes. With such a content of the dust fraction in the flux a favorable complex of mechanical properties of the samples cut from the welded plates is achieved.

3. Optimal content of liquid glass in the flux, which allows a favorable complex of mechanical properties to be obtained, is (20-30 %). This flux is recommended for deposition and welding elements of mechanized support.

4. It is proposed to introduce a carbon-fluorine-containing additive FD-UFS in the amount of 2-8 % to reduce the level of contamination of the weld metal with nonmetallic oxide inclusions and increase

IOP Conf. Series: Earth and Environmental Science **84** (2017) 012019 doi:10.1088/1755-1315/84/1/012019

the mechanical properties of the weld seam. The addition of FD-UFS additive reduces the level of contamination with nonmetallic inclusions, reducing their size and quantity.

References

[1] Puchol R Q et al 2009 *Welding Int.* **23(2)** 132–40
[2] Crespo A C et al 2007 *Welding Int.* **21(7)** 502–11
[3] Crespo A C et al 2009 *Welding Int.* **23(2)** 120–31
[4] Golovko V V and Potapov N N 2011 *Welding Int.* **25(11)** 889– 93
[5] Volobuev Yu S et al 2012 *Welding Int.* **26(8)** 649–53
[6] Volobuev Yu S, Surkov A V et al 2010 *Welding Int.* **24(4)** 298–300
[7] Potapov N N and Kurlanov S A 1987 *Welding Int.* **1(10)** 951–4
[8] Babushkin P L and Persits V Yu 1991 *Welding Int.* **5(9)** 741–2
[9] Pavlov I V and Oleinichenko K A 1995 *Welding Int.* **9(4)** 329–32
[10] Chigarev V V and Kosenko A A 1994 *Welding Int.* **8(10)** 808–9
[11] Kurlanov S A et al 1993 *Welding Int.* **7(1)** 65–8
[12] Bublik O V and Chamov S V 2010 *Welding Int.* **24(9)** 730–3
[13] Gur'ev S V et al 2012 *Welding Int.* **26(8)** 646–8
[14] Parshin S G 2012 *Welding Int.* **26(10)** 800–4
[15] Cruz-Crespo A et al 2005 *Welding Int.* **19(7)** 544–51
[16] Barmin L N et al 1989 *Welding Int.* **3(2)** 109–11
[17] Kazakov Yu V et al 1991 *Welding Int.* **5(3)** 202–5
[18] Potapov N N et al 2009 *Welding International* **23(10)** 800–3
[19] Kozyrev N A, Kryukov R E et al 2016 IOP *Conf. Series: Materials Sci. and Eng.* **125** 012034
[20] Kozyrev N A, Kryukov R E et al 2016 IOP *Conf. Series: Materials Sci. and Eng.* **150** 012032
[21] Kryukov N E, Kryukov E N, Kozyrev N A, Kryukov R E and Kozyreva O A 2016 *Flux for Welding Patent of the Russian Federation* 2576717, MPK8 V23 K35/362 publ. 03.10.2016 Bul. No. 7
[22] Kryukov N E, Kryukov E N, Kozyrev N A, Kryukov R E and Kozyreva O 2016 *Flux for Welding Patent of the Russian Federation* 2579412, MPK8 V23 K35/362 publ. 04.10.2016 Bul. No. 10

IOP Conf. Series: Earth and Environmental Science **84** (2017) 012020 doi:10.1088/1755-1315/84/1/012020

Enhancing down-the-hole air hammer capacity in directional drilling

V I Klishin[1], V V Timonin[2], D I Kokoulin[2], S E Alekseev[2] and B Kubanychbek[2]

[1]Institute of Coal of the Federal Research Centre of Coal and Coal Chemistry, Siberian Branch of the Russian Academy of Sciences, 10 Leningradsky Prospect, Kemerovo, 650065, Russia,
[2]Chinakal Institute of Mining, Siberian Branch of the Russian Academy of Sciences, 54 Krasnyi pr., Novosibirsk, 630091, Russia

E-mail: klishinvi@icc.kemsc.ru

Abstract. The authors discuss the issue connected with drilling trajectory deviation and present the technique of rotary–percussion drilling with a down-the-hole air hammer. The article describes pilot testing of the air hammer drill PNB76 in Berezovskaya Mine. The ways of improving the air hammer drill are identified, and the basic diagram and R&D test data are given.

1. Introduction

One of the problems in hole-making in coal mines is to ensure straight-line drilling. A shortage of rotary drilling rigs is limited application in hard rocks [1]. Of special concern is drilling in hard rocks for the purpose of directional hydraulic fracturing in backs [2]. Furthermore, drilled holes often miss a targeted excavation and are lost, which worsens the drilling performance. With any method used, drilling trajectory is generally deviated due to the increased axial pressure exerted on a drilling assembly.

It is possible to ensure pre-set path of hole-making with the method of directional drilling [3, 4]; on the other hand, it needs extra expenses for positioning devices and tools meant to change travel trajectory of the drill head are required. A promising approach to straight-line hole-making is rotary–percussion drilling using down-the-hole air hammer drills. As compared with the other techniques, this method allows minimum pressing force between the drill head and the hole bottom, as well as low speed and minor torque transmitted to the drilling assembly, which enables minimized distortion of hole trajectory and ensures straight-line drilling [5].

2. Experimental research procedure

It is advisable to reduce the load on the drilling assembly even more with the down-the-hole air hammer drills having minimum diameter for the given drill rig. In this case, forces induced under percussive interaction between the drill head and the hole bottom are applied closer to the center of the hole bottom, which mitigates bending moments transmitted to the drilling assembly. Creation of a small-diameter hammer drill is a complex technological problem in as much as the cross dimensions of down-the-hole air hammer drills are strictly limited in this case and the possibility to generate acting faces of the wanted size is small. The scheme developed with the aim to enlarge acting faces of an air hammer drill contains a number of chambers and the striking part area subjected to pressure is larger than the area of cross section of the housing cavity [6, 7].

Furthermore, straight-line drilling needs more uniform distribution of impact load at the hole bottom. The latter is achieved when the striking part and the tool collide at the center and the impact load is uniformly distributed in these parts. The central collision reduces lateral oscillation of the air hammer drill, which is beneficial for the straight-line drilling performance.

Based on the considerations presented above, at the Institute of Mining, a small-dimension air hammer drill PNB76 was designed [8, 9, 10]. Under order of Berezovskaya Mine, Spetsgidravlika LLC manufactured three PNB76 machines with the drill bits KNB76 (figure 1) for operation with the drill rig SBR-400 [11].

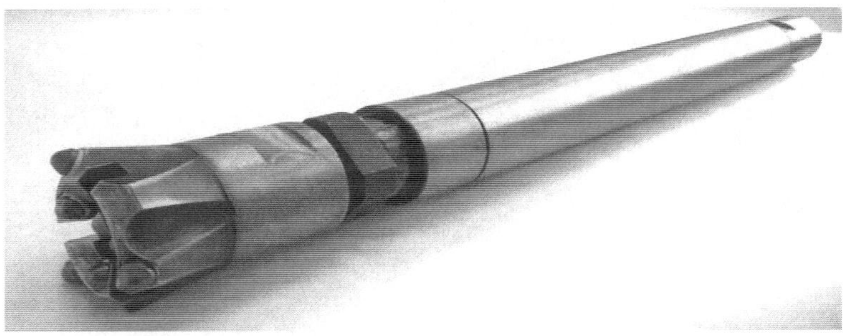

Figure 1. R&D model of the air hammer drill PNB76 with the drill bit KNB76.

After testing of serviceability in a mine, the air hammer drills were delivered to Berezovskaya Mine, Kemerovo Region, in autumn 2016. The commercial testing involved full-bottom process hole drilling with a diameter of 76 mm in rocks having strength of 90–120 MPa. The hole was intended to bypass water from an excavation to a lower-level drive. Drilling used the drill rig SBR-400 with the drill rods having diameter of 50 mm.

Drilling involved two methods, namely:

- rotary drilling with cutting bits;
- drilling with the down-the-hole air hammer drill equipped with a drill bit with sphero-cylindrical tungsten–carbide inserts.

3. Results and discussion

With the first method applied, the cutting bits failed within the first meters of drilling. Drill bits designed at Gornyi Instrument, Sibselmash and by specialists of Berezovskaya Mine failed too.

Drilling under the same conditions but using the down-the-hole air hammer drill PNB76 (figure 2) showed stable operation of the machine with the reliable lock mode. The power source for the air hammer drill was diesel compressor of an explosion-proof design. The compressed air pressure was 0.4 MPa at the flow rate of 6.0 m^3/s. The 76 mm-diameter hole drilling rate in rocks having strength of 110–120 MPa made 70 mm/min. The drill rig SBR-400 operated in the regime of moderation. The fluid pressure was not higher than 5.0 MPa in the pressure line and not higher than 3.5 MPa in the feed line, which conformed with the driving torque not more than 300 Nm at the drilling assembly and the pulldown of the latter not more than 300 kg. Thus, the air hammer drill implemented the main drilling functions, and the detrimental effect of high torsional moments, which contributed to drilling deviation from a preset trajectory, was reduced.

The process hole between excavations was drilled by the air hammer drill in three shifts. The hole was 40 m long, and it preserved the preset trajectory.

The further hole drilling with the air hammer drill PNB76 was terminated due to failure of the bits KNB76. The cause of the failure of the bits was that the hammer drill operating at the low air pressure developed impact energy that appeared to be lower than the energy required for rock destruction, and the bits suffered from high-rate wear as a result. Unlike edge bits, bit inserts need higher loading. From practical knowledge of efficient rock breakage, the load on bit inserts is to be not less than 10 J. Inconformity with this standard is the cause of poor performance of drilling equipment.

Figure 2. Actual drilling by the drill PNB76 and the drill rig SBR-400 in Berezovskaya Mine, Kemerovo Region.

Based on the full-scale test data, some conclusions had been drawn, namely:

- directional straight-line drilling in hard rocks is feasible with the down-the-hole rotary–percussion air hammer drills;
- in order to ensure efficient drilling of holes with a diameter of 76 mm, it is required to increase impact energy developed by the proposed air hammer drill by 1.5–2 times based on improvement of its design and build-up of feed line pressure to 0.8–1.0 MPa;
- the new design of a bit with the tungsten–carbide edge inserts will enable efficient utilization of the method of rotary–percussion drilling;
- regular sharpening of the edge bits is required when dulling area reaches 3 mm.

In accordance with the conclusions drawn and based on the full-scale tests in Berezovskaya Mine, the Institute of Mining implemented design improvement for the air hammer drill PNB76 and engineered a new drill bit.

Aiming to increase energy of the air hammer drill at the pressure to 0.6 MPa, the air distribution system is equipped with an elastic valve intended to build-up pressure in the power stroke chamber [12. 13]. The layout of the improved design air hammer drill PNB76 is demonstrated in figure 3.

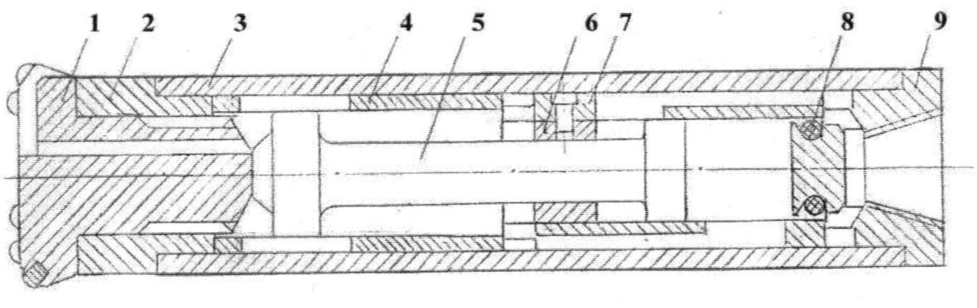

Figure 3. Improved layout of PNB76 hammer: 1 – drill bit; 2 – axle box; 3 – housing; 4 – front casing 5 – striking part; 6 – C-ring; 7 – back casing; 8 – elastic valve; 9 – adapter.

This design allows the air hammer drill to operate at the standard pressure of 0.6 MP and under an increased pressure of 1.2 MPs, which enhances the machine generality. This approach makes it possible to update both small-diameter and large-diameter hammer drills. The design of the edge drill bit is shown in figure 4.

The R&D model of the modernized air hammer drill PNB76U was tested in a laboratory (figure 5).

During the tests, diagrams of pressure in the chambers of PNB76U air hammer drill equipped with an elastic valve were recorded at the air pressure of 0.6 MPa. These diagrams are presented in figure 6 in comparison with the pressure diagrams recorded for the air hammer drill PNB76 without the elastic valve.

Figure 4. Edge drill bit.

Figure 5. Tests of PNB76U hammer on the vertical test bench GD-251.

Apparently, pressure in the power stroke chamber at the end of the stroke is considerably higher in the air hammer drill PNB76U than in PNB76. This is the source to increase average pressure in the power stroke chamber and, accordingly, to enhance energy performance of the machine.

It is seen in the pressure diagram of PNB76U air hammer drill that the pressure difference between the feed line and the power stroke chamber is sufficient for timely actuation of the elastic valve, which ensures reliable start and operation of the machine. It is noteworthy that the time of the working cycle becomes shorter, which increases the impact frequency and the air hammer drill capacity.

After processing the diagrams using the known procedure [14], the main energy parameters of the two machines are summarized in table 1 below.

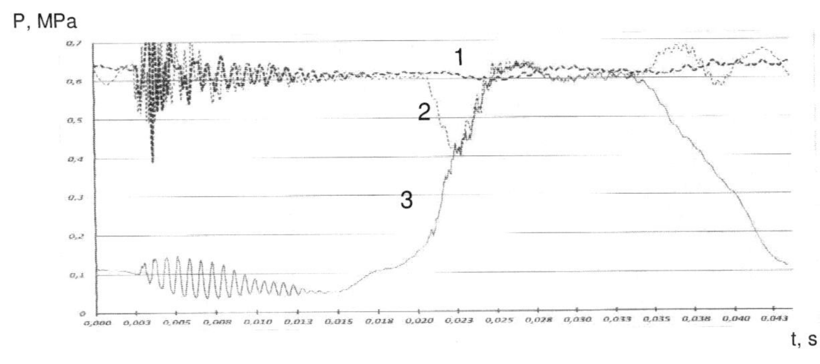

(a)

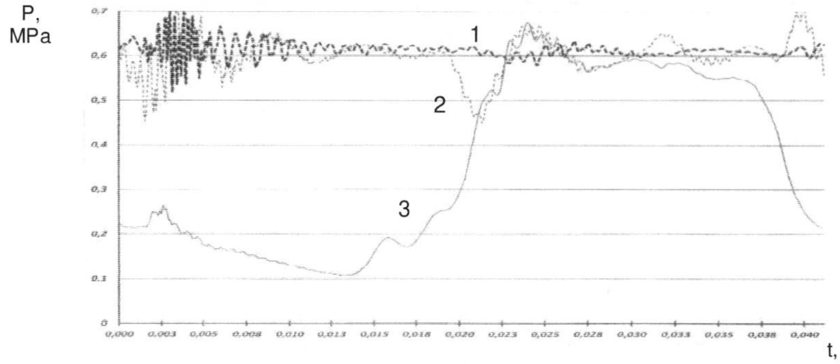

(b)

Figure 6. Pressure in the chambers of (a) PNB76 hammer drill without the valve and (b) PNB76U hammer drill with the valve: 1 – feed line; 2 – idle stroke chamber; 3 – power stroke chamber.

Table 1. Parameters of air hammer drills.

Parameters	PNB76 (without valve)	PNB76U (with valve)
Outer diameter of housing, mm	63.5	
Length (without drill bit), mm	675	
Weight	1.45	
Air pressure, MPa	0.6	
Impact energy, J	40	64.8
Impact frequency, min^{-}	1371	1456
Impact capacity, W	914	1573

After the tests, trial drilling using the drill rig NKR100M was undertaken in a block of granite having strength of 140 MPa (figure 7). Penetration rate made 100 mm/min.

Figure 7. Trial drilling in granite with PNB76U air hammer drill and NKR100M drill rig.

4. Conclusion

Thus, the basic diagram of the air hammer drill PNB76 allows including the elastic valve to enhance energy performance at the same capacity of the machine, namely, closed-type operation with the full exhaust of spent air toward the drill hole bottom, which improves removal of drill cuttings and enhances the hammer drill durability. During operation of the air hammer drill at the air pressure of 0.6 MPa, the unit impact energy is increased by 62 %, impact frequency—by 6 % and the impact capacity – by 72 %. These enhancements enable efficient operation of the air hammer drill both under standard and increased pressure of compressed air.

IOP Conf. Series: Earth and Environmental Science **84** (2017) 012020 doi:10.1088/1755-1315/84/1/012020

References

[1] Klishin V I, Kokoulin D I and Fokin Yu S 2007 *Ugol* **4(972)** 25–8

[2] Klishin V I, Nikolsky A M, Opruk G Yu, Neverov A A and Neverov S A 2008 *Ugol* **11(991)** 12–7

[3] Hungerford F, Ren T and Aziz N 2013 *Int. J. Min. Sci. Technol.* **23** 543–53

[4] Kondratenko A S, Timonin V V and Patutin A V 2016 *J. Min. Sci.* **52** **1** 129–34

[5] Timonin V V, Kokoulin D I, Alekseev S E and Kubanychbek B 2016

[6] *Proc. Conf. of Chinakal Inst. Mining, SB RAS on Problems of Development of Mining Sciences and Mining Industry* (Novosibirsk: IGD SO RAN)

[7] Repin A A, Alekseev S E, Timonin V V and Karpov V N 2015 *Proc. XXII Int. Conf. Miner's Week-2015* pp 475–82

[8] Alekseev S E, Timonin V V, Kokoulin D I, Shakhtorin I O and Kybanychbek B 2015 *J. Fundament. Appl. Min. Sci.* **2** 187–93

[9] Alekseev S E 1997 *Down-the-hole Pneumatic Impact Device* RF Patent No. 2090730 Bull. No. 26

[10] Repin A A, Alekseev S E and Pyatnin G A 2009 *Done-the-hole Air Hammer Drill* RF Patent No. 2343266 Bull. No. 1

[11] Repin A A, Alekseev S E and Karpov V N 2012 *Down-the-hole Air Hammer Drill* RF Useful Model 121854 Bull. No. 31

[12] Klishin V I, Kokoulin D I, Kubanychbek B and Gurtenko A P 2010 *J. Min. Sci.* **46(4)** 411–5

[13] Gaun V A 1981 *Pneumatic Percussion Device* USSR Auhtor's Certificate No. 848 615 Bull No. 27

[14] Petreev A M and Primychkin A Yu 2016 *J. Min. Sci.* **52(1)** 135–45

[15] Esin N N 1965 *Procedure to Test and Adjust Air Hammers* (Novosibirsk: SO RAN)

Ensuring stability of maintained goaf by means of directional hydraulic fracturing (DHF)

V I Klishin, G Yu Opruk and A S. Teleguz

Institute of Coal of the Federal Research Centre of Coal and Coal Chemistry, Siberian Branch of the Russian Academy of Sciences, 10 Leningradsky Prospect, Kemerovo, 650065, Russia

E-mail: klishinvi@icc.kemsc.ru

Abstract. The article describes how directional hydraulic fracturing (DHF) of roof rock was used while the longwall face was operating and airway for the isolated methane-air mixture drainage from the goaf was preserved. The authors give the reasons for using DHF in mining and geological conditions of Esaulskaya mine. They describe the sequence of the performed operations aimed at weakening roof rocks and also list the special equipment used during the process. Based on the results of the performed operations, the authors make a conclusion concerning the effectiveness of this work.

1. Introduction

Industrial experimental work aimed at implementation of directional hydraulic fracturing of roof in different process flows is taking place in Kuzbass [1, 2].

Longwall face 26-25bis in Esaulskaya mine has been extracted under extremely complex conditions, namely in the pillar between previously extracted pillars 26-23 and 26-25. The sizes of the protective pillars of airway and belt road do not exceed 6.0 m. Mining-and-geological and mining-and-technical parameters of the working area 26-25bis are given in table 1.

The main roof is represented by fine grain and coarse grain siltstone with thin interbeds of sandstone. The change of main roof types in controlled by the change of lithologic types combinations. In the second part of the longwall face coarse grain siltstone prevailed. This shows that the main roof is formed by the tight rock. Hardness factor according to Prof. Protodiakonov scale for fine grain siltstone equals 6, for coarse grain siltstone it is 7, for sandstone – 9. Estimated spacing of the initial main roof caving is 109.0 m and the secondary – 35.0 m.

Thus, the presence of extremely complex mining and geological conditions of extraction indicated the possibility of sufficient roof hanging when the longwall face is working which leads to excessive loads on props, reinforcement of the preserved part of airway with the possibility to create critical stress on the marginal part of the working. To reduce the negative effect of the excessive pressure of the roof tight rock, the specialists decided to use directional hydraulic fracturing [3, 4, 5].

According to the project documents, ventilation of mine workings during coal extraction from longwall face 26-25bis was performed using combined ventilation circuit with isolated methane-air mixture drainage from the goaf along the maintained part of airway 26-25bis and workings of gas drainage system. The width of protective pillars was 6 m. The main goal during the extraction of longwall face 26-25bis deposits was to preserve a part of the airway 26-25bis for isolated methane-air mixture drainage from the goaf after the longwall face advance.

Table 1. Mining-and-geological and mining-and-technical parameters of the working area 26-25bis.

Parameter	UM	Value
Longwall face index		26-25bis
Seam		26a
Coal grade		G
Operating ash content	%	30.7
Mining depth	m	520-550
Seam thickness	m	1.85-1.92
Extraction height	m	2.15
The angle of incidence of the reservoir, from-to, - along the longwall face - along the extraction column	degree	0-2 1-8
Length of working area	m	1140
Length of working face	m	120

2. Results and discussion

The specialists of Institute of Coal of FIC UUKh SB RAS and mine technical workers offered a technical plan to perform a directional hydraulic fracturing of the roof to reduce the strength of rock pressure in the preserved part of the airway 26-25bis in the langwall face 26-25bis (figure 1), which made it possible to "cut" of the roofing bracket over the airway 26-25bis behind the langwall face and save the airway during column extraction in langwall face 26-25bis [6, 7].

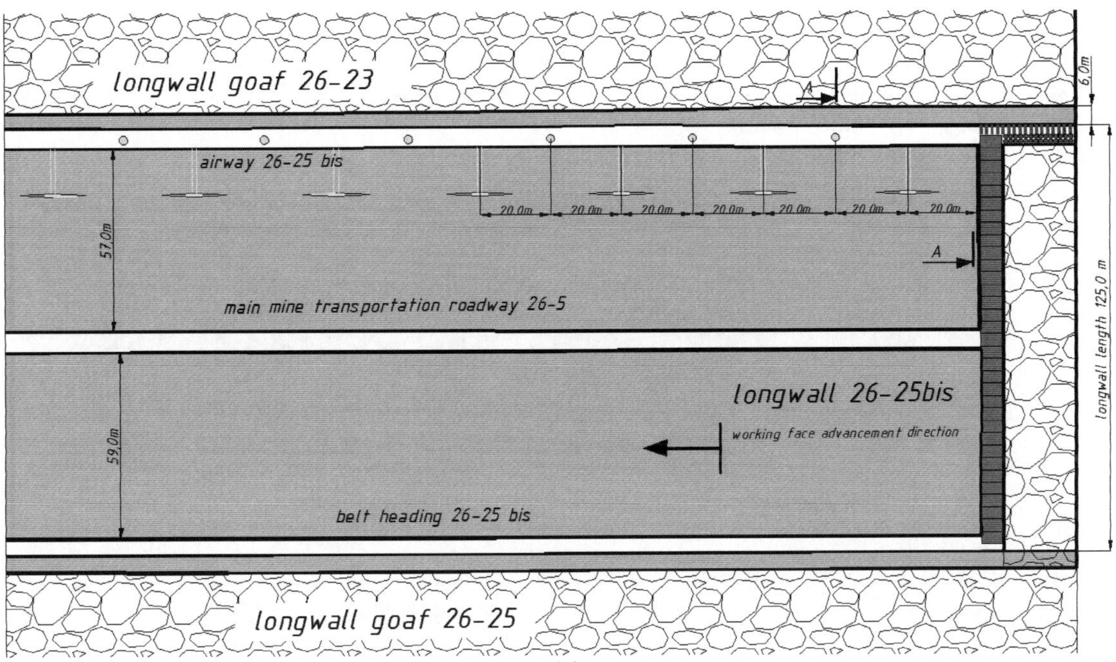

(a)

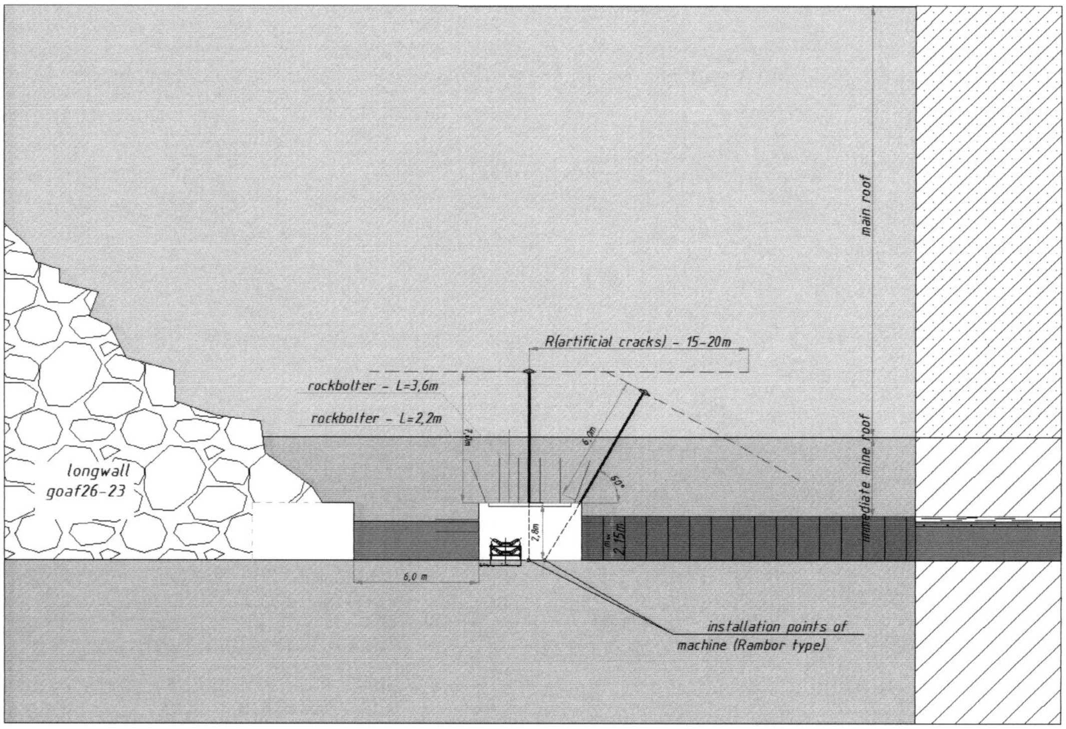

(b)

Figure 1. DHF engineering diagram: a) boreholes layout necessary for DHF implementation; b) vertical scheme of holes layout for DHF implementation in airway 26-25bis (crosscut A-A).

The engineering procedures involved in performing DHF were established using both standard general use equipment and special one. The boreholes and initiating crevices were made with pneumatic drilling roof bolt setter. Rock boring bits (46 mm in diameter) were used for drilling. Initiating crevices were made with mechanical crevice initiating devices (figure 2), which were fixed on the steel stems of the drill instead of boring bit [8].

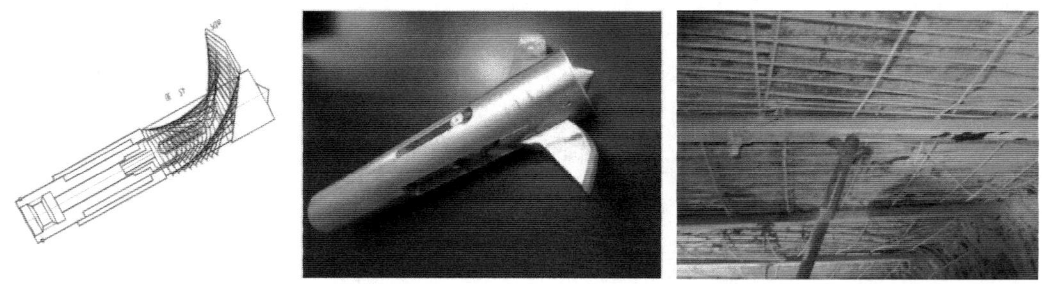

Figure 2. Crevice initiating device SHCHM – 45/1.

Sealing of initiating crevice zone was performed by means of hydraulic sealer "Taurus" type (figure 3). Special set of high-pressure pipes was used to deliver sealing agent to the borehole bottom. Compression unit was connected with the pipe ends sticking from the borehole through special adapter (collector) by means of high-pressure flexible hoses. Fluid was forced into the initiating crevice zone by means of high-pressure unit of mechanical complex.

Control over the drilling effectiveness and DHF process was performed by means of video endoscope. As a result of these measures it was found out that the boreholes were drilled on the project depth and there is an initiating crevice in each borehole opening (figure 4).

First of all, DHF is implemented in vertical boreholes when artificial crevices spread along the formation and then in the slant holes when the crevices develop at an angle to the formation. The artificial crevice will come into contact between the main roof and immediate mine roof and further spread along the contacting part (figure 5). When the liquid is pumped into the borehole, the DHF starts 5-10 sec after the emulsion is supplied into the system. There is sharp pressure drop on manometer near the initiating crevice.

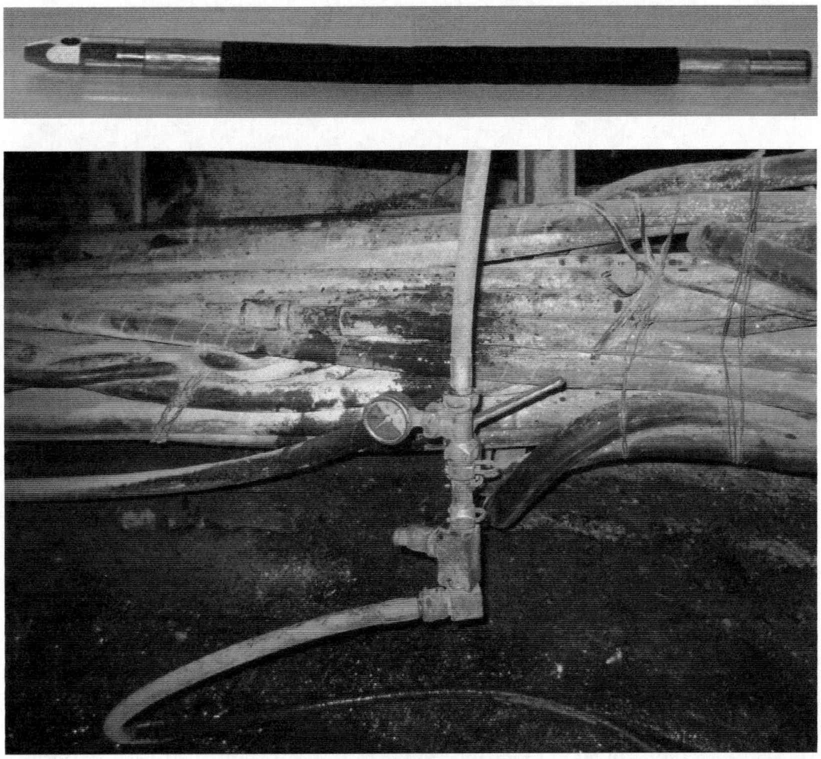

Figure 3. Sealer “Taurus”.

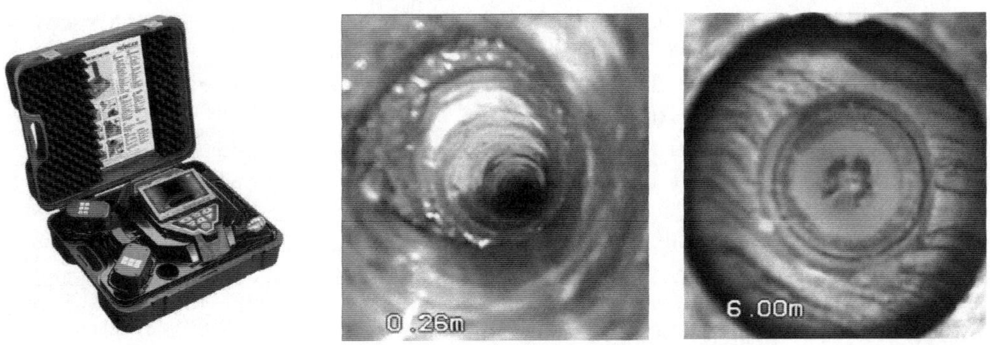

(a) (b) (c)

Figure 4. Borehole with initiating crevice before DHY (a – video endoscope, b – hole collar, c – borehole bottom).

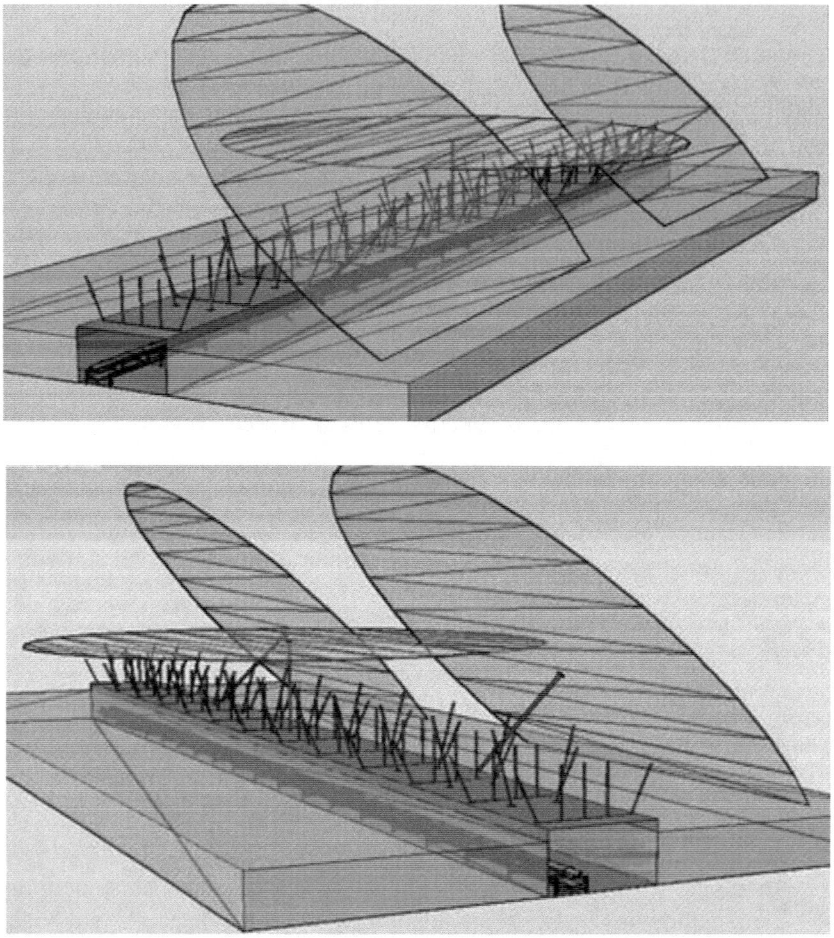

Figure 5. 3-D model of DHF planes formation for underground pressure control.

The authors suggested to use not only visual inspection to monitor the state of DHF parameters through rupture boreholes, but also seismic acoustic rapid test for danger detection. To do that, they installed structure-born noise sensors manufactured by "MarcoGmbH" on supports No.29 and No.31. They were used for data collection and further analysis of seismic acoustic impulses coming from the rock mass. Structure-born noise (vibration) sensors for using in mines have holding magnets which help fix them on any metal surfaces. The sensor detects the frequencies within the range 2.5 – 25000 Hz (period resonance ~50 kHz) [9]. The sensors were placed in fixation hole where the roof beam is fixed to the front row of hydraulic props (figure 6).

System operation principles: structure-born noise sensor detects audible and inaudible sounds (infrasonic vibrations), coming off the rock mass or the part of longwall located near the face. Impulses are registered according to the set parameters. Then the data is transmitted to the data collection gauge, which has a plug-in microSD card. After that the impulses are analyzed by means of special software. Frequency seismic record appears and then data is selected. Depending on the type of contact between the roof beam and roofing, the signal coming from sensor №1 (red) and sensor №2 (green) (figure 7) can have different impulses.

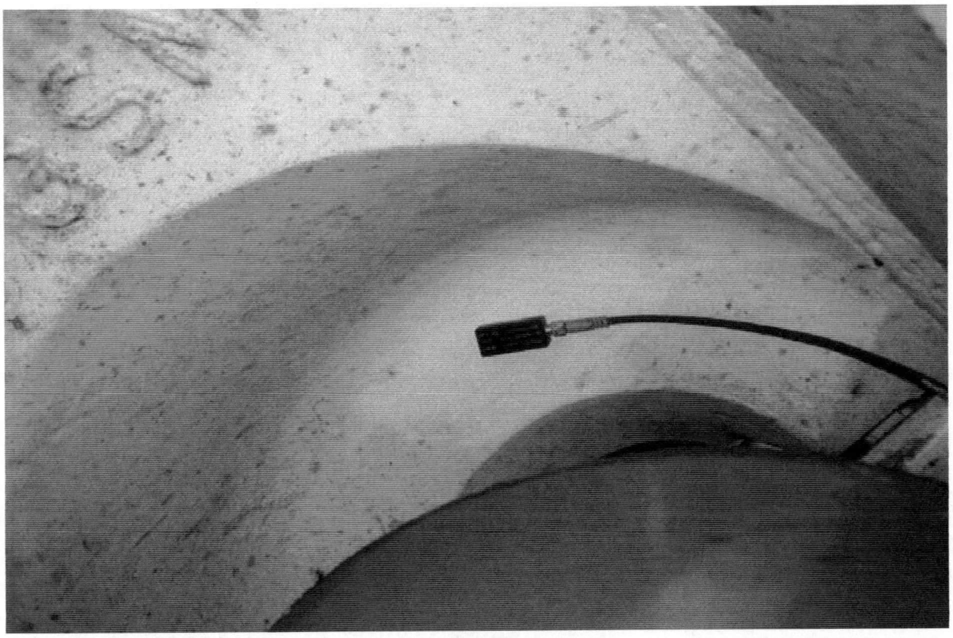

Figure 6. Structure-born noise sensors placement in the roof support.

Received audible and inaudible impulses (infrasonic vibrations) also show that conveyor belt in the longwall and extracting machine are working. This will be seismic acoustic activity rapid analysis of audible impulses only. It is necessary to consider only peak readings of the structure-born noise sensor.

As a result of that work, the authors got a seismic record with audible and inaudible impulses (figure 7). Arrows show the audible impulses.

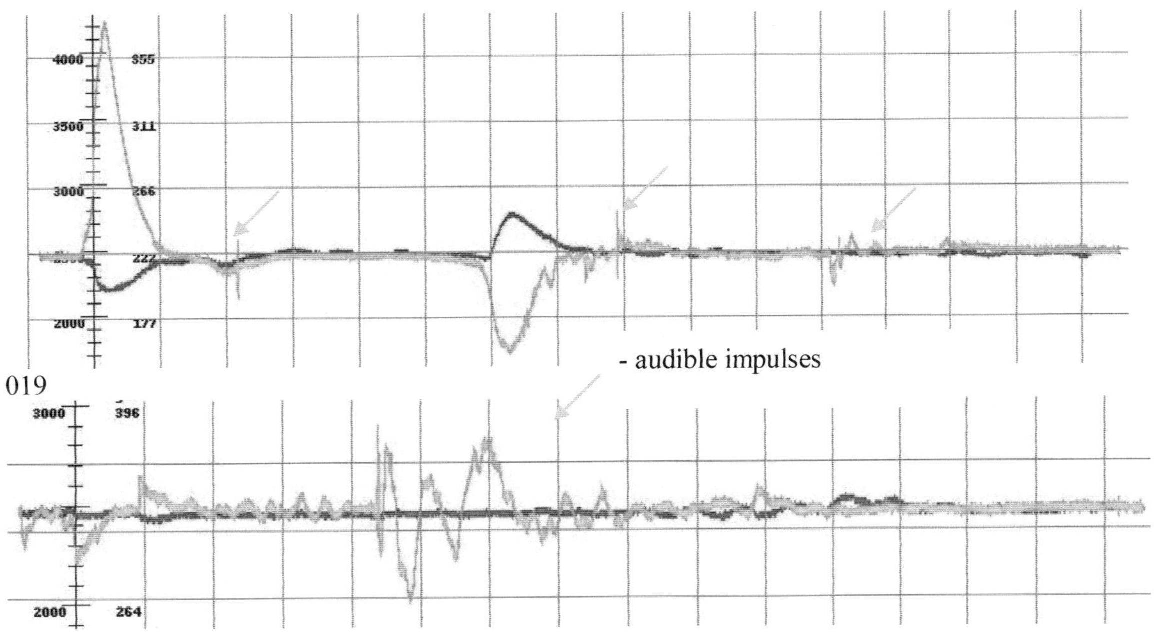

Figure 7. Seismic impulses.

The seismic acoustic rapid danger analysis method allowed to determine the main criteria for rock pressure dynamic displays. Two main criteria were determined for the given conditions, namely quantity and accumulation of impulses within some time period. While using impulse accumulation criterion it was found out that the accumulation process takes place with deceleration, which indicates safe state of the rock mass.

3. Conclusion

Thus the DHF scheme used for roof rock rupture to form short roof bracket after the advancement of the langwall face as a part of the sequence of measures developed to extract coal deposits in extremely complex mining and geological conditions let reduce pressure on props of the reinforcement of the preserved airway part. This guaranteed necessary (according to the calculations) methane-air mixture volume drainage during the extraction of the whole extraction column (figure 8).

Figure 8. The state of the preserved part of the airway 26-25bis.

Acknowledgements

The research was funded by the Russian Scientific Fund Grant (Project №17-17-01143).

References

[1] Klishin V I 2002 *Adjustment of Powered Roof Support to Dynamic Loading Conditions* (Novosibirsk: Nauka) p 200
[2] Klishin V I, Nikolsky A M, Golubev Yu G, Neverov A A and Neverov SA 2008 *Trade Magazine "Siberian Coal in XXI Century"* (Kemerovo) **7(11)** 40–1
[3] Klishin V I, Lakontsev Yu M, Sazhin P V, Klishin S V and Nikolsky A M 2008 *Proc. Sci. Conf. on Fundamental Challenges of Technogenic Geologic Environment Formation* (Novosibirsk: Coal Mining Institute of SB RAS) pp 123–7
[4] Klishin V I, Buchatsky V M and Konovalov L M 2007 *Ugol* **6** 40–3
[5] Klishin V I, Nikolsky A M, Opruk G Yu, Neverov A A and Neverov S A 2008 *Ugol* **11** 12–7
[6] Klishin V I, Opruk G Yu, Senturev A V and Nikolaev A V 2015 *Ugol* **11** 12–6
[7] *Instruction on the Choice of the Parameters and Methods of Roof Weakening in the Extraction area 1991* (Russian Research Institute of Mining Geomechanics and Survey) p 102
[8] Kurlenia M V, Klishin V I and Kokoulin D I 2013 *Crevice initiating device* Patent RF No 129148 Bulletin 17
[9] Klishin V I, Opruk G Yu and Vessel A O 2016 *Science Intensive Technologies in Mineral Resources Extraction and Usage* **3** 27–33

Identification of mine rescue equipment reduction gears technical condition

B L Gerike[1, 2], V I Klishin[1] and E G Kuzin[2]

[1]Institute of Coal of the Federal Research Centre of Coal and Coal Chemistry, Siberian Branch of the Russian Academy of Sciences, 10 Leningradsky Prospect, Kemerovo, 650065, Russia
[2]T. F. Gorbachev Kuzbass State Technical University, 28 Vesennyaya Street, Kemerovo, 650000, Russia

E-mail: gbl_42@mail.ru

Abstract. The article presents the reasons for adopting intelligent service of mine belt conveyer drives concerning evaluation of their technical condition based on the diagnostic techniques instead of regular preventative maintenance. The article reveals the diagnostic results of belt conveyer drive reduction gears condition taking into account the parameters of lubricating oil, vibration and temperature. Usage of a complex approach to evaluate technical conditions allows reliability of the forecast to be improved, which makes it possible not only to prevent accidental breakdowns and eliminate unscheduled downtime, but also to bring sufficient economic benefits through reduction of the term and scope of work during overhauls.

1. Introduction

Tough working conditions of mining machines and high level of dynamic load lead to decrease in equipment service life. Quantitative estimation of equipment reliability concerning one of the parameters (resource) has become widely spread in all industries. Technical diagnosis is one of the most important methods of reliability enhancement in the operation conditions. The aims of the technical diagnosis are the following [1 - 3]:

- technical condition control (whether the parameters of the equipment comply with the technical documentation objectives) and determination on the basis of this data of the technical condition type at the given moment;
- determination of the place and cause of equipment failure (malfunction);
- forecast of technical condition.
- According to NPAOP-10.0-3.01-90 "Safe Operation of Face Machines, Complexes and Aggregates Standards" the following requirements to the technical diagnosis are determined:
- mining machines and equipment have to comprise diagnostic software subsystem which will implement safe control over technical condition via parameters indication and measuring on machines, automatic mechanisms, electrical, hydraulic and pneumatic systems, lubrication systems and also bearing assemblies;
- the diagnostic software subsystem, as a rule, should consist of the following: non-destructive control of the units technical condition, determination of sudden and parametric failures of mining machines and their systems, detection of gradual failures by means of forecasting of monitored parameters changes, continuous and periodic technical condition monitoring.

- All parts of the mining machines and equipment, depending on service conditions and type of wearing-out process, can be divided into the following groups:
- digger teeth and cutting edges of excavator buckets; prop bases, cribs, conveyer worms, coal cutter bars, bulldozer blades; tracks, apron rolls, pins, bushings, driving gears track sprockets of crawler track machines, etc. Service life of these parts depends on the abrasive wear;
- the second group includes parts with screws and grooves, tooth couplings, rolling bearing housings on shafts, machines, surfaces of gear drives, etc. Their service life is determined by mechanical wear of the surfaces of these parts;
- the third group consists of the parts of internal combustion engines of dumpers, bulldozers, scrapers, mine diesel locomotives, machines for thermal rock destruction and flame-jet drilling, etc. Their service life is limited by molecular chemical and corrosion mechanical wear;
- the fourth group includes rolling and sleeve-type bearings, shocks, springs, rods, connecting-rod bolts, etc. Their service life depends on metal fatigue limit.

2. Problem description

At present moment there are many belt conveyers working in coal mining companies [4]. The performance of Kuzbass coal mining industry depends on their operational condition. Soon it is expected that there will be an increase in power supply capacity and technical provision of belt conveyers, their performance and the rock mass transportation length. Partial controlled-velocity electric drives are becoming more widely spread in the industry [5, 6].

Increasing coal extraction volume in fully-mechanized working faces and better work environment safety conditions lead to the necessity of reliable transportation systems development. High efficiency and uninterrupted operation as well as decrease in energy consumption comprise the main goal which is set for the producers of mine belt conveyer continuous flow lines. Another but not less important goal is to reduce costs of maintenance and repair [7, 8]. To ensure fault-free performance of belt conveyer as long as possible it is necessary to determine the causes why its different parts fail [3, 9].

Analysis of equipment outages which were due to belt conveyer reduction gear failure reveals that they account for 7.4% - 18.2 % of failures and on average it is 12 %. Besides average time needed for repair is 24 - 48 hours. It is necessary to point out that the most frequent cause of equipment failures is belt break (up to 50 %), and on average it takes 1.5-2 hours to eliminate the failure. Consequently, identification of real technical condition of mine belt conveyers is a present-day objective.

3. Results and discussion

Vibration control method has proven to be a good way to control technical condition of mechanical equipment [3, 8, 10]. Vibration diagnosis is used to control current condition of equipment, identify of possible faults, evaluate residual operation time, determine dates and scope of repair work. Analysis of Russian and foreign experience concerning control over technical condition over systems with rotational movement of power assemblies shows that control over equipment condition concerning vibration parameters is the most efficient one (up to 77 %) in detecting possible faults [11]. Besides, when other functional diagnostics methods such as oil spectrum analysis [12] and heat monitoring [13] are used with this method the accuracy of identification why the fault has appeared increases up to 95%.

Functional diagnosis from the viewpoint of ensuring safe operation of mining machines has to play a key role in R&D areas, production and control over quality of technological equipment (figure 1). And if such forms of technological equipment service which consider real condition of equipment are more and more spread in coal mining and ore mining industries [2, 3], in the sphere of such equipment manufacturing in spite of GOST ISO 9000-2011 standards implementation, new forms of quality control are still not used.

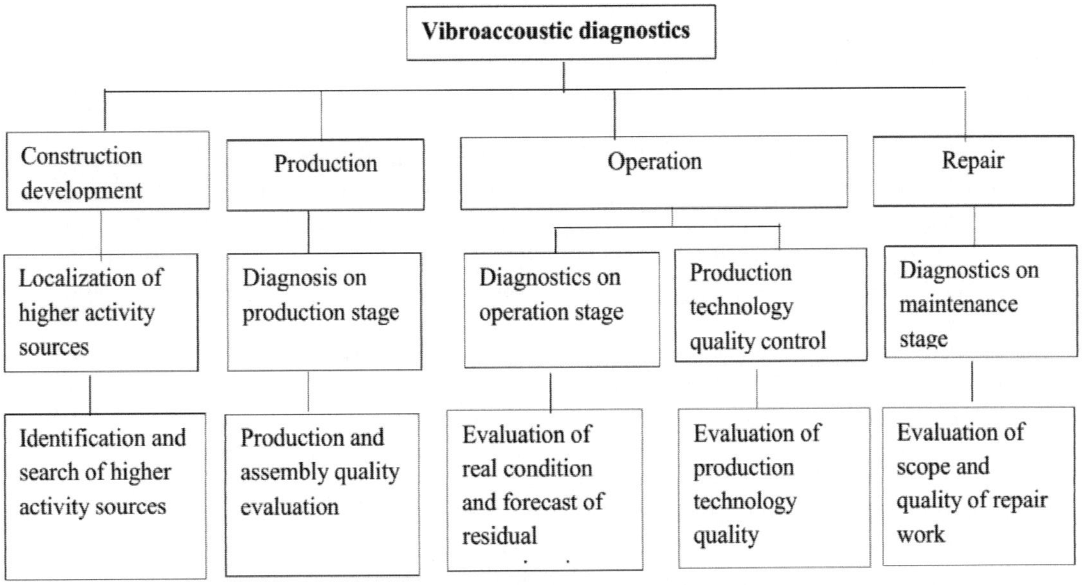

Figure 1. Spheres of using vibration diagnostics on different stages of machineoperation life.

The defects which appear at the moment in reduction of gear production can be divided into defects which appear when its parts are manufactured during reduction gear assembly (figure 2) [14, 15].

Full analysis of technical condition of a reduction gear after its assembly and testing on test bed will allow not only manufacturing defects to be identified and localized, but will also eliminate the possibility of low-quality equipment delivery to the customer. Besides, the obtained data can form the basis for further development of automatic quality control systems.

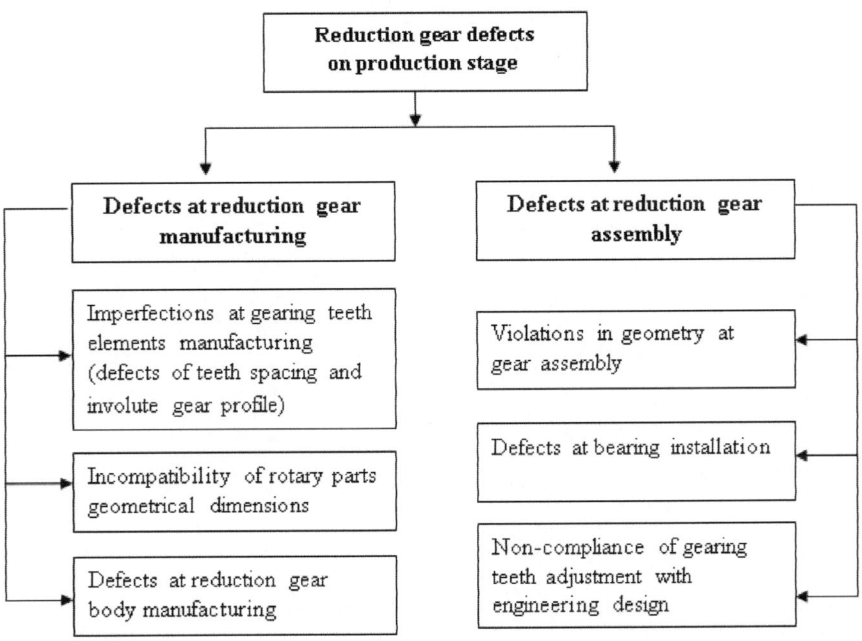

Figure 2. Defects appearing in reduction gears during manufacturing.

Analysis of vibration control methods allows us to make a conclusion that it is reasonable to use the support mask method to automate the control over manufactured products to be used in coal mining industry. This method is based on the assumption that defects which appear as a result of assembly work generate vibration in some set frequency bands with particular correlation between the controlled parameters.

Support mask method allows the width of frequency band, its position and values of assessment criteria, which are compared with current values at random, to be determined. Analyzing the changes of the controlled parameter within the frequency band (number of frequency bands can vary from 6 to 30), one evaluates and forecasts the condition of equipment [16].

Frequency ranges of spectrum mask (band width) usually take the values due to the following conditions [17]:

- "high energy" spectrum component that indicates that there is imbalance or misalignment – $(0.5...1.5)\times f_r$ and $(1.5...2.5)\times f_r$;
- "low energy" spectrum component that indicates that there are defects connected with rolling bearing – $(7.5...15.5)\times f_r$;
- $(2.5...10.5)\times f_r$— general failure of system rigidity;
- first mid-range frequency band $(3... 15)\times f_r$;
- second mid-range frequency band $(15...40)\times f_r$;
- first high-range frequency band 40×fr...20 kHz;
- $(0.1...0.9)\times f_r$ – to detect the defects of an oil wedge of rolling bearings;
- $(n\pm1)\times f_r$ – to indicate the damage on couplings.

Usage of modern technologies that allow control over equipment technical condition to be automated makes it possible to implement individual approach to each machine produced while evaluating its technical condition and to establish the threshold level for initial, operational and marginal state.

Figure 3a gives an example of vibration signal spectrum in 1 control point in reduction gear RKTS-400 produced by JSC "Anzhermash" (figure 4). Figure 3b presents an example of average spectrum mask. All measurements were made with vibration analyzer "Corvet" and processed using software Safe Plant, developed by scientific development and production center "Diatech". On the given figure frequency range (2; 3000 Hz) is divided into 27 bands. Each band is numbered according to mean square value of vibration velocity V_{CKZ}, determined at direct and reverse rotation of the output shaft.

a *b*

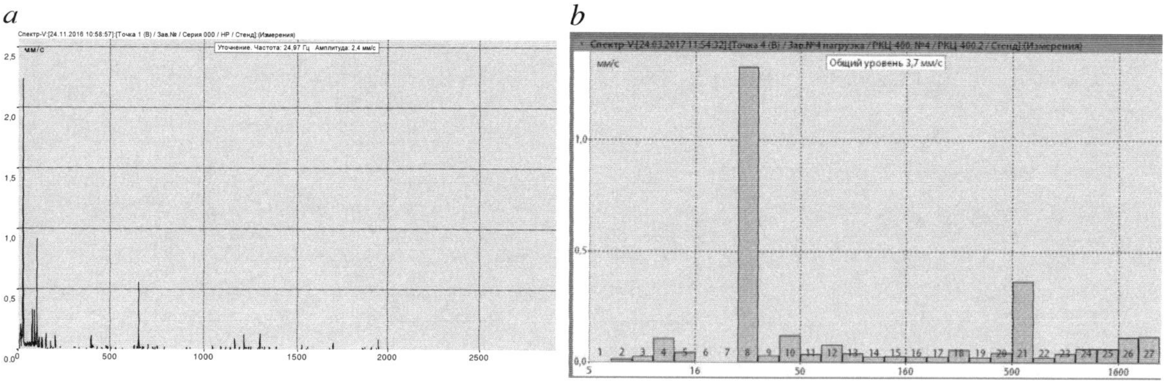

Figure 3. Vibration signal spectrum (a) and average spectrum mask of the vibration signal (b).

Complex approach which presupposes evaluation of technical condition of mine belt conveyer drives 3 LL1600 (transportation length L=850 m, technical performance Q=3500 ton per hour, belt speed v=0-4 m/sec) was implemented on industrial site (Taldinskaya Zapadnaya mine of JSC "SUEK-Kuzbass") in compliance with parameters determined for lubrication oil, vibration and heat control.

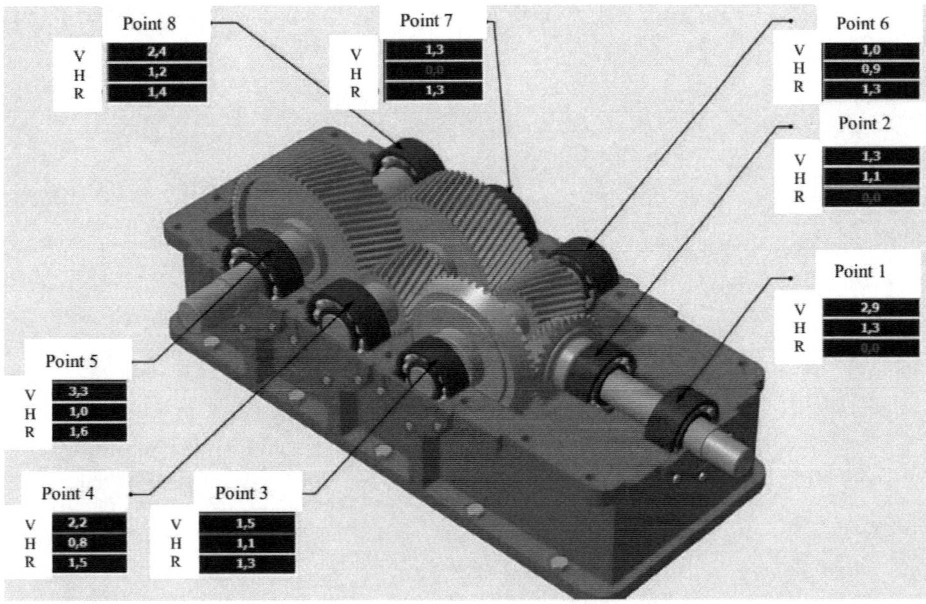

Figure 4. Vibration measurement control points in reduction gear RKTS-400

Figure 5 presents conveyer 3LL1600 design and shows the placement of reduction gears with conventional names P1 – P3.

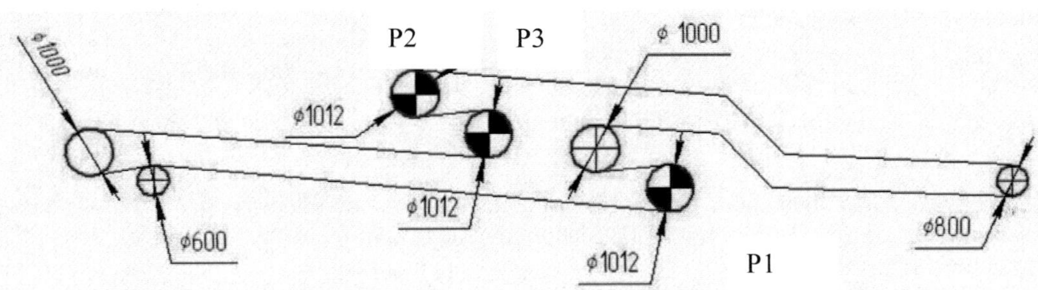

Figure 5. Conveyer 3LL1600 design and drives P1, P2 and P3 placement.

Reduction gears "Moventas Santasalo" (conventional names P1, P2, P3) are used in conveyer drives:

- Cone cylindrical type D3RST82XO (analogue RKTS-400);
- Gear ratio, i= 20,6128;
- Design mechanical power of the reduction gear at service factor FS=1 $P_{\text{m.design}}$=995 kW;
- Design thermal power of the reduction gear at service factor FS=1 and ambient temperature t_{amb}=20 °C $P_{\text{t.nom}}$=779 kW;
- Oil temperature limit t_{M}=90 °C;
- Adjusted engine power P=500 kW;
- Frequency of high speed shaft rotations n=1500, rotations per minute (25 Hz).

After conveyer commissioning vibration parameters were measured for 2 months depending on the load and conveyer belt speed (see figure 6, 7). Information about the load and speed was recorded upon the indications of frequency conversion station CHPSAH displays, which is designed to be used for continuous control of speed and rotation torque on mono- and multirotor belt conveyer drives.

General vibration level was monitored by vibration analyzer "Agat-M" and analyzed using software "Dominant".

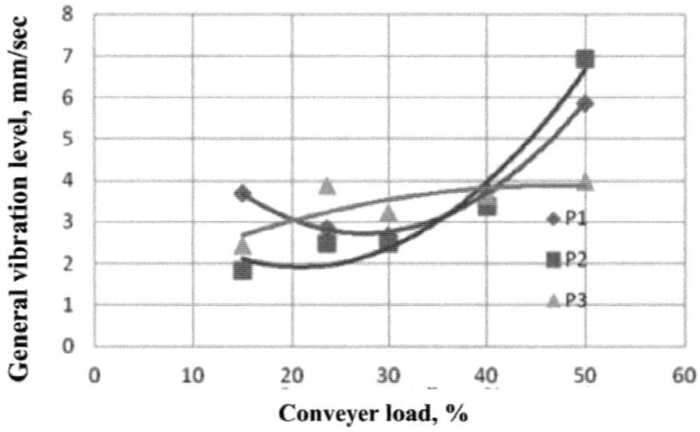

Figure 6. Dependence of general vibration level within the range 100-2000 Hz on conveyer load.

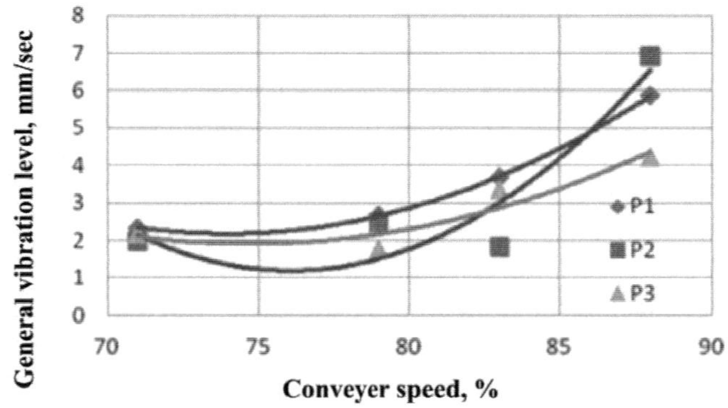

Figure 7. Dependence of general vibration level within the range 100-2000 Hz on conveyer belt speed.

The peculiarity of drives with adjustable rotation speed is dependence of vibration levels on rotation frequency of the driving engine. On debugging stage minimum vibration levels are indicated when load level is 25-30%.

Vibration load analysis was conducted simultaneously with analysis of operating oil parameters depending on the belt conveyer operating time.

Tables 1-3 give examples of wear products accumulation in reduction gears P1 – P3 oil and maximum acceptable values of their content. Their analysis shows that almost in all samples silicon content exceeds maximum acceptable level and indicates that the packings used in the reduction gears are not working well enough.

Table 1. Wear products in oil of reduction gear P1, gram/tonne.

Element	Maximum acceptable value	Sample				
		29.08.14	19.02.15	12.03.15	06.07.15	20.11.15
Fe	200	40.34	160.12	204.8	171.52	326.32
Si	35	37.64	33.88	44.55	39.83	43.29
Cu	150	5.53	0.48	8.96	1.65	2.16
Al	7	1.52	1.54	2.38	2.00	2.09
Cr	5	0.68	1.30	1.07	1.15	1.62
Pb	-	2.55	3.45	4.23	2.48	3.44
Sn	-	5.78	9.34	7.90	6.98	10.52

IOP Conf. Series: Earth and Environmental Science **84** (2017) 012022 doi:10.1088/1755-1315/84/1/012022

Table 2. Wear products in oil of reduction gear P2, gram /tonne

Maximum acceptable value	Sample				
	29.08.14	19.02.15	12.03.15	06.07.15	20.11.15
200	26.85	30.5853	32.3602	32.50	55.04
35	33.11	31.0649	37.1218	39.42	39.64
150	7.07	2.8362	2.1811	6.42	14.75
7	1.10	1.2960	1.7622	1.79	1.67
5	0.95	0.5097	0.7743	1.03	1.34
-	2.63	2.5013	2.8376	4.14	2.60
-	7.94	6.0796	7.0574	5.68	7.65

Table 3. Wear products in oil of reduction gear P2, gram /tonne.

Element	Maximum acceptable value	Sample				
		29.08.14	19.02.15	12.03.15	06.07.15	20.11.15
Fe	200	73.17	314.64	384.1	322.45	876.71
Si	35	36.07	34.49	38.66	40.81	40.77
Cu	150	13.67	11.98	15.7125	9.81	30.63
Al	7	1.43	1.76	2.4578	2.50	2.48
Cr	5	0.87	2.54	3.1097	2.19	10.50
Pb	-	1.75	3.69	1.7229	3.46	2.28
Sn	-	6.89	10.99	12.88	13.31	27.97

Summarized data concerning temperature, viscosity and general vibration level is given in table 4.

Table 4. Values of temperature, viscosity of oil and general vibration level.

Reduction gear No	Equipment body temperature, °C		Kinematic viscosity, mm²/s at 40°C, according to GOST 6258-85	Operational viscosity, mm²/sec at oil maximum temperature T_{max}, °C	General vibration level, mm/sec
	T_{max}	T_{avg}			
P1	47.0	35.6	345	290	5.9
P2	50.7	42.0	353	270	7.2
P3	44.2	37.4	329.7	300	4.2

Analyzing thermoplastic records, taken at operational load (figure 8), one can come to the conclusion that the most sufficient heating takes place in the lower part of the output shaft cover. Besides, temperatures of reduction gears differ. Based on the results of these measurements one can obtain operational viscosity of oil.

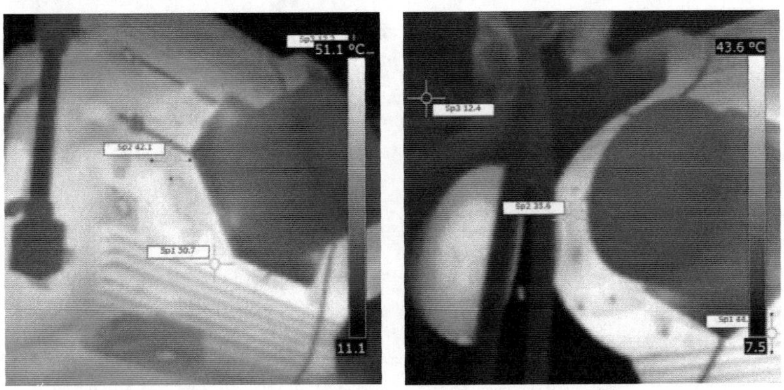

(a) (b)

Figure 8. Thermoplastic records of belt conveyer reduction gears:

(a) reduction gear P2,

(b) reduction gear P3.

Figure 9 presents the dependence of kinematic viscosity change on running time corrected taking body temperature and drive operation time into account.

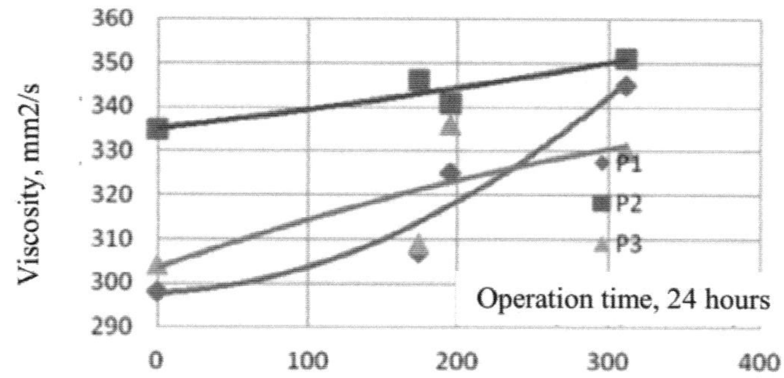

Figure 9. Operational changes in oil kinematic viscosity in relation to drives operation time.

Comparing the results of complex study when technical condition of reduction gears working on mine belt conveyer 3LL1600 were considered one can confirm the following:

- according to the general vibration level, technical condition of drive P1 is considered as satisfactory one [GOST ISO 10816-1-97. Machine state control based on the vibration of non-rotating part measuring. Part 1. General Requirements];
- technical condition of reduction gear P2 is considered as acceptable (according to general vibration level, high temperature of body and low operational viscosity of oil). Rated value of oil viscosity reaches the highest point in reduction gear P2 that indicates that there is evaporation of fractions with low boiling points;
- according to the general vibration level and its spectrum content (figure 10), technical condition of drive P3 is estimated as marginal acceptable. In vibration spectrum one can notice evident wear of high speed shaft bearings elements. This is confirmed by considerable quantity of mechanical impurities.

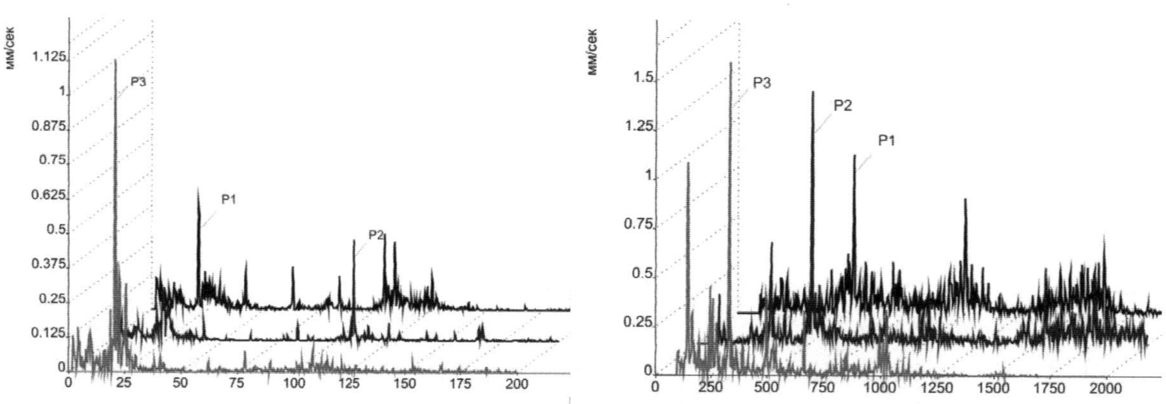

(a) (b)

Figure 10. Comparison of vibration spectra in low frequency (2…200 Hz) (a) and medium frequency (200…2000 Hz) (b) ranges.

Reduction gear producer "Moventas Santasalo" recommends to make the first oil change in 800 – 1000 working hours, and then every 10 000 hours of work or once a year. In fact the first oil change did not take place. In 5 000 working hours there was oil refilling.oil condition at present moment is unsatisfactory (apart from reduction gear P2). Chief mechanic department of mine administration has received the recommendations that oil in reduction gears P1 and P3 has to be changed and high speed shaft bearing should be replaced.

4. Conclusions

The suggested approach to the standardization of parameters of mechanical vibrations can be used in practical work when a company develops standards for vibration of the produced equipment to include in the datasheet.

Development of multiple spectrum masks for wide standard range of mining equipment is one of the conditions of high quality mining equipment production by mining machinery manufacturers. It also makes possible to switch to new ways of mining machines maintenance and repair.

Results of belt conveyer 3LL1600 drives technical condition evaluation (using complex method based on vibration acoustic signals parameters monitoring), analyzing emission spectrum content of bearing oil and heat monitoring of major nodes of rolling bearing allow to trace the changes of reduction gear elements technical condition depending on its load and operating speed. The suggested approach allows not only the accuracy of mining equipment reduction gears technical condition to be determined, but also activity aimed at development of methodological regulatory basis for construction of prediction models to be organized that will help observe changes in technical condition based on sufficient statistic data about the development of some belt conveyer reduction gear defects.

References

[1] Ansari F Uhr P and Fathi M 2014 *International Journal of Services, Economics and Management* **6**
[2] Ierace S and Cavalieri S 2013 *Management and Production Engineering Review* **4**
[3] Shafiei-Monfared S and Jenab K 2012 *IEEE Transactions on Engineering Management* **59**
[4] Hammershmidt A 2015 *Coal* **5** 14–5
[5] Almgren T, Andréasson N, Palmgre M, Patriksson M, Strömberg A B, Wojciechowsk A and Önnheim M 2012 *Proc. 4th World Conf. on Production & Operations Management and the 19th Int. Annual EurOMA Conf.* (University of Amsterdam)
[6] Silva F J S et al 2016 *Journal of Quality in Maintenance Engineering* **22**
[7] Kovalev V, Khoreshok A, Gerike B and Meshkov A 2016 *One Estimation of the Technical Condition of the Selective Headers Regarding* (The conference was funded by RFBR according to the research project # 16-05-20506) (Kemerovo, Russia) pp 291–6
[8] Gupta G and Mishra R P 2016 *J. of Quality in Maintenance Eng.* **22**
[9] Dehghanian P and Aminifar F 2013 *IEEE Transactions on Power Delivery* **28**
[10] Rudolff L, Arghir M, Bonneau O, Guingo S, Chemla G and Rehard E 2012 *Jornal of Engineering for Gas Turbines and Power* **134** (18)
[11] Kelly S Graham 2013 *Advanced Vibration Analysis (Dekker Mechanical Engineering)* p 637
[12] Prabhakar D and Raj V P 2013 *Int. J. Sci. & Tech. Research* **2**
[13] Gustafson A, Schunnesson H, Galar D and Kumar U 2013 *Int. J. of Mining, Reclamation and Env.* **27**
[14] Golmakani H R and Morteza P 2014 *J. of Quality in Maintenance Eng.* **20**
[15] Al-Chalabi H S et al 2015 *J. of Quality in Maintenance Eng.* **21**
[16] Hartman J C and Tan C H 2014 *The Engineering Economist* **59**
[17] Anil Rana 2016 *J. of Quality in Maintenance Eng.* **22**

Calculation of optimal modes for electric-contact welding of rails of mine haulage tracks

R A Shevchenko, N A Kozyrev, A A Usoltsev, R E Kriukov and P E Shishkin

Siberian State Industrial University, 42 Kirova Street, Novokuznetsk, 654007, Russia

E-mail: kozyrev_na@mtsp.sibsiu.ru

Abstract. The choice of thermal regime is based on the exclusion of formation of quenching structures (martensite and bainite), causing additional stresses and cracks which lead to the destruction of rails. After welded joint upset and cooling at the time of reaching the required temperature it is proposed to perform quasi-isothermal exposure by passing pulses of alternating current through the welded joint. The method for calculating the quasi-isothermal exposure is described that depends on the strength of the welding current and different rails section. It is suggested that after welding the rails during quenching, a quasi-isothermal holding is carried out in the temperature range of the formation of the fine-dispersed structure by passing pulses of alternating electric current through the welded joint maintaining this temperature until the end of the transformation. It is shown that the use of quasi-isothermal exposure at a chosen temperature of 600 – 650 °C makes it possible to obtain a finely dispersed structure of the welded seam of rails of mine haulage tracks without additional heat treatment.

1. Introduction

At present, the method of pulsed flash welding has become the most widespread in contact welding of rails. This method of contact welding is the most efficient and technological in comparison with continuous flush welding. During contact welding of rails, as well as during other types of welding, heating occurs and continuous cooling of metal in the heat-affected zone (HAZ). Depending on the steel chemical composition the technological process of welding is selected among the existing methods of welding: continuous or pulsed flash welding determining the linear value and temperature fields in the HAZ welded joint [1, 2]. The choice of thermal regime is based on the exclusion of formation of quenching structures (martensite and bainite) causing additional stresses and cracks which lead to rail destruction [3]. In connection with this [4, 5], the development of such regimes for rails made of chromium steel is of particular importance.

It is known that an increase in chromium content of steel causes a shift to the right of C-shaped curves of the beginning (H) and end (K) of austenite decay on the diagram of isothermal austenite decay (Figure 1) leading to a decrease in critical cooling rate at which austenite is transformed into martensite [6, 7]. At a rapid heating of the welded joint, which is provided by the method of pulsed flash welding and subsequent intensive cooling in the HAZ, a high-strength layer with a martensite structure is formed in the place of microvolumes with an increased content of chromium, nickel and carbon. Martensite sections [3] play the role of stress concentrators and lead to the formation of defects in welded joints (development of fatigue cracks in the head, neck and the base of the rail with a brittle fracture).

This problem in the process of track construction during rails welding is solved by mandatory heat treatment of the welded joint. Heat treatment is performed with induction units that increase the costs. This disadvantage in practice is suggested to be corrected by combining continuous and pulsed flash welding methods, changing the heating intensity during welding and controlling the cooling rate [4]. At the same time, partial usage of the method of continuous flash welding, can cause welding defects.

2. Results and discussion

This paper focuses on the alternative way of problem solution, allowing any of the discussed methods to be used. After welded joints upset and cooling at the time of reaching the required temperature it is proposed to perform quasi-isothermal exposure by passing pulses of alternating electric current through the welded joint. To avoid the formation of quenching structures the exposure temperature is selected based on the obtaining of a fine-grained structure of the weld metal (figure 1). The exposure time is determined by the incubation period of the formation of the required structure and is controlled by the number of current pulses.

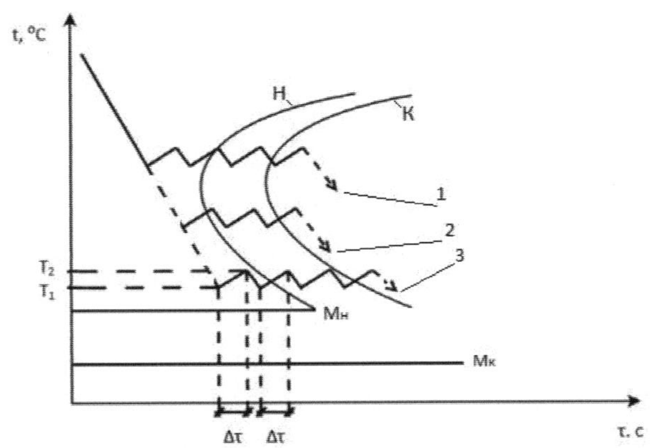

Figure 1. Scheme of isothermal austenite decay with different quasi-isothermal modes of samples exposure after welding for obtaining of perlite (1), sorbitol (2) or bainite (3) structures.

During the steel cooling in the austenite state (from high-temperatures range region), the austenite transformation occurs only after its supercooling below the eutectoid temperature Ar_1 = 727 °C, which is explained by the change in the free energy of alloys phases and structures upon heating and cooling. At low temperatures perlite has less amount of free energy as compared to austenite, so with steel supercooling the austenite transforms in a plate-like perlite (ferritic-cementite mixture). The greater the degree of austenite supercooling, the finer the ferritic-cementite mixture is. The resulting pearlite structures depend on temperature. With a small degree of austenite supercooling in the temperature range 727 – 650 °C perlite is obtained. With a greater degree of supercooling in the temperature range of 650 – 600 ° C sorbitol is formed after austenite transformation (a finer ferritic-cementite mixture is finer than perlite). At even higher degrees of supercooling in the temperature range 600 – 500 °C, troostite (a finer plate ferritic-cementite mixture in comparison with sorbitol) is obtained [8].

In order to judge about the degree of austenite supercooling, the procedure for calculating the cooling of the welded joint was used [9]. It was assumed that with the upset ($t = t_w$), the heated metal of each rod is removed at length d. In this case, for the initial temperature distribution at the cooling stage ($t > t_w$) we take the known solution $T(x, t_w)$ at the end of the heating. Placing the beginning of the X-axis in the center of the weld we obtain,

$$T_{(x,t)} = \frac{1}{2}\cdot(T_b - T_0)\cdot exp\left(-\frac{v\cdot d}{a} - \frac{x^2}{4\cdot a\cdot(t-t_w)}\right)\cdot\left[exp\left(\frac{[2\cdot v\cdot(t-t_w)-x]^2}{4\cdot a\cdot(t-t_w)}\right) \times \Phi^*\left(\frac{2\cdot v\cdot(t-t_w)-x}{\sqrt{4\cdot a\cdot(t-t_w)}}\right) + exp\left(\frac{[2\cdot v\cdot(t-t_w)+x]^2}{4\cdot a\cdot(t-t_w)}\right)\cdot\Phi^*\left(\frac{2\cdot v\cdot(t-t_w)+x}{\sqrt{4\cdot a\cdot(t-t_w)}}\right)\right]; \quad (1)$$

$$\Phi^*(u) = 1 - \frac{2}{\sqrt{\pi}}\int_0^u exp(x^{-2})dx, \quad (2)$$

IOP Conf. Series: Earth and Environmental Science **84** (2017) 012023 doi:10.1088/1755-1315/84/1/012023

where T(x, t) is the function of temperature dependence on time and distance from the center of the weld, °C; *x* is the distance from the heat source, mm; v – flash rate, mm/s; d is the value of the upset, mm; t – total welding and cooling time, s; t_w is the welding time, s; λ – thermal conductivity, J/(cm·s·°C); T_b is the temperature of the liquid metal in the spark gap, °C; T_0 is the ambient temperature, °C; a is the coefficient of temperature conductivity, cm²/s; Φ* (u) is the error function.

The heating time was calculated from the data of [10, 11]. The amount of heat required to heat the sample from T_1 to T_2 was calculated by the formula:

$$Q = mc(T_2 - T_1), \quad (3)$$

where Q is the amount of heat, J; m is the mass of the heated volume of metal, kg; c – specific heat capacity of metal, J/(kg °C); T_1 – initial temperature of metal, °C; T_2 – final temperature of metal, °C.

The heating of the samples is described by the Joule-Lenz law:

$$Q = I^2 R \Delta\tau, \quad (4)$$

where I is the current passing through the sample A; R is the resistance of metal samplet, Ohm; Δτ – current transit time, s.

Using formulas (3) and (4), the necessary time was determined for heating the sample from temperature T_1 to T_2 at a given current I:

$$I^2 R \Delta\tau = mc(T_2 - T_1); \quad (5)$$

$$\Delta\tau = \frac{mc(T_2 - T_1)}{I^2 R}. \quad (6)$$

The mass of the heated volume of metal was calculated by the formula:

$$m = Sl\rho, \quad (7)$$

where S – the cross-sectional area of the sample (rail), mm²; *l* is the distance between the electrodes of the welding machine, mm; ρ is the density of steel, kg/mm³.

The resistance of the heated section was calculated by the formula:

$$R = \rho_e \frac{l}{S}, \quad (8)$$

where ρ_e is the electrical resistivity of steel, Ohm·mm²/mm.

Quasi-isothermal exposure can be carried out by passing pulses of alternating electric current through a welded joint from the welding machine which performs the welding of the product.

Based on the set parameters: T_1 = 600 °C; T_2 = 650 °C; c = 565 J/(kg·°C); *l* = 200 mm; ρ = 7.85·10⁻⁶ kg/mm³; ρ_e = 0.25 Ohm·mm₂/m to calculate the required heating time for a welded joint of different cross-section at a temperature of 600 – 650 °C the calculation of the heating time at different current strengths using the above formulas was carried out and a nomograph was constructed (figure2).

To calculate the cooling after heating, provided that the temperature was distributed evenly to a certain extent over the sample, Newton's heat transfer law was used:

$$\frac{dQ}{d\tau} = \alpha A(T_0 - T), \quad (9)$$

where α is the heat transfer coefficient, W/(mm^2·°C); A is the surface area of the body through which heat is transferred, mm^2; T – body temperature, °C; T_0 is the ambient temperature, °C.

Since Q = CT, where C is the heat capacity of the body, the differential equation can be written as:

$$\frac{dT}{d\tau} = \frac{\alpha A}{c}(T_0 - T) \tag{10}$$

The solution of this equation has the form:

$$T(\tau) = T_0 + (T_2 - T_0)e^{-\left(\frac{\alpha A}{c}\cdot\tau\right)}, \tag{11}$$

where T_2 is the body temperature after heating, ° C.

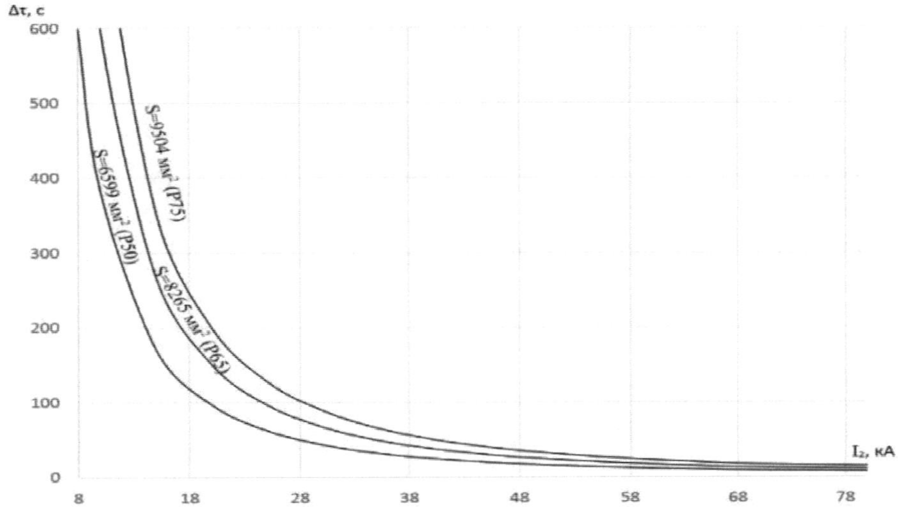

Figure 2. Dependence of heating time Δτ ($T_1 - T_2$) on the current across the secondary winding of transformer at different cross-sectional areas S of the heated samples.

When calculating the quasi-isothermal exposure for 3 pulses each 0.5 seconds long at α = 100 W/(m^2·°C); c = 266.76 J/°C; A = 300 mm^2; T_2 = 650 °C; T_0 = 20 °C; l = 200 mm; I = 11700 A, we plotted the exposure curves (figure 3).

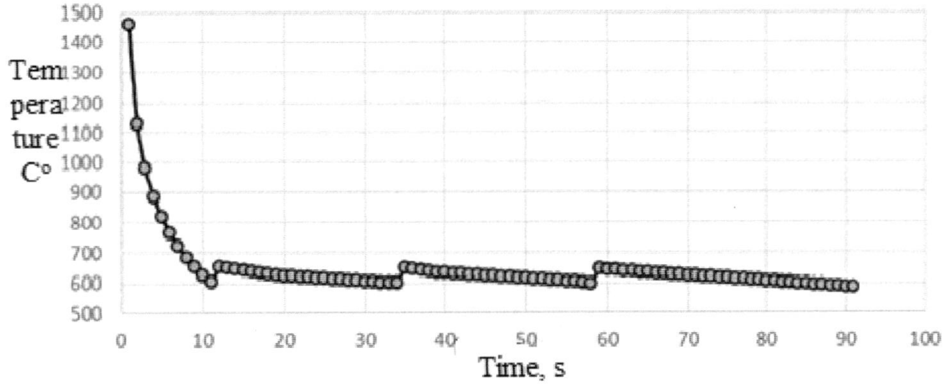

Figure 3. Calculated graph of quasi-isothermal exposure.

The obtained results of calculations formed the basis for the development of a new technology for welding rails without the use of post-welding heat treatment, which is very important for mines characterized by increased combustion hazard.

3. Conclusions

1. The calculation and construction of a nomograph for determining the necessary heating time for a welded joint of rails of different cross-sections depending on the welding current strength was carried out.

2. After welding the rails during cooling it is proposed to conduct a quasi-isothermal exposure in the temperature range of the formation of a finely dispersed structure by passing pulses of alternating electric current through a welded joint maintaining this temperature until the end of the transformation.

3. It is shown that the use of quasi-isothermal exposure at a temperature from 600 to 650 °C makes it possible to obtain a finely dispersed structure of the welded seam of rails of mine haulage tracks without additional heat treatment.

4. Refernces

[1] Mitsuru F et al 2015 *JFE Technical Report* **20** 159–63
[2] Saita K et al 2013 *Nippon Steel & Sumitomo Metal Tech. Rep.* **105** 84–92
[3] Kozyrev N A et al 2006 *Railway Rails From Electric Steel* (Novokuznetsk: EvrazHolding, Novokuznetsk Metallurgical Combine) p 388
[4] Shur E A and Rezanov V A 2012 *Vestnik VNIIZhT* **3** 20–2
[5] Gavrilov V V 2012 *Metro and Tunnels* **1** 42–3
[6] Gulyaev A P and Gulyaev A A 2012 *Metal Science: Textbook for Universities* (Moscow: ID Alliance) p 644
[7] Shevchenko R E 2016 *Proc. of All-Russian. Sci. Conf. for Students, Graduate students and Young Scientists* (Novokuznetsk: SibSIU) pp 196–8
[8] Skugorova L P 1989 *Materials for Construction of Gas and Oil Pipelines and Storage Facilities* (Moscow: Nedra) p 344
[9] Karkhin V et al 2008 *Welding* **1** 12 –7
[10] Lansberg G S 1972 *Elementary Textbook of Physics* (Moscow: Nauka) vol 1 p 565
[11] Danilov I A and Ivanov I A 1998 *General Electrical Engineering with the Basics of Electronics* (Moscow: Higher School) p 752

IOP Conf. Series: Earth and Environmental Science **84** (2017) 012024 doi:10.1088/1755-1315/84/1/012024

Use of barium-strontium carbonatite for flux welding and surfacing of mining machines

R E Kryukov, N A Kozyrev and A A Usoltsev

Siberian State Industrial University, 42 Kirova Street, Novokuznetsk, 654007, Russia

E-mail: kozyrev_na@mtsp.sibsiu.ru

Abstract. The results of application of barium-strontium carbonatite for modifying and refining iron-carbon alloys, used for welding and surfacing in ore mining and smelting industry, are generalized. The technology of manufacturing a flux additive containing 70 % of barium-strontium carbonatite and 30 % of liquid glass is proposed. Several compositions of welding fluxes based on silicomanganese slag were tested. The flux additive was introduced in an amount of 1, 3, 5 %. Technological features of welding with the application of the examined fluxes are determined. X-ray spectral analysis of the chemical composition of examined fluxes, slag crusts and weld metal was carried out, as well as metallographic investigations of welded joints. The principal possibility of applying barium-strontium carbonatite as a refining and gas-protective additive for welding fluxes is shown. The use of barium-strontium carbonatite reduces the contamination of the weld seam with nonmetallic inclusions: non-deforming silicates, spot oxides and brittle silicates, and increases the desulfurizing capacity of welding fluxes.

1. Introduction

The issues of development of new flux compositions and the use of relatively inexpensive natural materials in the form of additives to welding fluxes that allow purposefully the physical and chemical state of the metal melt to be controlled in order to improve the qualitative characteristics of the weld are paid much attention both in Russia and abroad [1 - 12]. In metallurgy one of the promising areas is the development of technologies that allow steel with barium and strontium from oxide materials to be modified and refined skipping the stages of ligature production [13 - 15].

As a material for such technologies the most promising one is the deposit of complex carbonate ores containing calcium, barium and strontium in the northeast of Irkutsk region. The ores have the following mineral composition: 70-80% strontium-calcium-barium carbonate, 10 % potassium feldspar, 10-20 % pyroxene. The production of barium-strontium modifier BSK-2 according to TU 1717-001-75073896–2005 was developed by “NPK Metalltekhnoprom” OJSC. The modifier is intended for the production of steel, cast iron and non-ferrous alloys, as well as for welding electrode coatings [16 - 19]. The barium-strontium modifier contains 13.0-19.0% BaO, 3.5 – 7.5 % SrO, 17.5 – 25.5 % CaO, 19.8 – 29.8 % SiO2, 0.7 – 1.1 % MgO, 2.5 – 3.5 % K_2O, 1.0 – 2.0 % Na2O, 1.5 – 6.5 % Fe_2O_3, 0 – 0.4 % MnO, 1.9 – 3.9 % Al_2O_3, 0.7 – 1.1 % TiO_2, 16.0-20.0 % CO_2.

The results of X-ray phase analysis [19] showed that the main compounds that make up the barium-strontium modifier are BaCaO $(CO_3)_2$, calcite $CaCO_3$, calciostrontianite $CaSr(CO_3)_2$, dolomite MgCO3 and siderite $FeCO_3$. The differential-thermal analysis determined that up to 1223 K the dissociation of dolomite, calcite, baritocalcite and calciostrontianite occurs. This means that at temperatures of welding processes (1873-1923 K) it is possible to use this material with the decomposition of carbonate constituents at a temperature of 1273 K according to the reactions [13]:

$$MgCO_3 \rightarrow MgO + CO_2\uparrow, \quad (1)$$
$$CaCO_3 \rightarrow CaO + CO_2\uparrow, \quad (2)$$
$$BaCa(CO_3)_2 \rightarrow BaCO_3 + CaO + CO_2\uparrow, \quad (3)$$
$$CaSr(CO_3)_2 \rightarrow SrCO_3 + CaO + CO_2\uparrow \quad (4)$$

with the subsequent dissociation of barium and strontium carbonates according to the reactions:

$$BaCO_3 \xrightarrow{>1273K} BaO + CO_2\uparrow, \quad (5)$$
$$SrCO_3 \xrightarrow{1373K} SrO + CO_2\uparrow. \quad (6)$$

In [20, 21] to determine the conditions for the reduction of barium and strontium from oxides thermodynamic modeling methods implemented in the Terra software complex were used.

Based on the study of decomposition kinetics of barytocalcite and calciostrontianite the authors in believe [22, 23] that the process proceeds in two stages, as a result of which CaO, BaO, and SrO are formed and carbon dioxide (CO_2) is released. The release of CO_2 during barium and strontium carbonates dissociation creates a protective atmosphere for the welded products, thereby protecting the molten metal from oxidation and improving the quality of the weld. At the same time the protection is due to high-temperature decomposition of carbonates with the formation of CO2.

Calculations [24, 25] show that during decomposition of 1 kg of CaCO3 (n.c.) 0.224 m^3 of CO_2 is formed, 1 kg of $MgCO_3$ – 0.267 m^3, 1 kg of $FeCO_3$ – 0.192 m^3, 1 kg of $MnCO_3$ – 0.194 m^3, 1 kg of Na_2CO_3 – 0.211 m^3. Accordingly, with the decomposition of 1 kg of $BaCO_3$ (n.c.), 0.114 m_3 of CO_2 is formed, 1 kg of $SrCO_3$ – 0.152 m^3 of CO_2. When heated to the temperature of welding processes 1800 °C (2073 K) the gas expansion in 7.6 times takes place. Apparently, without taking into account the costs of carbonates decomposition, the most optimal is the use of $MgCO_3$, $CaCO_3$, BaCa $(CO_3)_2$ and $CaSr(CO_3)_2$ as components that make it possible to obtain the greatest amount of CO_2 during decomposition of 1 kg of material.

The carbonates dissociation contained in barium-strontium carbonatite (MgO, CaO, BaO, SrO) increases the basicity of slag and, apparently, contributes to the refining of the weld metal from nonmetallic inclusions. It is believed thatbarium and strontium, being modifiers, affect the metallic structure of the welded joint [19].

In [26] it is shown that it is possible to use carbonatite as an additive to welding fluxes. Further studies concerning the effect of the additive of barium-strontium carbonatite, when introduced into welding fluxes, on the quality of obtained welds are given in this paper.

2. Methods of research

In a series of experiments performed in the laboratory conditions various compositions of welding fluxes were prepared and investigated.

Welding flux No. 1 (flux additive) was made by mixing barium-strontium carbonatite with liquid glass in a ratio of 70% and 30%, respectively. After exposure at room temperature it was dried in an oven at a temperature of 300 °C, then cooled, crushed and sieved with fractionation 0.45-2.5 mm. Flux No. 2 – flux based on silicomanganese slag of fraction 0.45-2.5 mm, fluxes No. 3, 4, 5 – silicomanganese slag of fraction 0.45-2.5 mm mixed with flux additive (barium-strontium carbonatite with liquid glass in a ratio of 70% to 30%) in amounts of 1; 3 and 5%, respectively.

Butt welding was performed without a double bevel on samples with measures 500×75 mm, thickness 16 mm, from sheet steel 09G2S. The process was carried out using a S-08GA wire with a diameter 4 mm by ASAW-1250 welding tractor in the following modes: I = 700 A; U = 30 V; V = 35 m/h.

3. Results and research

Chemical compositions of fluxes, slag crusts and weld metal are given in tables 1, 2 and 3. As it can be seen from the tables there is an insignificant recovery of barium and strontium from the material, and a change in the chemical composition of the weld indicates an increase in the degree of desulfurization with the use of a flux additive.

Table 1. Chemical composition of flux, %.

Sample No.	Mass fraction of elements,%												
	MnO	SiO_2	CaO	MgO	Al_2O_3	FeO	F	Na_2O	K_2O	BaO	SrO	S	P
1	0.87	32.52	12.15	1.02	3.54	9.96	-	2.41	2.48	14.24	4.86	0.25	0.21
2	9.58	50.26	18.51	7.98	10.55	1.54	0.38	0.41	0.61	отс.	отс.	0.13	0.05
3	8.11	47.64	29.79	6.19	6.97	0.31	0.21	0.27	отс.	0.25	0.10	0.15	0.01
4	7.96	48.28	28.36	6.15	7.05	0.42	0.33	0.43	0.03	0.61	0.21	0.15	0.02
5	8.19	48.44	28.86	5.08	7.04	0.54	0.31	0.43	0.04	0.65	0.25	0.15	0.02

Table 2. Chemical composition of slag crusts.

Sample No.	Mass fraction of elements,%												
	FeO	MnO	CaO	SiO_2	Al_2O_3	MgO	Na_2O	K_2O	BaO	SrO	F	S	P
1	2.42	4.92	18.81	26.21	14.84	18.11	0.70	0.38	0.51	0.10	-	0.15	0.04
2	2.11	8.05	23.83	46.92	10.27	6.90	0.38	0.66	отс.	отс.	0.74	0.13	0.01
3	1.76	8.21	29.34	46.74	7.17	5.79	0.28	отс.	0.25	0.10	0.23	0.15	0.01
4	2.45	8.42	27.66	46.90	6.92	6.00	0.39	0.04	0.59	0.24	0.27	0.15	0.02
5	2.70	7.93	27.36	46.94	6.87	5.94	0.47	0.12	0.90	0.32	0.32	0.15	0.02

Table 3. Chemical composition of welded joints.

Sample No.	Mass fraction of elements,%											
	C	Si	Mn	Cr	Ni	Cu	Nb	Al	S	P	Sr	Ba
1	0.11	0.13	0.53	0.02	0.05	0.10	0.002	-	0.030	0.021	0.0004	0.004
2	0.09	0.71	1.51	0.03	0.10	0.11	0.014	0.023	0.019	0.013	отс.	отс.
3	0.08	0.54	1.38	0.02	0.04	0.06	0.016	0.012	0.011	0.009	0.0004	0.0012
4	0.08	0.51	1.33	0.02	0.03	0.06	0.015	0.021	0.010	0.008	0.0004	0.0021
5	0.09	0.57	1.41	0.02	0.03	0.04	0.020	0.020	0.008	0.009	0.0004	0.0033

Visual quality control of the welded joints of samples No. 1, prepared with the use of flux additives only, showed unsatisfactory quality – uneven shape of the seam, with poor separation of the slag crust. The remaining samples are of satisfactory quality.

Samples were cut from the welded plates and X-ray spectral analysis of the weld metal composition, metallographic examinations of the weld metal were performed, the mechanical properties were identified – they are given in table 5. In connection with the unsatisfactory quality of the welded bead on sample No. 1 no mechanical properties were determined.

The metallographic study was carried out on polished sections without etching with OLYMPUS GX-51 optical microscope with magnification × 100. The results of the analysis for the presence of nonmetallic inclusions in the weld zone, carried out in accordance with GOST 1778-70, are shown in figure 1 and table 4. The points of nonmetallic inclusions were established when examining the entire area of welded seams of non-etched polished sections. In each field of vision the dimensions were determined separately for each type of nonmetallic inclusions by comparison with reference scales.

Studies indicate a reduction in the contamination of the weld metal by non-deforming silicates and the absence of brittle silicates.

The microstructure was studied with OLYMPUSGX-51 optical microscope in a bright field in the magnification range × 500 after etching the sample surface in a 4% nitric acid solution. The grain size was determined according to GOST 5639-82 at magnification × 100. The microstructures of the weld metal are shown in figure 2. The weld metal is characterized by the formation of a ferrite-perlite structure of vismanthtiten orientation. In this case in the samples containing strontium and barium a slight decrease in the grain size is observed. The grain size in the welded sample with a flux that does not contain barium-strontium carbonatite is No. 4 on the grain scale, and the grain size in the welded samples (No. 3, No. 4, No. 5) with the flux containing barium-strontium carbonatite is No. 4, No. 5.

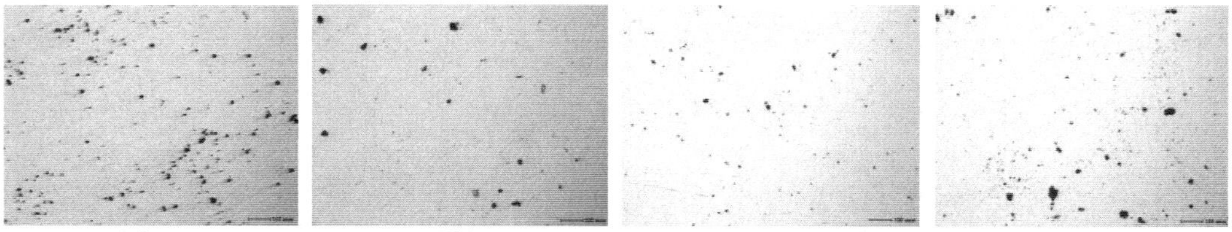

(a) (b) (c) (d)

Figure 1. Non-metallic inclusions in the weld zone of samples: a) sample 3; b) sample 4; c) sample 5; d) sample 6.

Table 4. Non-metallic inclusions in the weld zone.

Sample No.	Non-metallic inclusions, point		
	Non-deforming silicates	Brittle silicates	Spit oxides
3	4b; 3b; 4a	3b	1a
4	4b; 3a	-	1a; 2a
5	4b; 2b; 2a	-	1a; 2a
6	4b; 3a; 4a	-	1a; 2a; 3a

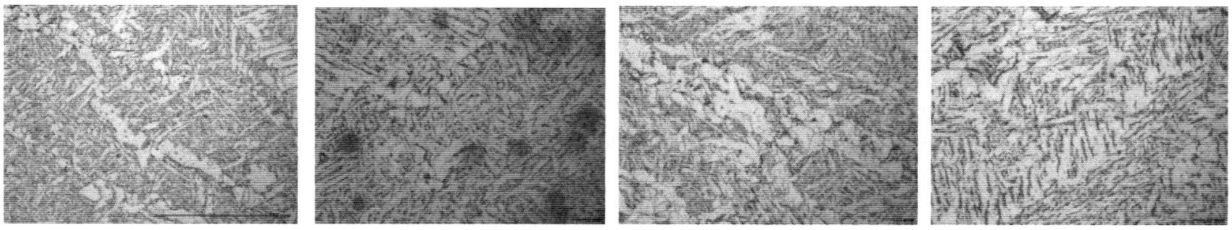

(a) (b) (c) (d)

Figure 2. Welded joints microstructure of samples with the addition of barium-strontium carbonatite, %: a) 0; b) 1; c) 3; d) 5.

Samples were cut from the welded plates and mechanical properties were determined. As a result of studying the mechanical properties of samples cut from welded plates results were obtained indicating an increase in the values of the impact hardness and and yield stress (figure 3, 4).

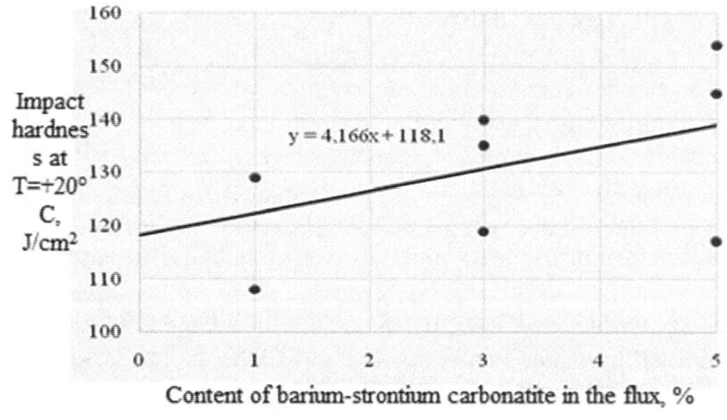

Figure 3. Influence of the content of barium-strontium carbonatite in the flux on impact hardness.

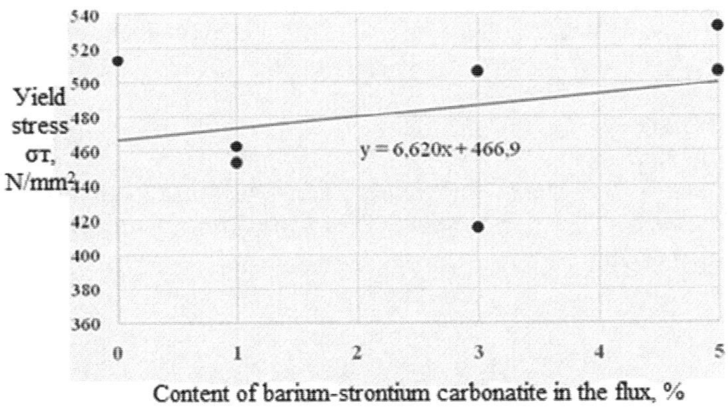

Figure 4. Influence of the content of barium-strontium carbonatite in the flux on yield stress.

4. Conclusions

1. The principal possibility of using barium-strontium carbonatite as a refining and gas-protective additive for welding fluxes is shown.

2. The use of barium-strontium carbonatite allows contamination of the weld seam with nonmetallic inclusions (non-deforming silicates, spot oxides and brittle silicates) to be reduced and the desulfurizing capacity of welding fluxes to be increased.

3. Introduction of barium-strontium carbonatite in the amount of 5% into the flux from silicomanganese slag forms a ferrite-perlite structure of the weld metal of vismanthtiten orientation, while the grain size from No. 4 to No. 4, No. 5 is slightly reduced.

4. The impact hardness and yield stress of the samples are increased by introduction of barium-strontium carbonatite into the flux composition.

References

[1] Crespo A C et al 2009 *Welding Int.* **23(2)** 120–31
[2] Golovko V V and Potapov N N 2011 *Welding Int.* **25(11)** 889–93
[3] Puchol R Q et al 2009 *Welding Int.* **23(2)** 132–40
[4] Crespo A C et al 2007 *Welding Int.* **21(7)** 502–11
[5] Volobuev Yu S 2012 *Welding Int.* **26(8)** 649–53
[6] Volobuev Yu S 2010 *Welding Int.* **24(4)** 298–300
[7] Potapov N N 1987 *Welding Int.* **1(10)** 951–4
[8] Babushkin P L 1991 *Welding Int.* **5(9)** 741–2
[9] Pavlov I V 1995 *Welding Int.* **9(4)** 329–32
[10] Chigarev V V 1994 *Welding Int.* **8(10)** 808–9
[11] Kurlanov S A 1993 *Welding Int.* **7(1)** 65–8
[12] Bublik O V 2010 *Welding Int.* **24(9)** 730–3
[13] Deryabin A A et al 2007 *Steel* **11** 134–41
[14] Deryabin A A et al 2008 *Electrometallurgy* **6** 35–8
[15] Ryabchikov I V et al 1983 *Ferroalloys with Rare-earth and Alkaline-earth Metals* (Moscow: Metallurgy) p 272
[16] Grigoriev Yu V and Ryabchikov I V 2005 *Ferrous Metallurgy* **7** 3–5
[17] Chernyak S S and Romen B M 196 *High-Manganese Steel in Precious Metal Industry* (Irkutsk: Irkutsk State University) p 377
[18] Ivakin V L et al 2004 *A New Technology for Improving the Quality of Metals and Zlloys with Barium-cesium Carbonate* (Irkutsk: Irkutsk State University) p123
[19] Rozhikhina I D et al 2015 *Ferrous Metallurgy* **58(15)** 871–5

[20] Vatolin I A et al 1994 *Thermodynamic Modeling in High-temperature Inorganic Systems* (Moscow: Metallurgy) p 352
[21] Trusov B G 2005 *III Int. Symp. on Combustion and Plasma Chemistry* (Almaty: Kazak University) 52–7
[22] Volobueva Yu S 2012 *Welding Int.* **26(8)** 649–53
[23] Golovko V V 2011 *Welding Int.* **25(11)** 889–93
[24] Kryukov R E et al 2014 *Ferrous metallurgy* **10** 25–8
[25] Kryukov R E et al 2015 *IOP Conference Series: Materials Science and Engineering* **91** 012016
[26] Lipatova U I et al 2016 *Proc. All-Russian Sci. Conf. of Students, Graduate Students and Young Scientists* (Novokuznetsk: SibSIU) vol 3 pp 266–71

Analysis of rail welding methods for mine rail access with the use of modern technologies

A A Usoltsev, R A Shevchenko, N A Kozyrev, R E Kriukov and P E Shishkin

Siberian State Industrial University, 42 Kirova Street, Novokuznetsk, 654007, Russia

E-mail: kozyrev_na@mtsp.sibsiu.ru

Abstract. Welded joint zones are weak sections of the railway track for all traffic cases (in the case of high-speed traffic and heavy traffic). In the paper advantages and disadvantages of the basic ways of rails welding, which are widely used today, are considered: electrocontact and aluminothermic. Carefully selected mode of differentially thermally strengthened rails string will allow the process of correction after heat treatment to be minimized and internal residual compressive stresses to be kept.

Particular attention should be paid to the method of rails welding, in which after rails welding during their cooling it is offered to perform quasi-isothermal exposure in the temperature range of fine structure formation by passing pulses of alternating electric current through the welded joint maintaining this temperature until the end of the transformation. The use of quasi-isothermal exposure at a temperature of 600 – 650 °C makes it possible to obtain a finely dispersed structure of the welded seam of rails without additional heat treatment.

1. Introduction

Rail access are an important part of industrial transport, which means a set of transport devices and facilities that serve the needs of an enterprise in the transportation of finished products, raw materials and semi-finished products. Also, the intensity of cargo flows puts forward new requirements to the upper structure of track, in particular, to the main element – rails. At present, both on the Russian railways and abroad the jointed construction of tracks is neglected.

One of the main drawbacks of the jointed track is the presence of a joint. The development of technologies that make it possible to provide a continuous welded railway track is an actual direction at present [1 - 6]. It should be taken into account that the operation of the railway access in the country takes place in difficult climatic and operational conditions. Thus, the use of modern technologies for construction and repair of access roads will increase the quality of the road.

2. Results and discussion

To the advantages of a continuous welded rails can be attributed [7, 8]:

- a 30-40% reduction in the costs spent on the current road maintenance and improvement of traffic safety, construction reliability;
- a 8-10% reduction in the main resistivity to traffic and, therefore, saving of fuel and electricity for traction, which is very important in the context of a continuous increase in energy prices;
- increase in the service life of the track upper structure due to a smaller than in the jointed track of rails damage (cracks in the edges of bolt holes, head dislocation, crushing and saddles). Thus, failures of continuous welded rail strings due to defects (contact fatigue and in the

joints) arise 1.8-2.0 times less often than of jointed track, and without taking into account balancing intervals in 3-4 times;

- reduction in the amount of works on the road correction (up to 25-30%) associated with sinks in the joints, especially works on the elimination of splashes which with the increase of axial loads become a big problem;
- decrease in the intensity of lateral wear of the outer rail line in the curves and, accordingly, the damage to the rails for this reason in 1.5-1.6 times;
- reduction of the need to clean crushed-stone ballast along the coal-ore routes in 1.5-2.0 times;
- reduction of metal costs for track joint fastening (up to 4.5 t-km);
- reduction of expenses for repair of running parts of wagons and locomotives;
- increase the comfort of passengers;
- increase in the work reliability of electric track chains of automatic blocking.

Important advantage of the continuous welded construction of the upper structure of the track is the fact that it allows the reinforced concrete sub-base to be used, which increases the stability margin, resistance to longitudinal and transverse displacements of rails and provides equal rigidity of the track along the length. In addition, the use of reinforced concrete sleepers reduces the consumption of commercial timber.

Welded joint zones are weak sections of the track for all traffic cases (in the case of high-speed traffic and heavy traffic). Removals due to the defects of welded joints the total number of rails removal reach 30%, while the total length of the joint zone is not more than 2% of the length of the rail. The reasons for this are: a change in the homogeneity of microstructure in the zones of weld and the thermal effect; creation of an unfavorable diagram of internal residual stresses; creation of conditions during welding for the formation of internal defects, which are stress concentrators and weaken the rail section with a welded seam; the deformation of the rail in the weld zone with the subsequent formation of "saddlings" during operation.

The creation of rails with a resource of 1500 – 2000 million tonnes is possible only with a comprehensive optimization of metallurgical quality, the structure of the metal matrix, the residual stresses and straightness curves [9, 10]. High metallurgical quality of rails is due primarily to the absence of aggregates and separate large brittle-decomposed oxide nonmetallic inclusions in them, which is ensured by the low content of aluminum (less than 0.004%) and oxygen (total – less than 20 ppm and bind in high alumina inclusions – less than 10 ppm).

The main disadvantages, which reduce the resource of all these rails in operation, remain:

- presence of residual stresses in the head of rails, which are induced there by the last technological operation in the production stream – by cold correction on roller machines;
- creation of soft areas with reduced resistance to wear and saddling in the heat affected zones after welding and subsequent local induction heat treatment of welded joints, which results in unevenness in these places and increased dynamic impact of wheels of rolling stock;
- a noticeable decrease in the impact hardness, fracture toughness and critical size of fatigue cracks during quenching from rolling heating as compared to quenching with a separate recrystallization heating.

All these disadvantages can be overcome by designing and development of welded rail strings with length 800 m welded from solid-rolled rails of 100 m, followed by heat treatment by continuously sequential induction heating of the entire cross section of the rail and further differentiated cooling, aimed at receiving in the rail head of sorbite and troostosorbite structure, maximum possible uniform for all-rolled rail and for the welded joint without heat-affected zones. The obtained rail string with heat treatment should provide overhaul service life at least 1500 – 2000 million tonnes brutto, provided periodic preventive grinding or milling is performed.

Such an approach will ensure the maximal possible strength uniformity of welded joints and solid rolled rails with the production of long rails for a continuous welded track with equal wear resistance and saddling during operation, which will eliminate local unevenness of the continuous welded track during the lifetime of rails. Besides, this provides a favorable diagram of internal residual stresses that contribute to the generation of the greatest resistance to nucleation and distribution of contact fatigue cracks in the head, corrosion-fatigue cracks in the sole and longitudinal cracks in the neck of the rail, both in the rolling and welded parts of rail strings.

Correctly selected mode of differentially thermally strengthened rail strings will allow the process of straightening after heat treatment to be minimized and internal residual compressive stresses to be kept. It is known that the rails thermally strengthened by the technology of continuously-sequential through induction heating and subsequent differential cooling by compressed air from the side of the head and sole at the enterprises of SISCO (Canada) (currently the plant is closed) and Sozheray (Tata Steel, France), had residual compressive stresses in the head of 100-200 MPa, and in the sole 100-150 MPa. The rails thermally strengthened from rolling heat by differentiated quenching or from a separate heating by volumetric quenching in oil, as a result of cold straightening on rollers, have tensile stresses in the head and the sole up to 250-300 MPa [11].

In the construction, repair and maintenance of a continuous welded track, one of the key technologies is the welding of rails, for which the following methods have been developed and patented:

- pressure welding: electrocontact, gas-press, induction, by laser, friction, etc.;
- aluminothermic welding;
- electric arc welding: by electrodes, under a layer of flux, in the environment of shielding gases, electroslag, by powder wires, etc.

The method of pressure welding is based on the heating of the ends of rails to the temperature of plastic state (above 1000 °C) and squeezing them with a certain force (depending on the cross-sectional area and the physical-mechanical properties of the rails metal) [12]. The ends of rails can be heated by electric current – electrocontact method, by gas burners – gas press, by high frequency currents (inductors), laser, plasma, heat released during friction, etc. When welding by pressure there is no filler metal, i.e. the ends of rails are welded directly to each other.

The aluminothermic method is based on the production of metals and alloys by the reduction of their oxides in exothermic reaction with aluminum. The reaction proceeds with the release of a large amount of heat. The molten metal is poured into the gap between the connected rails [13].

The electric arc welding is based on melting of electrode arc by electrode metal (rod or wire) and filling with it a gap between the connected rails.

Aluminothermic and electric arc methods differ significantly from the methods of pressure welding because a welded joint with a width of 15-25 mm and more consists of a filler metal having a cast structure.

All methods of welding are characterized by the presence of a heat-affected zone (HAZ) – a modified structure of the base metal of rails directly adjacent to the weld. This leads to the occurrence of residual stresses and, as a consequence, to decrease in the strength characteristics of the welded joint. The width of the HAZ depends on the time of exposure of the base metal to high temperatures, weight of the filler metal, welding method and parameters.

On the welded joints performed by the methods of pressure welding, heat treatment (including differentiated, with hardening of the rail head in the HAZ) can be used to increase their mechanical properties [14].

Since during aluminothermite and electric arc welding there is an additive metal in the weld seam that is very different in terms of chemical composition and structure from the metal of the rails, the heat treatment of these joints does not give a tangible increase in their mechanical properties.

Among these methods of rail welding by pressure the electrocontact method (EM) is most widely used – in 95% of all cases. In Russia on the railways EM is only applied, by which annually about 600 thousand joints of rails are performed by rail-welding trains and up to 50 thousand joints by moving rail-welding machines [15]. Consider the advantages and disadvantages of existing EM technologies.

Advantages of EM:

- high quality of the welded joint;

- the presence of monitoring system in the welding machine that allows the deviation of welding parameters to be controlled;
- high mechanization and automation of works (in stationary conditions);
- high performance of the process.

Disadvantages of EM:

- impossibility to weld joints in the area of turnouts;
- expensive equipment;
- need for long schedule gaps in the movement of trains during welding operations.

At present, the method of pulsed flash welding has become most widespread in contact welding of rails. This method of contact welding is the most economical and technological in comparison with continuous flash welding. During contact welding of rails and welding by other methods, heating and continuous cooling of the metal in the HAZ occurs. Depending on the chemical composition of steel, the welding process is selected using the existing methods of flash welding: continuous or pulsed, which determine the linear magnitude and temperature fields in the HAZ of welded joint [16, 17].

The choice of thermal regime is based on the exclusion of the formation of quenching structures (martensite and bainite), causing additional stresses and cracks, which lead to destruction of rails [8]. In connection with this [18], the development of such regimes for rails of high-speed railways made of chrome steel is of particular importance.

However, when using the continuous welded construction of the track top structure, a number of issues remains unsolved, in particular, the issues of increasing the strength of welded joints and the HAZ influence require further work as the number of dangerous defects in this area is 13-15% of the total number of defects on the rail string.

This problem in the process of construction continuous welded tracks is solved by the obligatory thermal treatment of the welded joint. Heat treatment is performed by induction units, which increases the costs. This disadvantage in practice is suggested to be corrected by combining continuous and pulsed flash welding methods, changing the heating intensity during welding and adjusting the cooling rate [18]. At the same time, partially using the method of continuous flash welding, welding defects may appear.

It is proposed in [19] after welding the rails during their cooling to perform a quasi-isothermal exposure in the temperature range of the formation of fine structure by passing pulses of alternating electric current through the welded joint maintaining this temperature until the end of the transformation. Quasi-isothermal exposure at a temperature of 600 – 650 °C makes it possible to obtain a finely dispersed structure of the welded seam of railway rails without additional heat treatment.

Until recently, the main method of rails welding was electrocontact method ES. However, due to the impossibility of rails welding in the turnout zones by contact machines, since 1995 the aluminothermic rails welding (ATRW) has been used on the railways of Russia, and at present the issues related to the use of ATRW not only in the zone of turnouts but also on the running line are being solved.

Advantages of ATRW:

- mobility;
- small duration of schedule gaps in the movement of trains;
- welding process does not consume electricity.

Disadvantages of ATRW:

- there is no possibility to monitor and control the welding process;
- a wide and uncontrolled zone of thermal influence;
- the influence of mixture quality on the quality of welded joints;
- a strong dependence of the welding quality on the experience of a welder.

In order to understand the processes occurring in ATRW, it is necessary to understand the basic principles of this welding method. In general, the technology is described in [20 - 24]. The termite, used for welding rails, is made of iron scale – wastes of steel production and metal aluminum. They are crushed to obtain grains of 0.1 to 2.5 mm in diameter. Primary aluminum should contain at least 98-99% of pure aluminum. Before grinding, scale should be well burned in order to remove moisture and oil from it. Dust from the grinded scale and aluminum is removed by an air separator. Scale in its chemical composition is different. The smaller content of silicon in the scale, the better it is for a welding thermit. The content in the scale of oxygen should not be below 25%. Iron with oxygen can form three oxides: FeO – iron oxide; Fe_2O_3 – ferric oxide and Fe_3O_4 – black of iron. Under normal conditions, the constituents of the thermit mixture do not interact, but if the thermit mixture is heated to 1100 – 1200 °C a chemical reaction begins between its components. As a result of this reaction, metallic aluminum is oxidized and converted to alumina, and the scale is reduced to iron. This reaction is accompanied by a large release of heat spontaneously and uncontrollably. The thermit mixture contains: 23.7% of aluminum and 76.3% of scale. When combusted 1.0 kg of thermit mixture releases 3188.22 kJ of heat, which makes it possible to obtain a temperature of molten metal 2700 – 3000 °C.

Scale and aluminum may contain various impurities. Therefore, the percentage of aluminum and scale in the thermit mixture is calculated taking into account the purity of aluminum and the oxygen content in the scale. The practice of thermit welding showed that the scale in the thermit mixture should be (7-8)% higher than the calculated content. Then the termite metal is denser and welds better with the metal of the rail.

To use effectively the heat generated during the reaction and increase in the metal yield, finely grinded pieces of steel – wastes of nail-making production are added to the thermit mixture during manufacture for the welding of rails. In the process of flash, steel increases the yield of the thermit metal and lowers the initial temperature of the thermit reaction products. Depending on the weight of the thermit batch, these wastes are added to the thermit mixture from 12 to 20% of the estimated yield of the thermit metal. To improve the mechanical properties of the weld metal, ferroalloys are introduced into the thermit mixture, in most cases ferromanganese, ferrosilicon, ferromolybdenum, ferrotitanium, ferrovanadium. Ferroalloys cause the production of thermit metal with mechanical properties close to the metal of the welded rails. Thus, the chemical composition of the metal formed from the thermit mixture is far from welded rail steel with all the consequences. In addition, the thermit component itself is a source of oxides of both exogenous –unreacted iron oxide and endogenous – formed during oxidation-reduction reactions. It should also be noted that slag inclusions and gases that are not always formed during the reaction time manage to float out of the reaction zone. As a result, the welding zone becomes contaminated with non-metallic oxide inclusions and various micropores, which are the centers of crack initiation.

For the welding of rails, an intermediate casting method can now be used in which only molten metal without slag is used from the products of thermit reaction increasing the quality. However, ATRW welding is a joint obtained by casting processes in which non-metallic oxide inclusions, pores, cavities, g bleed, slag inclusions, internal cracks and micropores are possible.

Great problems arise during ATRW welding in winter. At low temperatures the cooling rate increases and as a result crystallization of the molten metal of the weld pool accelerates, the released gases and slag particles do not have time to float, saturating the metal with pores and slag inclusions. Increased heat removal from the heated metal and increase in the content of gases in it contribute to the formation of cracks in the seam and the HAZ. If the storage rules are not observed, a certain amount of moisture is possible in the materials, which contributes to the saturation of the weld metal with hydrogen. [25] shows the results of operational tests of welded joints welded by various technologies. It is shown that preheating is the main technological operation that influences the development of a necessary structure and provides the required quality of the welded joint. Experiments have shown that if the heating is insufficient or the ambient temperature is low, the temperature of the rails ends drops sharply, which leads to an increase in the rates of cooling of the weld metal and the weld zone. This contributes to the formation of quenching structures, increased brittleness and the formation of microcracks, which reduce the strength of the joints. For example, when welding rails at an ambient temperature of 15 °C, the cooling rates at the peripheral sections of rail base are almost twice as high as in the head.

In [26] experiments on ATRW welding at negative temperatures are presented. The ambient temperature has a negative effect on the quality of welds. The results of experiments obtained at different welding temperatures do not meet standard requirements.

It should be noted that experience of the operation of welded joints of ATRW on the railway network and testing on the Experimental Ring VNIIZhT show that their quality is somewhat inferior to the quality of the joint produced by EM, which is conditioned by the welding method itself. Therefore, the main companies that carry out ATRW of railroads on the railway network ("Snaga", "Welding Surfacing Company", "GT-Aluminothermic Welding", "Railtech", "Welding Technologies") set up this method solely for welding rails within turnout, on bridges, overpasses, in tunnels, when it is difficult to organize the necessary windows in the traffic schedule for the use of ATRW.

In [27] the calculation is made, according to which the cost of electrocontact welding is more expensive than aluminothermic welding of rails by 41.8%. However, when considering the terms of guarantee of the joints welded by the EM and the ATRW according to the quantity of the cargo passed over them (for EM according to the RZD standards 1.08.002 – 2009 for rails of the type R75 and R65 – 150 million tonnes brutto, for rails of type R50 – 120 million tonnes brutto: for ATRW according to TU 0921 – 127 – 01124323 – 2005 for rails of type R75 and R65 – 120 million tonnes brutto: for rails of type R50 – 100 million tonnes brutto) the cost of a joint for each million tonne brutto changes. When EM is used: for rails of type R75 and R65 – 81.33 rub., for rails of type R50 – 101.67 rub. When ATRW is used: for rails of type R75 and R65 – 59.17 rub., for rails of type R50 – 71 rub. Thus, the present costs of EM is more expensive than ATRW by 27.25% when welding rails of type R75, R65 and by 30.17% when welding rails of type R50. Moreover, it should be pointed out that the analysis of data on the removal of 897 defective welded joints of ATRW on the rail network in 2009 and 9 months of 2010 revealed that 89% of the ATSW joints (710 pcs.) did not work during the warranty period. This also reduces the difference between the cost of EM and ATRW. The cost of ATRW increases with the further joints operation. The current periodicity of monitoring the ATRW joints (not later than 6 months from the moment of welding, then at least once a year), most of which are operated with safety pads, are higher than the periodicity of monitoring of joints welded by electrocontact method, operated without safety pads (at least once a year in the first two years after welding, then at least once every two years) [28].

3. Conclusion

From the data given, it follows that welding ATRW today has not proved itself to be a reliable and qualitative method of connecting the rails, the economic component is related to the quality characteristics and safety of transportation by rail. For railway accesses at mining and processing enterprises it is recommended to weld the rails by the electrocontact method.

It should be noted that both methods of welding are continuously developing, in particular, the German company "Elektro-Thermit GmbH & Co. KG" declares about the achieved successes in improvement of quality of a joint made by ATRW. Despite the good advertising component, mass use and introduction of this technology for rails welding should necessarily be preceded by large-scale studies, laboratory and testing tests, operational tests on the sections of railways with different geometries, density of freight traffic and climatic conditions.

Refernces

[1] Meade B 1997 *Welding J.* **76(9)** 47–52
[2] Kargin V A, Tikhomirova L B, Abramov A D and Galai M S 2014 *Welding Int.* **28(3)** 245–7
[3] Yamamoto R, Komizu Y and Fukada Y 2014 *Welding Int.* **28(7)** 510–20
[4] Karimine K, Uchino K and Okamura M 1997 *Welding Int.* **11(6)** 452–61
[5] Kuchuk-Yatsenko et al 2008 *Welding Int.* **22(5)** 338–41
[6] Irving B 1997 *Welding J.* **22(9)** 33–37
[7] Klimenko L V 2004 *Supplement to the Magazine "The World of Transport"* **1** 88–93

[8] Kozyrev N A et al 2006 *Railway Rails from Electric Steel* (Novokuznetsk: EvrazHolding, Novokuznetsk Metallurgical Combine) p 388
[9] Shur E A 2008 *Rail Transport* **2** 41– 5
[10] Shur E A 2006 *Influence of the Properties of Metal Matrix on the Operational Stability of Rails* (Ekaterinburg: UIM) 37–63
[11] Zolotarsky A F 1976 *Thermally strengthened rails* (Moscow: Transport) p 264
[12] Kuchuk-Yatsenko S I et al 2010 *Welding Int.* **24(6)** 455–61
[13] Wegrzyn J and Maxur M 1992 *Welding Int.* **6(1)** 5–8
[14] Genkin Z 2005 *Welding Int.* **19(2)** 160–4
[15] Kalashnikov E A and Korolev Yu A 2015 *Railway and Track Economy* **8** 2–6
[16] Mitsuru F et al 2015 *JFE Tech. Rep.* **20** 159–63
[17] Saita K et al 2013 *Nippon Steel & Sumitomo Metal Tech. Rep.* **105** 84–92
[18] Shur E A and Rezanov E A 2012 *Vestnik VNIIZhT* **3** 20–2
[19] Shevchenko R A 2016 *Proc. All-Russian Sci. Conf. "Science and Youth: Problems, Searches, Solutions"* (Novokuznetsk: SibSIU) 20(III) pp 259–61
[20] Voronin N N et al 2008 *Technology of Aluminothermic Welding of Rails* (Moscow: MIIT) p 117
[21] Voronin N N et al 2013 *Aluminothermic Welding of* (Moscow: Educational and Methodological Center for Education in Railway Transport) p 195
[22] Sergejevs D and Mikhaylovs S 2008 *Transport Problems* **3** 33–7
[23] Karguin V A et al 2015 *Welding Int.* **29(2)** 155–7
[24] Lee F T 2006 *Welding J.* **85(1)** 24
[25] Voronina O N 2014 *Development of Rails Constructions, their Butt Joints and Processing Technologies Cand. Thesis* (Moscow) p 228
[26] Voronin N N et al 2012 *The World of Transport* **4** 56–9
[27] Velichko D V 2013 *Proc. Int. Sci. Conf "Actual Problems of Modern Science"* 93–6
[28] Rukavchuk Yu P et al 2011 *Railway and Track Economy* **4** 26–7

Mathematical models of functioning and allocation indicators of road-transport complex resources in the fuel and raw materials region

V A Buyvis, A V Novichikhin and M V Temlyantsev

Siberian State Industrial University, 42 Kirova Street, Novokuznetsk, 654007, Russia

E-mail: buyvis_va@mail.ru

Abstract. A number of features of coal industry functioning was determined for the conditions of Kemerovo region, and the specifics of planning and organization of coal transportation were revealed. The analysis of indicators of motor and railway types of transport in the process of coal transportation was executed. The necessity of improving the tools of coal products transportation in the modern conditions is substantiated. Specific features of functioning of a road-transport complex in the fuel and raw material region (on the example of Kemerovo region) are determined. The modern scientific and applied problems of functioning and allocation of the road-transport complex resources are identified. To justify the management decisions on the development and improvement of road-transport complex a set of indicators are proposed: infrastructural, transportation performance, operating, social and economic. Mathematical models of indicators are recommended for formulation and justification of decisions made during operational and strategic planning of development, evaluation and development of algorithms of functioning and allocation of road-transport sector in Kemerovo region in the future.

1. Introduction

At present, the basic industries of the Russian Federation in accordance with the structure of gross added value are: transport and communications – 7.8% and extraction of minerals – 9.4%, while the share of coal from the total amount of extracted energy products is 13.9% [1].

One of the main centers of coal industry in the Russian Federation is the Kuznetsk coal basin, which is located on the territory of Kemerovo region. Currently, more than 56% of the country coal is mined in the region, as well as about 80% of all coking coal, and total geological coal reserves are estimated at 319 billion tonnes [1, 2].

The coal industry shows positive dynamics of coal production in the Russian Federation and Kemerovo region [3] and according to forecasts in future periods the growth of production will remain [1, 2]. The volumes of coal production in the Russian Federation and Kemerovo region from 1998 to 2020 are shown in figure 1.

More than 85% of coal from Kemerovo region is exported, thus, the transportation distances on average exceeds 4.500 km, and the transport component in the cost of coal for the consumer is more than 45% [4]. Rail transport is mainly used for coal transportation to remote points of destination in Russia, coal in the transportation structure is more than 26% [3]. Motor transport for the coal transportation is used mainly for short distances – the average delivery range is 45 km, but their volumes exceed the railroad volumes [4].

Figure 1. Volumes of coal mining in the Russian Federation and Kemerovo Region [1 - 3].

Table 1 presents the dynamics in the volumes of products transportation by road and rail transport in the Russian Federation for the period 2000-2020. From 2010 there has been a steady growth in indicators, showing the overcoming of the crisis phenomena of 2008.

Table 1. Volumes of products transportation by road and rail transport in the Russian Federation for the period 2000-2020 [4].

Type of transport	Volumes of transportation, million tonnes / year				
	2000	2005	2010	2015	2020 (prediction)
Public railroad	1046.8	1273.3	1312	1380	1558.3
Railway industry	3140.4	3486.6	3272.2	3757.1	4355.5
Motor transport of general use	550	612	498.3	571.8	628.8
Automotive industry	5328	6073	4738.1	6091.5	7140.6

Thus, at the present time there is no doubt that necessity of improving the tools of coal products transportation in the Russian Federation is urgent.

2. Methods of research

With an intensive growth in the share of volumes of motor transportation (table 1), the load on the infrastructure significantly increases, with no significant improvement in the level of its development and technical capabilities [4]. In this regard, we shall consider the problematic aspects of motor transportation process.

At present, almost 27% of the road network is operated with exceedance of the normative load, while the local road network is not enough developed [5]. In Kemerovo region, 26.73% of public roads do not meet the regulatory requirements for transport performance indicators, including 66.36%

of regional or inter-municipal roads and 11.32% of local roads [6, 7]. Acceleration of motorization of the country has not yet led to a corresponding increase in the volume of construction, reconstruction and repair of the road network, and repair of roads in recent years has even somewhat decreased. For example, with the increase in the length of public roads by 15% over the past 10 years, the fleet of vehicles has increased by more than 1.7 times [8]. The axial load of 11.5 tonnes was observed on the sections of federal roads with a total length of 2613 km (5.2% of the total network length), with a requirement of 22 000 km, ten-tonnes load – on 11.9 thousand km (23.8% of the total network length) at a demand of 26 thousand km respectively [8, 9].

To overcome the above-mentioned problem situations it is necessary to solve the following modern scientific and applied problems of functioning and allocation of road-transport complex resources:

- specification and improvement of the best practices in functioning of a road-transport complex in the fuel and raw material region;
- improvement of the system of planning functioning and allocation of the road-transport complex resources taking into account regional and industry-specific features of the fuel and raw material region [10];
- improvement of the tools for allocation of road-transport complex resources based on the development of flexible systems of incentives and change management;
- development of recommendations for improving the regulatory framework of the road-transport complex in the fuel and raw material region.

3. Results and discussion

To solve these scientific and applied problems we propose a set of math models of functioning and allocation indicators of road-transport complex resources as a tool for justifying management decisions and development of projects for future periods:

1. Infrastructure indicator Ind_1 – it is an indicator characterizing the length of transport communication sections, on which there are limitations of traffic and carrying capacities due to noncompliance with regulatory requirements:

$$Ind_1 = \sum_i \sum_j \sum_k (L_{\mathrm{TC}}^{ijk} + L_{\mathrm{AL}}^{ijk} + L_{\mathrm{OV}}^{ijk} + L_{\mathrm{RS}}^{ijk} + L_{\mathrm{A}}^{ijk}), \tag{1}$$

where i – organization servicing the road, $i \in [1; I]$; j – type of transportation (transit, technological, local, etc.), $j \in [1; J]$; k – administrative area, $k \in [1; K]$; $L_{TC}^{ijk}, L_{AL}^{ijk}, L_{OV}^{ijk}, L_{RS}^{ijk}, L_{A}^{ijk}$ – length of motor roads that do not meet the regulatory requirements for traffic capacity, axial loads, spaces for oversized vehicles, the type of road surface, and the absence of a year-round car traffic respectively, km.

2. Indicator of transportation performance Ind_2 – characterizes the volume of traffic on the reserve routes due to noncompliance with regulatory requirements for of motor roads, on which the main routes by transport-operational indicators are laid:

$$Ind_2 = \sum_i \sum_j \sum_k (Q_{ijk}^{\mathrm{TC}} \cdot L_{TC\,ijk}^{\mathrm{AD}} + Q_{ijk}^{\mathrm{AL}} \cdot L_{AL\,ijk}^{\mathrm{AD}} + Q_{ijk}^{\mathrm{OV}} \cdot L_{OV\,ijk}^{\mathrm{AD}} + Q_{ijk}^{\mathrm{RS}} \cdot L_{RS\,ijk}^{\mathrm{AD}} + Q_{ijk}^{\mathrm{A}} \cdot L_{A\ \,ijk}^{\mathrm{AD}}), \tag{2}$$

where $Q_{ijk}^{\mathrm{TC}}, Q_{ijk}^{\mathrm{AL}}, Q_{ijk}^{\mathrm{OV}}, Q_{ijk}^{\mathrm{RS}}, Q_{ijk}^{\mathrm{A}}$ – the volume of traffic on the additional routes due to noncompliance with regulatory requirements for of motor roads, on which the main routes for traffic capacity, axial loads, spaces for oversized vehicles, type of road surface, absence of year-round motor traffic, t; $L_{\mathrm{TC}\,ijk}^{\mathrm{AD}}, L_{\mathrm{AL}\,ijk}^{\mathrm{AD}}, L_{\mathrm{OV}\,ijk}^{\mathrm{AD}}, L_{\mathrm{RS}\,ijk}^{\mathrm{AD}}, L_{\mathrm{A}\ \,ijk}^{\mathrm{AD}}$ – route length on the reserve trajectories due to noncompliance with regulatory requirements for the basic routes: traffic capacity, axial loads, spaces for oversized vehicles, type of road surface, absence of year-round motor communication, km;

3. The operating indicator Ind_3 – is an indicator characterizing the value of shipment delivered by motor transport within a set timeframe exceeding the standard (contractual) period:

$$Ind_3 = \sum_i \sum_j \sum_k (Q_{ijk}^{\mathrm{TC}} \cdot t_{\mathrm{TC}\ ijk}^{\mathrm{AD}} + Q_{ijk}^{\mathrm{AL}} \cdot t_{AL\ ijk}^{\mathrm{AD}} + Q_{ijk}^{\mathrm{OV}} \cdot t_{OV\ ijk}^{\mathrm{AD}} + Q_{ijk}^{\mathrm{RS}} \cdot t_{\mathrm{RS}\ ijk}^{\mathrm{AD}} + Q_{ijk}^{\mathrm{A}} \cdot t_{\mathrm{A}\ ijk}^{\mathrm{AD}}), \quad (3)$$

where $Q_{ijk}^{\mathrm{TC}}, Q_{ijk}^{\mathrm{AL}}, Q_{ijk}^{\mathrm{OV}}, Q_{ijk}^{\mathrm{RS}}, Q_{ijk}^{\mathrm{A}}$ – volume of transportations delivered by motor transport within a set timeframe exceeding the standard (contractual) period due to noncompliance of motor roads with the normative requirements for traffic capacity, axial loads, type of road surface, absence of a year-round car traffic, t; $t_{\mathrm{TC}\ ijk}^{\mathrm{AD}}, t_{\mathrm{AL}\ ijk}^{\mathrm{AD}}, t_{\mathrm{OV}\ ijk}^{\mathrm{AD}}, t_{\mathrm{RD}\ ijk}^{\mathrm{AD}}, t_{\mathrm{A}\ ijk}^{\mathrm{AD}}$ – additional time of traveling on reserve (bypass) roads due noncompliance of motor roads with the normative requirements for traffic capacity, axial loads, type of road surface, absence of a year-round car traffic, day.

4. Social indicator Ind_4 – characterizes the value of additional time people spend travelling by roads due noncompliance of motor roads with the normative requirements:

$$S_4 = \sum_i \sum_j \sum_k \sum_p (S_{ijkp}^{\mathrm{TC}} \cdot t_{\mathrm{TC}\ ijk}^{\mathrm{ET}} + S_{ijkp}^{\mathrm{AL}} \cdot t_{\mathrm{AL}\ ijk}^{\mathrm{ET}} + S_{ijkp}^{\mathrm{ET}} \cdot t_{\mathrm{OV}\ ijk}^{\mathrm{ET}} + S_{ijkp}^{\mathrm{RS}} \cdot t_{\mathrm{RS}\ ijk}^{\mathrm{ET}} + S_{ijkp}^{\mathrm{A}} \cdot t_{\mathrm{A}\ ijk}^{\mathrm{ET}}), \quad (4)$$

where p – category of population, pensioners, employees, workers $p \in [1; P]$; $S_{ijkp}^{\mathrm{TC}}, S_{ijkp}^{\mathrm{AL}}, S_{ijkp}^{\mathrm{OV}}, S_{ijkp}^{\mathrm{RS}}, S_{ijkp}^{\mathrm{A}}$ – number of people that have to spend extra time travelling due to noncompliance with regulatory requirements for of motor roads, on which the main routes for traffic capacity, axial loads, spaces for oversized vehicles, type of road surface, absence of year-round motor traffic, pers.; $t_{\mathrm{TC}\ ijk}^{\mathrm{ET}}, t_{\mathrm{AL}\ ijk}^{\mathrm{ET}}, t_{\mathrm{OV}\ ijk}^{\mathrm{ET}}, t_{\mathrm{RS}\ ijk}^{\mathrm{ET}}, t_{\mathrm{A}\ ijk}^{\mathrm{ET}}$ – time of the enforced travelling due noncompliance with regulatory requirements for of motor roads, on which the main routes for traffic capacity, axial loads, spaces for oversized vehicles, type of road surface, absence of year-round motor traffic, hour;

5. Economic indicator Ind_5 – characterizes the efficiency of investments into the system of the road-transport complex, as an indicator it is proposed to use the net present value (*NPV*). In the general form NPV is defined [11]:

$$NPV = \sum_{t=1}^{T} \frac{CF_t}{(1+E)^t} - \sum_{t=1}^{T} \frac{Inv_t}{(1+E)^t}, \quad (5)$$

where CF_t is the amount of moneys received in the period t, rub.; Inv_t – the amount of investments in the period t, rub.; E – discount rate; T is the calculation period, years.

4. Conclusion

The proposed set of indicators is recommended for justifying the management decisions in the operational and strategic planning of development, effectiveness evaluate and models building of functioning and allocation of the road-transport complex resources in Kemerovo region in the future.

References

[1] *Official gazette of the Russian Federation* 2014 **27** Article 3786

[2] *Order of the Government of the Russian Federation "Energy Strategy of Russia for the Period till 2030"* No. 1715-r from November 13, 2009 http://minenergo.gov.ru/node/1026.

[3] *Transport and Communications in Russia* 2016 (Moscow: Rosstat) p 112

[4] Bogdanova A 2015 Transportmetrica A: Transport Science **11(7)** 561–78

[5] *Decree of the Government of the Russian Federation "On Approval of the Transport Strategy of the Russian Federation for the Period till 2030"* No. 1734-r from November 22, 2008

http://docs.cntd.ru/document/902132678
[6] *Decree of the Administration Board of Kemerovo region "On Approval of the Regional Program "Development of the Network of Public Roads of the Kemerovo Region" for 2015-2022"* No. 212 from June 30, 2015 http://docs.cntd.ru/document/428628801
[7] Klementschitz R and Roider O 2015 *Transport Problems* **10(4)** 57–74
[8] *Branch Road Methodical Document* No. 2018-00-2005 2005 (Moscow: Rosavtodor) p 9
[9] *Road Economy of Russia in Figures and Facts 2003* (Moscow: Rosavtodor) p 325
[10] *Osorio C and Chong L 2015 Transportation Science* **49(3)** pp 623–36
[11] *Financial Management: Theory and Practice* 2010 ed E S Stoyanova (Moscow: Perspektiva) p 656

Cognitive simulation of incident risks in the structure of loading and transport enterprise

S V Shishkina[1], Yu D Pristupa[1], L D Pavlova[2] and V N Fryanov[2]

[1]United Production and Transport Department of Kuzbass, LLC, 13 Kirova Avenue, Leninsk-Kuznetsky, 652518, Russia

[2]Siberian State Industrial University, 42 Kirova Street, Novokuznetsk, 654007, Russia

E-mail: ld_pavlova@mail.ru

Abstract. Organizational and technical system of a manufacturing enterprise was identified, which includes three subsystems: main production, industrial and social infrastructure. Based on the results of cognitive modeling, significant system concepts were identified that reduce the risks of incidents. The internal control influences formed in accordance with level of competence of heads of services, departments, sections, dispatchers, acting on the basis of regulations, job profiles. The second concept influencing the enterprise management system is personnel, which is assessed by the compliance of competencies of crane operators, loader operators, slingers, loaders, and acceptance/delivery agents to job responsibilities and labor functions. At a low level of professional competencies, the personnel does not fully comply with job duties and labor functions, the risk of an incident is maximal. The application of cognitive modeling allows us to identify the essential elements that ensure stable functioning of the system as a whole.

1. Introduction

The loading and transport enterprise (LTE), the structure of which includes a complex of facilities and services providing loading and unloading services as well as cargo shipping expeditions for mines, open pit mines, processing plants, as well as current repair of railway tracks and adjustment of own diesel locomotives, can be regarded as organizational and technical system, which contains three subsystems [1, 2]: main production, production and social infrastructure.

The main production includes the production processes of enterprise, during which materials or raw materials are converted into products. In the context of LTE, such processes are loading and transportation of coal in wagons for sale to consumers.

The production infrastructure consists of a complex of shops, management units and services that provide the necessary conditions for the main activity of the enterprise. At the LTE the production infrastructure includes the rolling stock service, track service, material and technical supply department, communication section, production control and labor protection service, production and technical department, etc.

The social infrastructure unites divisions of the enterprise that ensure the sanitary and hygienic working conditions and the social and everyday needs of workers.

The LTE specific features are the fulfillment of two main processes by the service of freight and field operations: loading and transportation of coal in wagons for sale to consumers, and the remaining processes are auxiliary providing all the technological cycles [3, 4].

In accordance with the management theory of complex production structures, a loading and transport enterprise that performs both economic and social functions can be identified as an organizational and technical system (OTS).

The relevance of study of subsystem interaction patterns in the LTE structure follows from the results of analysis of its economic activity, which confirm a significant gap in technical, organizational and socio-economic levels of development between the OTS subsystems. At the facilities of production and social infrastructure, the level of mechanization and automation is lower, the share of manual labor is much higher, and the level of management is low. It should be noted that during optimization and structural changes at enterprises, first of all, the personnel of production and social infrastructure are reduced, turning it into a "bottleneck" of LTE, negatively affects the economy, leads to an increase in operating costs. Technical progress in the development of loading transportation equipment, organization of processing, coal loading and transportation of wagons requires adequate changes in the subdivisions of production and social infrastructure.

As an object of research the enterprise named United Production and Transport Department of Kuzbass, LLC, was accepted. The general scheme of subsystems interaction in LTE structure is given in [2, 4, 5]. To study the nature of interaction of elements in subsystems, a cognitive approach was adopted that predicts the development scenarios of the facility and making of management decisions taking into account the risks of incidents.

2. Methods of research

Based on the analysis of theory and experience of cognitive modeling [6 - 8], the following main stages of the analysis of interacting factors (concepts) and modeling of possible development scenarios of situation in LTE structure are substantiated [9 - 13]:

- identification of concepts and variables of their state of the research object;
- construction of a conceptual model using uncertain and unreliable information;
- construction of cognitive maps with identification of concepts in graph vertices and connections between concepts;
- determination of values of cause-effect relationships with consideration of state variables of the complex of factors interacting in technological process;
- analysis of cognitive models;
- formation of adjacency matrix;
- calculation of accessibility matrix to evaluate the cognitive map connectivity;
- impulse simulation;
- correction of models;
- making decisions to manage the object state.

3. Results and discussion

Concept is the basic notion for the group (class) of elements of the object under study. The name of the concepts and their states in each class depend on many factors. Within the framework of LTE organizational and technical system the following indicators are distinguished [14, 15]:

1) external influences: control and supervisory bodies, regulatory documents, delivery of wagons, coal to warehouses, etc.;

2) internal control actions: the level of competence of heads of services, departments, sections, dispatchers and other job responsibilities within the framework of delegated authority;

3) conformity of personnel competencies (crane drivers, loader operators, slingers, movers, acceptance/delivery agents, etc.) to job duties and labor functions [16];

4) loading and unloading operations;

5) seasonal changes in the elements of the environment in which the control object operates: extreme negative and positive air temperatures, precipitation, reduction in the volumes of coal shipment during spring-summer period, increased probability of spontaneous combustion of coal stored in storages, seasonal and weather restrictions due to environmental factors;

6) shunting operations;
7) wear of production funds;
8) state and conditions of non-public routes operation;
9) compliance of the actual activity of LTE with its license for loading and unloading activities with regard to hazardous goods transported by railway;
10) wagon transportation;
11) communication and signaling: levels of efficiency, sustainability, continuity, efficiency and reliability of transmitted information;
12) risk of incidents: spillage of oil products from railway tanks during transportation, fire of rolling stock, fire in coal storage, train derailment from railway tracks, etc.;
13) repair works;
14) material and technical support.

Identification of concepts and variables of their state of control object for development of cognitive models was carried out on the basis of expert assessments of specialists from United PTU of Kuzbass, LLC. The interrelation and the significance of the concepts connections in the cognitive map were estimated on the following scale: (+1.0) – affects strongly; (+0.5) – affects; (0.0) – does not affect; (-0.5) – affects negatively; (-1.0) – affects very negatively.

Based on the results of concepts identification a cognitive map and an adjacency matrix were constructed that include 14 interrelated concepts.

At the first stage cognitive modeling was carried out to determine the influence of a single concept on the stability of the system as a whole, and on the second stage – the effect of the system on the concept. The calculations were made when one of the concepts was changed within the range 0÷1 with step Δ_k=0.1. The sum of the effects of the ith concept on the remaining concepts and the system as a whole for each kth step of the change of ith concept was determined by the formula

$$W_i^k = \sum_1^N v_{jk},$$

where W_i^k – sum of the influences of the ith concept on the remaining concepts of the system at the kth step of change of the ith concept; k – initial (starting) value of the ith concept at the kth step of its change; v_{jk} – numerical value of the jth concept at the kth step of change of the ith concept.

The modeling and calculation results of sums of influence of the ith concept on the remaining concepts of the system are shown in Figure 1, from the graphs of which it follows that for values of concept *i* within the range $0<v_i<0.2$, its effect significantly and gradually decreases, and at $v_i \approx 0.5$, the system becomes stable. At $v_i>0.7$ instability arises in the system due to the strong influence of the concept *i*.

The most significant influence on the system is provided by internal control actions (Figure 1), formed in accordance with the level of competence of heads of services, departments, sections, dispatchers acting on the basis of regulations, job descriptions. According to the graphs, the technological and organizational control actions at the concept value $v_{i-2}<0.3$ have the strongest impact on the state of the system as a whole, additional control actions allow the risk of an incident in the system to be reduced and the system to be brought to an equilibrium state. The system functions normally at the values of the concept $0.4<v_{i=2}<0.7$. At $v_{i=2}>0.7$ additional control actions on the system are not required, as they will lead to an increase in the risk of incidents.

The second strongest impact on the LTE management system is personnel (Figure 1), which is assessed by the compliance of the competencies of crane drivers, loader operators, slingers, loaders, and acceptance/delivery agents, etc., to job responsibilities and labor functions. At $v_{i=3}=0.1$ the competence of the staff does not fully correspond to the job description and labor functions, the risk of incidents is maximal. This is confirmed by the maximum risk of incident on the graph in Figure 1. It can be argued that at $v_{i=3}<0.1$, with sufficient competencies for performing production operations, the gradual adaptation of performers to the working conditions takes place in terms of acquiring skills and techniques for performing operations. The adaptation period is intensive, this fact is confirmed by the graph in figure 1, in which at $v_{i=3}>0.3$ the competencies are sufficient to perform production operation and provide functioning of the entire system in a stable mode. It should be noted that further increase in the competence of personnel at $v_{i=3}>0.7$ the risk of incidents does not increase.

The influence of other concepts on the risks of incidents in the system is smoothly reduced according to the graphs in figure 1.

Almost the same influence of two concepts on the system were revealed:

- external impacts $v_{i=1}$: supervisory authorities, regulatory documents, delivery of coal wagons to storages;
- communication and signaling $v_{i=11}$: the criterion is the risk of an incident due to unsatisfactory communication and signaling.

The influence of these concepts on the risk of an incident in the system decreases monotonically in the range $v_{i=1}=v_{i=11}<0.4$.

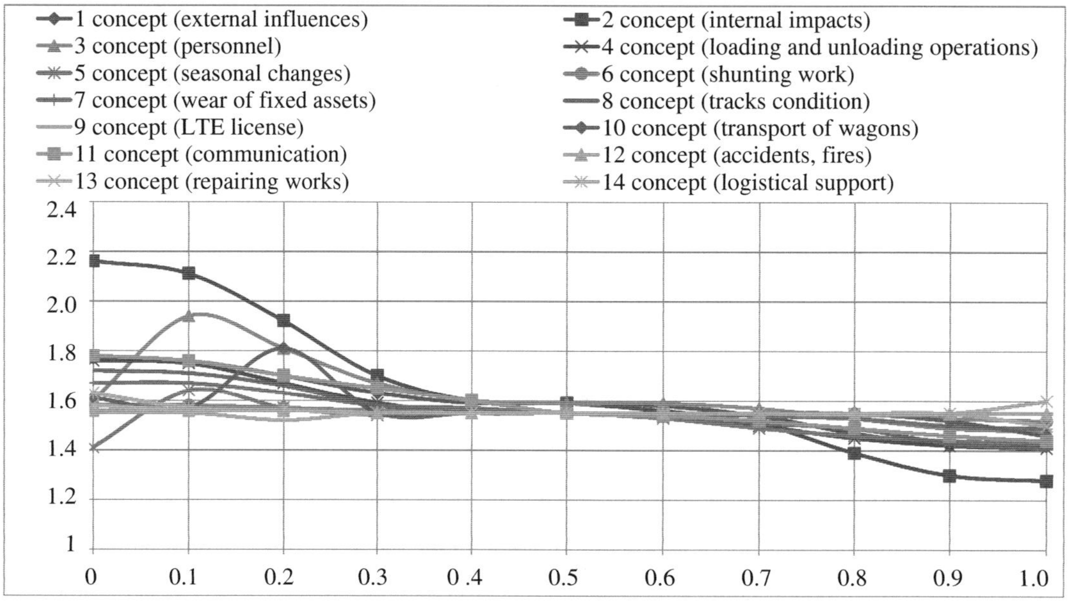

Figure 1. Graphs of variance of concepts action force on the system.

The developed algorithm for constructing functional dependencies between the relevant factors makes it possible to identify not only the influence of individual concepts on the management system, but also the influence of management system on individual concepts. The ranking of concepts by the strength of the system integral impact allows us to identify the concepts most relevant to the system (figure 2).

It should be noted that the graphs in figures 1 - 2 were obtained by varying the values of only one factor (concept) in the range from 0 to 1. In this case, the values of other factors are assumed to be the same, that is, $v_i=0.5$. However, the method of cognitive modeling allows different alternative versions of the state of individual concepts and the system as a whole to be considered.

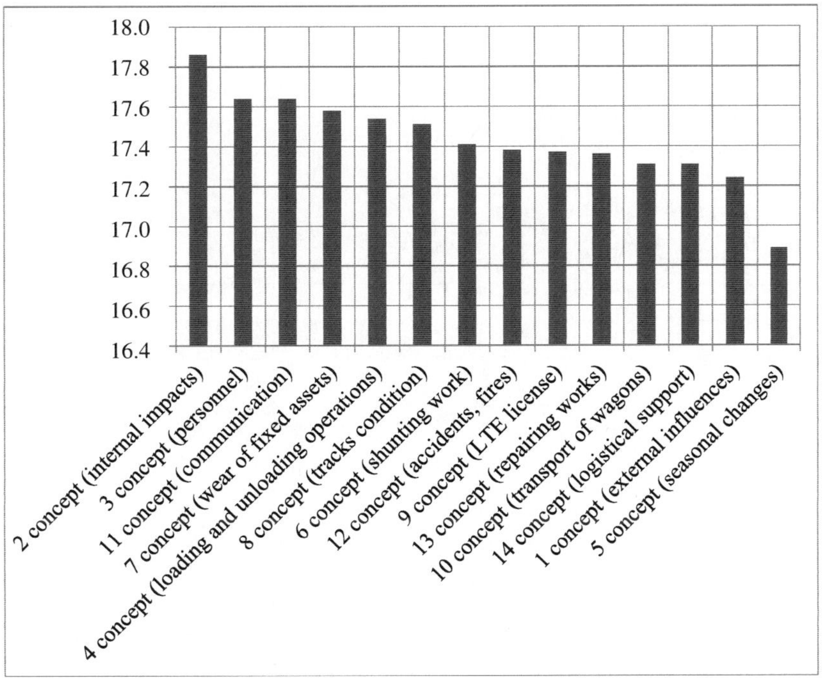

Figure 2. An ordered diagram of action force values of the system on individual concept.

There are possible variants of the worst system state, for example, $v_{i=1\div14}=0.1$ or the best one, for example $v_{i=1\div14}=0.9$.

The results of modeling and comparison of the basic variant at the values of all concepts $v_{i=1\div14}=0.5$ and variants at values of all concepts $v_{i=1\div14}=0.1$ and $v_{i=1\div14}=0.9$ are presented in table 1.

Table 1. Comparison of the forces of concepts influence on the system and influence of systems on the concepts for the three variants of simultaneous impact of all concepts on the system with $v_{i=1\div14}=0.1$; $v_{i=1\div14}=0.5$; $v_{i=1\div14}=0.9$.

Number of a concept	Action force of the ith concept on the system			Action force of the system on the ith concept		
	$v_{i=1\div14}=0.1$	$v_{i=1\div14}=0.5$	$v_{i=1\div14}=0.9$	$v_{i=1\div14}=0.1$	$v_{i=1\div14}=0.5$	$v_{i=1\div14}=0.9$
1	0.61	0.19	0.02	0.00	0.00	0.00
2	0.66	0.29	0.08	0.17	0.09	0.02
3	0.47	0.17	0.07	0.22	0.09	0.02
4	0.35	0.14	0.03	0.23	0.12	0.04
5	0.35	0.17	0.05	0.00	0.00	0.00
6	0.08	0.04	0.01	0.44	0.21	0.06
7	0.15	0.06	0.02	0.48	0.17	0.05
8	0.26	0.14	0.04	0.40	0.15	0.05
9	0.06	0.04	0.01	0.07	0.00	0.00
10	0.13	0.08	0.02	0.53	0.18	0.07
11	0.48	0.15	0.05	0.20	0.04	0.02
12	0.00	0.00	0.00	0.62	0.32	0.06
13	0.10	0.05	0.02	0.44	0.19	0.04
14	0.15	0.07	0.02	0.04	0.02	0.01

According to the results of the simulation given in table 1, a distribution histogram of influence forces of concepts on the system is constructed (figure 3) for the values of concepts $v_{i=1\div14}=0.1$; $v_{i=1\div14}=0.5$; $v_{i=1\div14}=0.9$ (the numbers of concepts correspond to the values shown in figures 1, 2).

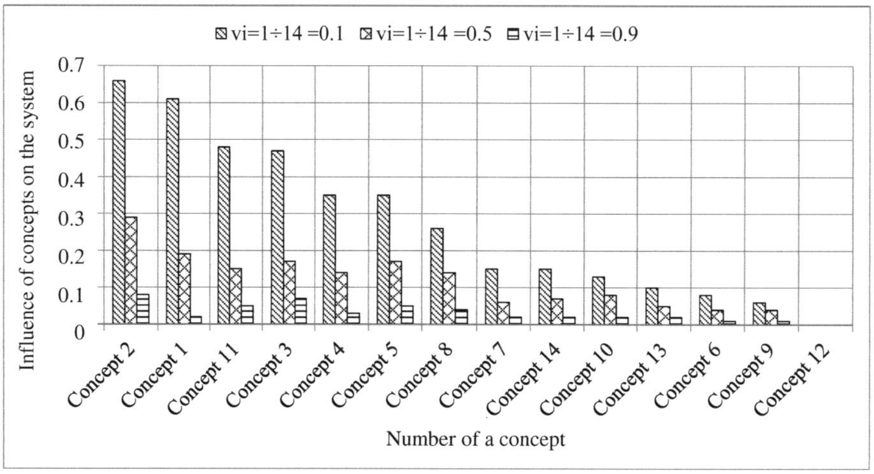

Figure 3. Distribution histogram of concepts influence on the system.

From table 1 and the graphs in figure 3 it follows that for critical values of concepts $v_{i=1\div14}$= 0.1 their influence on the system is ranked as follows (see Table 1): internal control actions (0.66), external influences (0.61), communication and signaling (0.48), compliance of personnel competencies (0.47), loading-unloading operations (0.35), seasonal changes (0.35). The influence of other concepts on the system is not significant in the range 0-0.26. In general, the state of the system at $v_{i=1\div14}$=0.1 should be considered as unstable.

At the influence on the concept system $v_{i=1\div14}$=0.9 their influence varies within the range 0-0.08, i.e. insignificantly (see table 1), the system state is stable, which corresponds to the graphs in figure 1. The boundary of the system transition to high stability is the condition $v_{i=1\div14}$=0.5.

Thus, based on the results of cognitive modeling, the most significant concepts that affect the stability of the system are identified.

4. Conclusions

1. It is established that the application of cognitive modeling methods for the analysis of complex production objects allows the essential elements of the system, that ensure the stable functioning of the system as a whole, to be identified.

2. The most significant impact on the system is achieved by the internal management actions of the heads of services and the compliance of the personnel competencies with job duties and performed labor functions.

3. In case when the personnel competences do not fully correspond to the duties and labor functions, the risk of incidents is maximal. In the process of acquiring additional skills and techniques for performing operations, the risk of incidents is reduced. Competencies sufficient to perform production operarions, ensure the functioning of the system in a stable mode.

References

[1] Kovalev V I, Octinin A T and Kovalev V I 2009 *Technology of Work Stations* (Moscow: UMTs ZhDT) vol 1 p 263

[2] Shishkina S V, Pristupa Yu D, Fryanov V N and Pavlova L D 2015 *Proc. Xth All-Russian Conf. on Automation Systems in Education, Science and Production* (Novokuznetsk: SibSIU) 2015 166–173

[3] Shishkina S V, Pristupa Yu D, Pavlova L D and Fryanov V N 2016 *Proc. IVth All-Russian Conf. on Modeling and Knowledge-based IT in Tech. and Socio-economic Systems* (Novokuznetsk: SibSIU) pp 8–14

IOP Conf. Series: Earth and Environmental Science **84** (2017) 012027 doi:10.1088/1755-1315/84/1/012027

[4] Pristupa Yu D, Shishkina S V, Smolin A V and Bytkar M Yu 2014 *Ugol* **12** 50–2
[5] Pristupa Yu D, Fryanov V N and Pavlova L D 2016 *IOP Conference Series: Earth and Environmental Science* **45** 012014
[6] Yamalov I U 2010 *Modeling of Management and Decision-making Processes in Emergency Situations* (Moscow: Laboratoriya Basovikh Znaniy) p 288
[7] Silov V B 1995 *Decision Making in Uncertain Situations* (Moscow: INPRO-RES) p 228
[8] Pegat A 2011 *Fuzzy Modeling and Control* (Moscow: Laboratoriya Basovikh Znaniy) p 798
[9] Meskon M H, Albert M and Hedouri F 1992 *Fundamentals of Management* (Moscow: Delo) p 704
[10] Burkov V N, Novikov D A and Shchepkin A V 2008 *Mechanisms for Management of Ecological and Economic Systems* (Moscow: Fizmatlit) p 244
[11] Gelovani V A et al 2001 *Intelligent Decision Support Systems in Emergency Situations Using Information on the State of Natural Environment* (Moscow: Editorial) p 304
[12] Arkhipova N I and Kulba V V 2008 *Control in Emergency Situations* (Moscow: RGGU) p 474
[13] Novikov D A 2005 *Theory of Organizational Systems Control* (Moscow: MPSI) p 584
[14] Emelyanov S V et al 2008 *Theory and Practice of Prediction in Control Systems* (Kemerovo: Kuzbassvuzizdat) p 487
[15] Reason J 2015 *Man: a Threat or a Guarantor of Safety (short version)* (Novokuznetsk: KemSU) p 151
[16] Labunsky L V 2003 *Development of Personnel Competences at a Mining Enterprise* (Ekaterinburg: UB RAS) p 230

IOP Conf. Series: Earth and Environmental Science **84** (2017) 012028 doi:10.1088/1755-1315/84/1/012028

Specific aspects of evaluation of control systems similarity

V F Evtushenko[1], V N Burkov[2], L P Myshlyaev[3] and G V Makarov[3]

[1]Siberian State Industrial University, 42 Kirova Street, Novokuznetsk, 654007, Russia
[2]Institute of Control Sciences of the Russian Academy of Sciences, 65 Profsoyuznaya Street, Moscow, 117997, Russia
[3]R&D Center of Control Systems LLC, 42 Kirova Street, Novokuznetsk, 654007, Russia

E-mail: evtushenko_vf@sibsiu.ru

Abstract. The paper discusses the problem of estimating the similarity of two or more control systems. In accordance with the statements made earlier it is indicated that such management systems should be characterized by a strict or approximate equality of the values of targeted indicators (criteria) of control effectiveness. Two main variants of procedures for estimating the control systems similarity are proposed. The first is based on the direct evaluation of the targeted indicators of control performance, the second – on the use of the so-called similarity relations, describing the conditions the fulfillment of which is valid for such systems. The procedures for evaluating the control systems similarity with the help of special *similarity relations* are associated with considerably less expenses. However, they are formed for a relatively narrow class of automatic control systems and are basically special cases of similarity, which requires further research in this direction.

1. Introduction

One of the primary and important tasks for development of the traditional theory of similarity [1] for control systems is the problem of estimating the similarity of two or more systems. Its solution is based on the axiomatics of similarity of control systems, stated in the form of *Statements*, according to which such control systems should be characterized by a strict or approximate equality of values of the targets (criteria) of control effectiveness. The formula expression for this statement for the j-th and l-th control systems can be written in the form [2]

$$\left\| q_j^n\{(t-T),t\} - q_l^n\{(t-T),t\} \right\| \le \delta q_{jl}^{n*};\ j \ne l;\ j = \overline{1,J};\quad l = \overline{1,L}, \tag{1}$$

where $\|\ \|$ – the measure of proximity; $q_j^n\{(t-T),t\}$ and $q_l^n\{(t-T),t\}$ – the normalized targeted values of the performance efficiency in the specified time interval T, respectively, of the j-th and l-th control systems; δq_{jl}^{n*} – the value determined by the maximum permissible (predetermined) threshold of the proximity of indicators $q_j^n\{(t-T),t\}$ and $q_l^n\{(t-T),t\}$, under which the effectiveness of these systems is considered to be the same; J and L are the number of control systems.

The strict equality of indicators values q_j^n and q_l^n corresponds to the functioning of control systems in the absence of influence on their behavior of uncontrolled disturbances. While the approximate equality of indicators values, represented, in particular, in the form of inequality (1), reflects the functioning of control systems under uncertainty. The numerical value of the acceptable threshold δq_{jl}^{n*} depends on the specific values of the level of uncertainty, its characteristics and methods of their estimation [3, 4]. We shall note here that preliminary studies of dynamic properties and conditions for the functioning of full-scale control objects and their operating physical models [5, 6] have made it possible to conclude that almost all of them are subject to the influence of uncontrolled disturbances.

2. Research methods

Two variants of procedures for evaluating the control systems similarity can be distinguished. The first is the direct use of the relation (1), the practical realization of which requires a reliable estimation, and in the case of a non-stationary control object – continuous or periodic tracking of the values of the target indicators of control systems efficiency and their subsequent comparison with the corresponding value δq_{jl}^{n*}. It is also possible to estimate the control systems similarity by the method of correlated processes [7], by calculating and subsequently analyzing the correlation moments between the values of the performance indicators of the two control systems.

The second version of the procedure for estimating the control systems similarity is based on the use of the so-called *similarity relations* [5] describing the conditions that are valid for such systems. They represent a linear or nonlinear combination of known characteristics of the actions properties, objects and control parts of systems, such as statistical and fractal time series, parameters of dynamic characteristics of the transformation channels for these effects of objects and control algorithms. *Similarity relations* can be obtained, in particular, by analytical methods or by using numerical simulation and search optimization methods. The statement of the problem, the solution scheme, conditions for the effective application of these methods are given in [5]).

3. Application of methods

Practical application of the first variant requires the following basic operations:

- normalizing of the values of targeted indicators q_j^n and q_l^n in order to bring them into a single coordinate scale of change. Here we can use the known normalization operation [8] by means of the difference relation obtained at a given interval of time T of efficiency indicator of the system under consideration and its minimum possible value to the difference of the maximum $q_j^{\max}$ and minimum $q_j^{\min}$ possible values, i.e.

$$q_j^n\{(t-T),t\} = \frac{q_j\{(t-T),t\} - q_j^{\min}}{q_j^{\max} - q_j^{\min}}, \tag{2}$$

where $q\{(t-T),t\}$ – the current value of the targeted criterion in the time interval T; $q_j^{\min}$ and $q_j^{\max}$ – accordingly, the minimum and maximum possible values of the targeted criteria for the functioning of the j-th control system;

- coordination of input, output effects and state variables of the j-th and l-th control systems. The control objects of these systems can operate at different coordinate and time scales. Therefore, to ensure a single coordinate-time space for the operation of both control systems, it is necessary to reconcile the trajectories of changes in the input, output actions and state variables of their control objects [9].

Such coordination can be provided, in particular, by converting the values of the input $V_j\,(t)$, output $Y_j\,(t)$ effects and the state variables of the $S_j\,(t)$ j-th control object represented in the form

$Z_j(t) = \{V_j(t); Y_j(t); S_j(t)\}$, by means of the scaling function $F_M^Z\{\cdot\}$ to the scale of the changes in the values of the output actions of the l-th control object $Z_{jl}(t)$

$$Z_{jl}(t) = F_M^Z\{Z_j(t)\} \quad (3)$$

The values obtained as a result of the value transformation (3) $Z_{jl}(t)$ change in the same range of values and time scale as the values $Z_j(t)$ of the j-th control object. The execution of operations using relations (2) and (3) is a necessary condition for the realization of the first variant of the similarity estimation, regardless of the determination method of specific values for target indicators of the functioning control systems;

- determination of the target indicators values of the effectiveness of control systems, estimation of which can be performed by processing full-scale, model or full-scale and model [10] data, including recurrent procedures.

4. Classification of methods for similarity evaluation

The results of experience generalization concerning the evaluation and study of the control systems similarity [1, 2, 4, 5, 7] made it possible to form a finite set of methods for estimating the similarity of control systems (Figure 1), the differences between them for the first variant are determined by possible ways of estimating the values of their performance indicators.

The first version of the procedure for assessing the control system similarity, based on the direct use of targeted indicators of their efficient performance using the relation (1), is more universal from the point of view of its practical application. It can be used to estimate the similarity of almost any control systems having different structures and functioning in different conditions, if the values of the targets are evaluated on a finite time interval with the use of full-scale, model, full-scale and model data. However, in order to obtain adequate estimation results, it is necessary to have a large number of arrays with a large volume of reliable data computation of which is very time consuming. In addition, obtainment of reliable data in the existing control systems is associated with considerable difficulties and cannot always be realized.

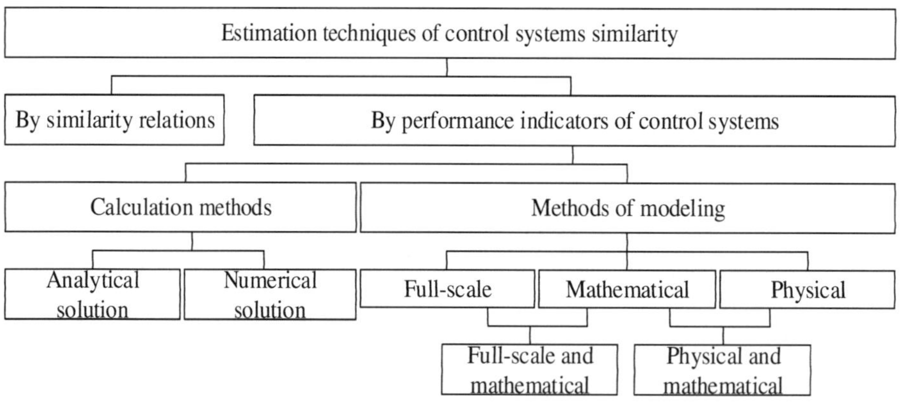

Figure 1. Classification of methods for evaluating the similarity of control systems.

The most preferable from the point of view of the cost of realization are the calculation methods for evaluating the control systems similarity, to which refer such methods that help to obtain an analytical solution of mathematical equations with an output to the target indicator of the functioning control systems. To implement the calculation method the initial information should be presented in the form of mathematical models of the channels for converting input effects and the effects themselves, including uncontrolled ones with fairly strict limitations on their structures and the range of parameters. It is possible to obtain a solution of the integral equation with an subintegral function comprising irrationality in the denominator only in very rare cases [3], in which on the area of effective application of the calculation method substantial limitations are imposed associated mainly with the structure and the range of variation of the values of model parameters transforming the channels of control objects, their input effects, control algorithms etc. Usually these restrictions are also very rarely implemented in practice. Area of the calculation methods can be somewhat extended in the cases when such numerical solution of integral equations is associated with a lower cost compared to the cost for the realization of the modelling methods of control systems.

The second version of the procedure for estimating the similarity of control systems, based on special *similarity relations* requires considerably less computing and time costs, and from this point of view it is more preferable. However, it, as well as analytical methods, is essentially limited in practical application. Such relationships are obtained only for a small class of control systems. In particular, they include control systems of positioning action with typical laws of regulation, the similarity relations for which and the scope of their application are given in [5].

5. Conclusions

The procedures for estimating the control systems similarity, based directly on the relation (1) are applicable for virtually any control systems of the same or similar structures. However, they require significant costs and time for reliable determination of the targeted indicators of their performance efficiency.

The procedures for evaluating the control systems similarity with the help of special *similarity relations* are associated with considerably less expenses. However, they are designed for a relatively narrow class of automatic control systems which requires further research in this direction.

Acknowledgments
The work was supported by the RFBR grant, project No. 15-07-01972

References

[1] Venikov V A 1976 *Theory of Similarity and Modeling* (Moscow: Higher School) p 479
[2] Myshlyaev L P et al 2012 *Steel in Translation* **42(12)** 823–4
[3] Rotach V Ya 1973 *Calculation of the Dynamics of Industrial Automatic Control Systems* (Moscow: Energia) p 439
[4] Barkovsky V V et al 1969 (Moscow: Mechanical Engineering) p 385
[5] Evtushenko V F et al 2016 *IOP Conference Series: Earth and Environmental Science* **45** 012010
[6] Tsiryapkina A V et al 2015 *Steel in Translation* **45 (12)** 943–8
[7] Pugachev V N 1973 *Combined Methods for Determining Probabilistic Characteristics* (Moscow: Sov. Radio) p 256
[8] Rykov A S 1999 *Methods of System Analysis: Optimization* (Moscow: Economika) p 255
[9] Krasovskii A A 1979 *Avtomatika i Telemekhanika* p 156–62
[10] Avdeev V P 1979 *Ferrous Metallurgy* **6** 131–4

IOP Conf. Series: Earth and Environmental Science **84** (2017) 012029 doi:10.1088/1755-1315/84/1/012029

Structures and efficiency areas of object control systems with various types of recycling

L P Myshlyaev[1], A V Tsyryapkina[2], V N Burkov[3] and K A Ivushkin[4]

[1]R&D Center of Control Systems LLC, 42 Kirova Street, Novokuznetsk, 654007, Russia
[2]Siberian State Industrial University, 42 Kirova Street, Novokuznetsk, 654007, Russia
[3]Institute of Control Sciences of the Russian Academy of Sciences, 65 Profsoyuznaya Street, Moscow, 117997, Russia
[4]Consolidated Company "Sibshakhtostroy", 9 Shosse Kuznetskoe, Novokuznetsk, 654034, Russia

E-mail: nicsu@ngs.ru

Abstract. The work is devoted to the study of automatic control systems (ACS) of objects with various types of recycling. The characteristic feature of such objects is the presence of a delay element in the coordinates. It is known that only with the help of recycling the use of raw materials and energy resources can be maximized. However, little attention has been paid to the control of such objects in the well-known publications, which gives the study a particular importance. The paper describes a classification of objects with recycling proposed by the authors with the identification of four classes of objects: recycling "by concentration", "by mass", "by parameters", combined recycling. ACS are synthesized for the first three classes of objects. The task of comparative analysis of the systems effectiveness with a model control law and the proposed synthesized systems is set, as well as the task of determination of areas of systems effective operation depending on the variable value of the ratio of the delay time in the recycling chain and in the direct circuit, the conclusions are drawn.

1. Introduction

Increase in the efficiency of technological processes of raw materials processing, environmental processes, and sometimes their feasibility is achieved through the introduction of positive feedbacks in controlled objects (objects with recycling). Examples of such objects include water-slime systems for mineral processing, pelletization of materials during the production of agglomerate in metallurgical production, socio-economic systems in the allocation of material and financial resources, in a variety of chemical processes [1, 2], in the formation of fractal structures of materials [3 - 6]. The introduction of positive feedbacks into controlled objects gives them qualitatively new properties, significantly changes the dynamics of behavior and complicates the control process.

2. Methods of research

Mathematical models of objects with a recycling in the state space in the general form are described by the expressions [7]:

$$\dot{X}(t) = A(t) \cdot X(t - \tau_x) + B(t) \cdot U(t - \tau_u) + C(t) \cdot W(t - \tau_w);$$
$$Y(t) = D(t) \cdot X\left(t - \tau_y\right) + F(t) \cdot E(t - \tau_e),$$

where *X, U, W, Y, E* – vectors of states, control, external influences, outputs and measurement errors; $\tau_x, \tau_u, \tau_w, \tau_y, \tau_e$ – corresponding delay times; $A(t), B(t), C(t), D(t), F(t)$ – matrices of the corresponding dimensions; *t* – continuous time.

Classification of objects with recycling, depending on the type of recycling and its effect on the controlled object is presented in table 1 [8].

Table 1. Classification of objects with recycling (with indication of the structure of the object model).

1 class	2 class
Object with recycling "by concentration". All material costs are fixed and the concentration of elements (temperature, etc.) of the material varies. The matrix $A(t)$ does not depend on the state $X(t)$ and $D(t) = 1$. 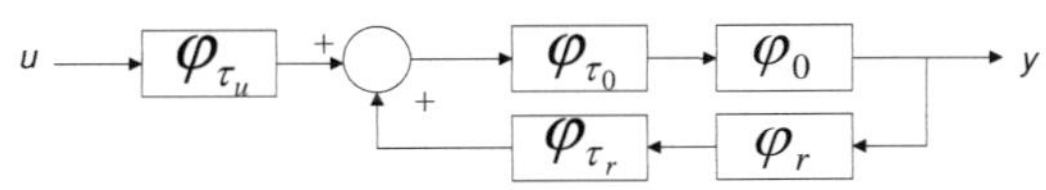	Object with recycling "by mass". For the output action a part of the material (finished product) arriving at the outlet of the object is taken and the other part of the material enters the recycling chain. The matrix $A(t)$ depends on the state $X(t)$ and $D(t) \neq 1$. 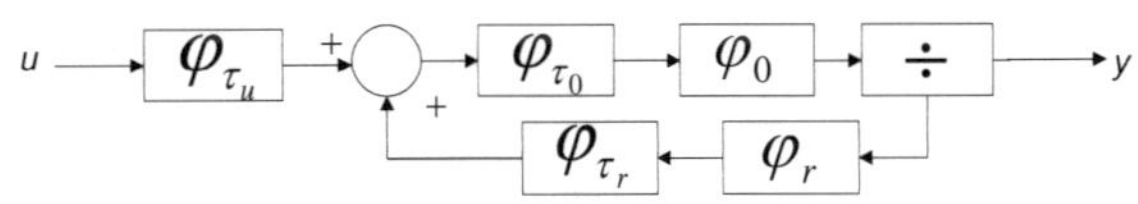
3 class	**4 class**
Object with recycling "by parameters". The recycling affects the parameters of their individual components. In this case the matrices $B(t)$ and $C(t)$ depend on the state $X(t)$. 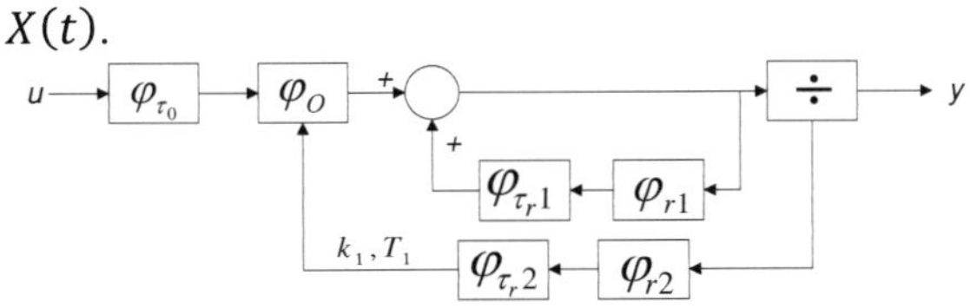	Object with a combined recycling. 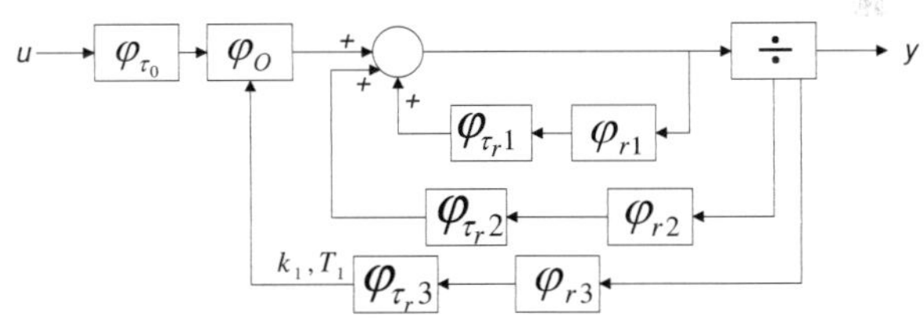

For the objects with recycling "by concentration", "by mass", "by parameters", the automatic control systems (ACS), presented in figure 1 [9 - 12], were synthesized.

For the obtained ACS and ACS with a model control laws a comparative efficiency analysis was carried out, and areas of effective system operation were determined depending on the ratio of the delay times in the recycling chain and in the direct chain.

3. Results and discussion

Formulation of the problem.

Given.

1. Structures of ACS objects with recycling "by concentration", "by mass", "by parameters" with a typical law of regulation f_R and basic structures of ACS.

2. Operators of ACS blocks

$\varphi_0(S) = \frac{k_0}{T_0 S+1}$, $\varphi_o^M(S) = \frac{k_o^M}{T_o^M S+1}$, $k_0 = k_o^M, T_0 = T_o^M$ – object with self-leveling in the direct chain;

$\varphi_r(S) = \frac{k_r}{T_r S+1}$, $\varphi_r^M(S) = \frac{k_r^M}{T_r^M S+1}$, $k_r = k_r^M, T_r = T_r^M$, where k_0, k_r – transmission factors, T_0, T_r – time constants; $f(S)$ – the proportional-integral control law, $f^e(S)$ is the extrapolation operator in the form of a "shift".

IOP Conf. Series: Earth and Environmental Science **84** (2017) 012029 doi:10.1088/1755-1315/84/1/012029

3. Variations of the ratio ${}^{\tau_r}/_{\tau_0}$ in the range [1; 45] for objects with recycling "by concentration" and "by mass" (at $\alpha = 0.6$), in the range [1; 2.5] – for objects with recycling "by parameters".

4. Limitations on the magnitude of the input action $u_2 \in [0; 100]$, affecting the value of the division block, for the ACS object with recycling "by parameters".

5. The efficiency criterion: $q(t) = \frac{1}{T}\int_{t-1}^{T} |y^*(\theta) - y(\theta)| d\theta$, where T – time of the transient process.

Required. To identify the area of ACS effective performance of objects with recycling of all classes.

The transition from the operators φ_0, φ_r, φ_{τ_u}, φ_{τ_0}, φ_{τ_r} to the recursive-difference form is made by the method of finite differences. The programming is carried out in the Microsoft Office Excel system. The results of numerical studies with variation of ratios ${}^{\tau_r}/_{\tau_0}$ for objects ACS with recycling "by concentration", "by mass", "by parameters" are presented in tables 2-4.

Table 2. Efficiency criterion for synthesized ACS with recycling "by concentration" and ACS with a model control law.

${}^{\tau_r}/_{\tau_0}$	Performance criterion, rel.u.	
	ACS with a model control law	Synthesized ACS
1	unstable	4.99
5	unstable	5.02
10	unstable	6.13
15	unstable	7.50
20	unstable	8.66
30	unstable	10.61
35	unstable	11.48
37	stability boundary	11.82
38	19.26	11.98
40	19.14	12.28
45	18.78	12.98

Table 3. Efficiency criterion for synthesized ACS with a recycling "by mass" and ACS with a model control law.

${}^{\tau_r}/_{\tau_0}$	Performance criterion, rel.u.	
	ATS with a model control law	Synthesized ACS
1	3.84	2.32
5	3.97	2.48
10	4.62	2.59
15	5.27	2.99
20	5.42	3.19
25	5.82	3.65
30	6.20	4.40
35	6.52	4.87
40	6.79	5.28
45	7.04	5.70

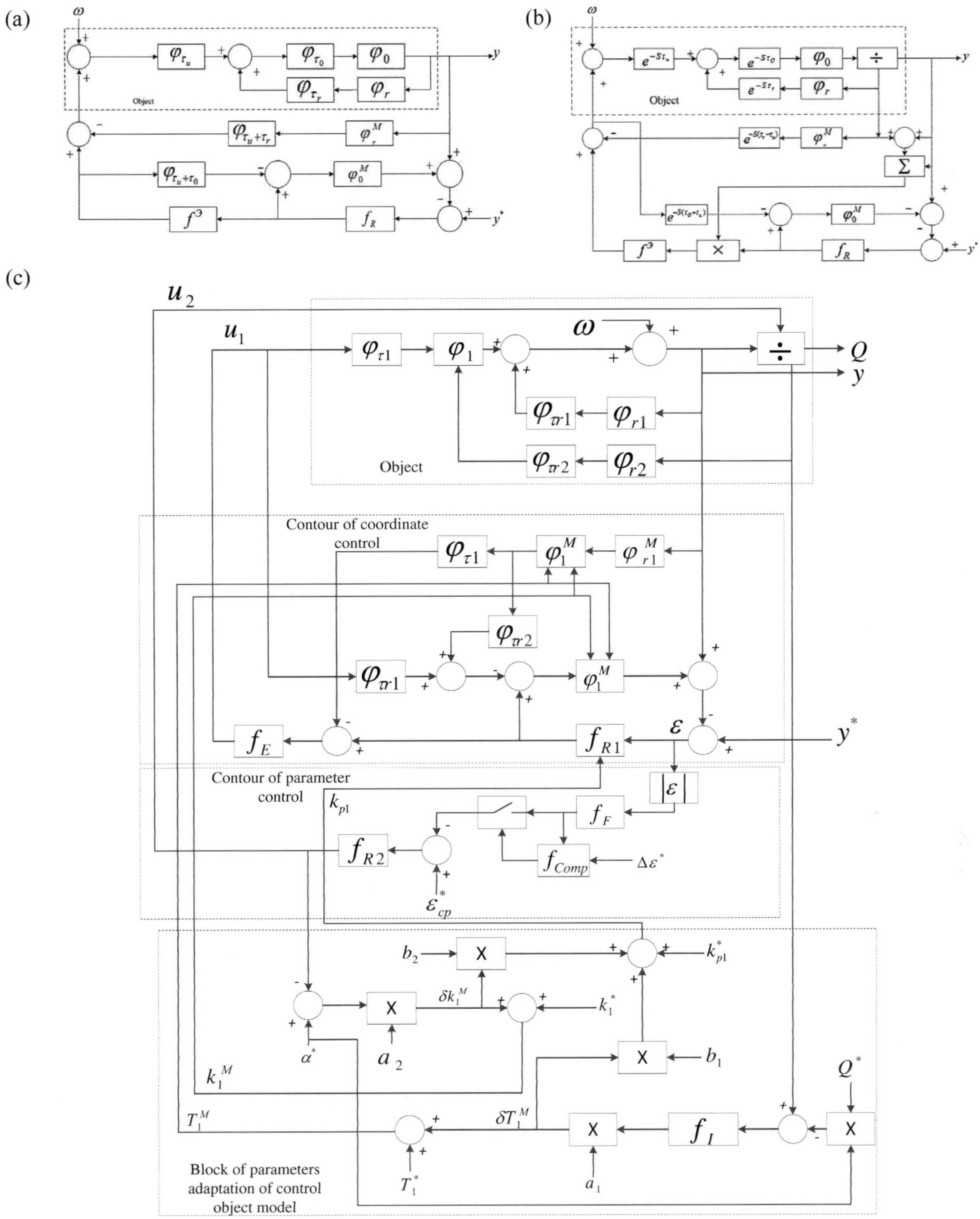

Figure 1. The structures of synthesized ACS of the object with recycling: (a) by concentration, (b) by mass, (c) by parameters.

Table 4. Performance criterion for a synthesized ACS with a recycling "by parameters"

ACS under the influence of external step actions according to disturbing input $\omega(t)$		ACS under the influence of external step actions according to controlling input $y^*(t)$	
τ_r/τ_0	Performance criterion, rel.u.	τ_r/τ_0	Performance criterion, rel.u.
1	3.708	1	5.248
1.25	3.824	1.25	5.310
1.5	3.954	1.5	5.369
1.75	3.943	1.75	5.429
2.0	4.521	2.0	5.512
2.1	Boundary of stability	2.1	Boundary of stability
2.25	unstable	2.25	unstable
2.5	unstable	2.5	unstable

4. Conclusions

Based on the results of problem solution the following conclusions are drawn:

1. The synthesized ACS of the object with the recycling "by concentration" remains stable for any ratios $\tau_r/\tau_0 \in [1; 45]$, while the ACS with a model control law becomes stable only when τ_r/τ_0 reaches value of 38.

2. The synthesized ACS and ACS with a model law of object regulation with recycling "by mass" are stable for any ratios $\tau_r/\tau_0 \in [1; 45]$.

3. ACS of object with recycling "by parameters" is stable for any values of the delay ratio in the straight chain and the recycling chain τ_r/τ_0 in the range [1; 2.1]; at $\tau_r/\tau_0 > 2.1$ the system becomes unstable.

4. For stable systems in the entire range of ratios τ_r/τ_0, the synthesized ACS of the object with a recycling exceeds the average modulus criterion ACS with a model control law not less than by 1.5 times.

5. Throughout the investigated range of ratios τ_r/τ_0, the time of the transient process of the synthesized ACS is less than the time of the ACS transient process with the model control law no less than 3 times.

The results of solution of this problem can be applied both to adjust the regulators in the creation of automation systems and to develop technological solutions at the design stage of units and process regulations, so that the design of aggregates of direct chain and in the recycling chain could meet the calculated conditions.

Acknowledgments

The work was carried according to the State Assignment of the Ministry of Education and Science of the Russian Federation No. 8.8611.2017 / 8.9.

References

[1] Uteush E V and Uteush Z V 1973 *Control of Grinding Aggregates* (Moscow: Mechanical Engineering) p 280
[2] Amelin A G and Yashke E V 1980 *Production of Sulfuric Acid* (Moscow: Higher School) p 245
[3] Paitgen H O et al 1993 *Beauty of Fractals. Images of Complex Dynamic Systems* (Moscow: Mir) p 176
[4] Ivanova V S et al 1994 *Synergetics and Fractals in Materials Science* (Moscow: Nauka) p 382.
[5] Starovackaya S N et al 2014 *Coll. of the Int. Seminar Papers* (Novokuznetsk: SibSIU) p 348
[6] Kiseleva M Y et al Smagin V I 2011 *Bulletin of TSU* **1** 5–12
[7] Rey U 1983 *Techniques of Technological Processes Control* (Moscow: Mir) p 368
[8] Tsiryapkina A V et al 2015 *Ferrous metallurgy* **12** 925–31
[9] Myshlyaev L P 1987 *Collection of Sci. Papers "Automatic control in the Automated Process*

Control System" (Moscow: Energoatomizdat) p 28–31

[10] Avdeev V P et al 1987 *Natural-mathematical Modeling in Control Systems: Textbook* (Kemerovo: KemSU) p 85

[11] Podchufarov Yu B 2002 *Physico-mathematical Modeling of Control Systems and Complexes* (Moscow: PH of Physics and Mathematics Literature) p 268

[12] Evtushenko V F et al 2016 *IOP Conference Series: Earth and Environmental Science* **45** 012010

IOP Conf. Series: Earth and Environmental Science **84** (2017) 012030 doi:10.1088/1755-1315/84/1/012030

Mechanical characteristics of a double-fed machine in asynchronous mode and prospects of its application in the electric drive of mining machines

V Yu Ostrovlyanchik, I Yu Popolzin, V A Kubarev and D A Marshev

Siberian State Industrial University, 42 Kirova Street, Novokuznetsk, 654007, Russia

E-mail: eidoline@yandex.ru

Abstract. The concept of a double-fed machine as an asynchronous motor with a phase rotor and a source of additional voltage is defined. Based on the analysis of a circuit replacing the double-fed machine, an expression is derived relating the moment, slip, amplitude and phase of additional voltage across the rotor. The conditions maximizing the moment with respect to amplitude and phase of additional voltage in the rotor circuit are also obtained, the phase surface of function of machine electromagnetic moment is constructed. The analysis of basic equation of electric drive motion in relation to electric drive of mine hoisting installations and the conclusion about the necessity of work in all four quadrants of coordinate plane "moment-slip" are made. Family of mechanical characteristics is constructed for a double-fed machine and its achievable speed control range in asynchronous mode is determined. Based on the type of mechanical characteristics and the calculated range of speed control, the conclusion is made about the suitability of using a dual-fed asynchronous machine for driving mine mechanisms with a small required speed control range and the need for organizing a combined operating mode for driving mine hoisting installations and other mechanisms with a large speed control range.

1. Introduction

Currently, most alternating current drives are built according to the scheme of asynchronous motors with a phase rotor and a rotor station. The use of these drive systems is due to their relative simplicity and good overload capacity. At the same time, the used electric drives with a phase rotor have low energy efficiency. These disadvantages are not typical for electric drive systems with frequency control of asynchronous motors, but the use of frequency control for existing asynchronous motors designed for an unchanged frequency of the supply voltage 50 Hz is associated with such problems as significant deterioration of machine mechanical characteristics at a underfrequency, steel overheating, reduction of efficiency rate and power factor [1]. Thus, to ensure acceptable energy and control characteristics of already existing motors with a phase rotor, it is optimal to use them in the dual-supply mode [2, 3]. The article is devoted to the study of characteristics of a dual-fed machine in an asynchronous mode and consideration of the expediency of using this scheme in electric drives of mine installations.

2. Methods of research

For the preparation of the present article, the following research methods were used: the method of a generalized two-phase electrical machine; method of equivalent circuits; method of partial derivatives; method of mathematical modeling.

3. Results and discussion

A double-fed machine (DFM) is a circuit of switching on an asynchronous motor with a phase rotor, in which an additional voltage is supplied to the rotor from an external power source, which usually is a frequency converter [4, 5]. The most important characteristics of a DFM are its mechanical characteristics. They determine the suitability of its application in the electric drive and the operating modes of the machine.

Depending on the method of generating the additional voltage, two modes of DFM operation are distinguished: asynchronous and synchronous. In asynchronous mode, the frequency of the additional voltage is equal to the frequency of rotor emf [6, 7], in synchronous mode – the frequency of additional voltage is set independently [8]. Let us consider the asynchronous operation mode of DFM.

To do this, let us turn to the T-shaped circuit of machine replacement [4, 9] (figure 1).

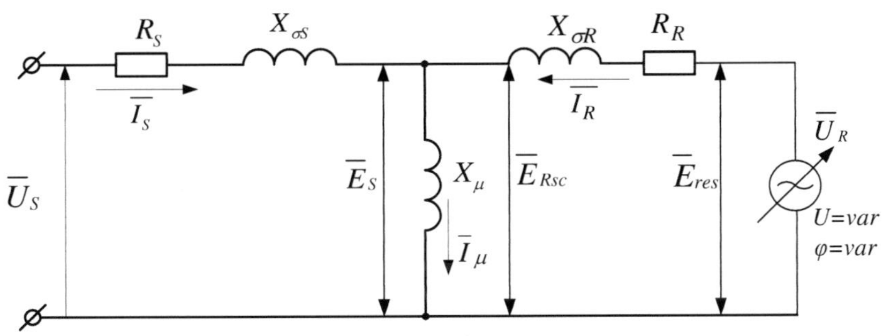

Figure 1. Circuit of DFM replacement in asynchronous mode.

The following designations are adopted on the equivalent circuit:

$\overline{U}_S$ – complex value of voltage across the stator; $\overline{I}_S$ – complex current value of the stator; $\overline{E}_S$ – complex value of emf induced by a rotating field in the stator; $X_{\sigma S}$ – intrinsic inductive resistance of stator scattering; I_μ – magnetizing current of the machine; X_μ – inductive resistance of the magnetizing circuit; $X_{\sigma R}$ – intristic inductive resistance of the rotor scattering, is determined from the expression $X_{\sigma R} = 2\pi f_S L_{\sigma R} \dfrac{\omega_0 - \omega}{\omega_0}$, where f_S – voltage frequency applied to the stator; R_S – active stator resistance; R_R – active rotor resistance; $\overline{E}_{res}$ – complex value of the total (resulting) emf acting in the rotor; $\overline{E}_{R_{sc}}$ – emf of the rotor in short-circuit mode; s – slip $s = \dfrac{\omega_0 - \omega}{\omega_0}$; ω_0 – angular rotation speed of the stator magnetic field; ω – angular rotation speed of the rotor; $\overline{U}_R$ – complex voltage value applied to the rotor.

In asynchronous mode, the frequency of this voltage is equal to the self-frequency of rotor emf: $f_R = sf_S$.

We shall write down the equations of electric equilibrium of the machine for T-shaped analogue circuit, expressing the self-emf of windings through the flux-linkages of windings, and the electromagnetic moment will be defined as a function of the vector product of flux-linkages of the stator and rotor windings:

$$\begin{cases} \overline{U}_S = j\omega_0 \overline{\Psi}_S + \overline{I}_S R_S, \\ \overline{U}_R = j\omega_0 s \overline{\Psi}_R + \overline{I}_R R_R, \\ M = -\frac{3}{2} z_P \frac{L_\mu}{L_S L_R - L_\mu^2} \left[\overline{\Psi}_S \times \overline{\Psi}_R \right] \end{cases} \quad (1)$$

where $\overline{\Psi}_S$ – complex value of stator flux-linkage; $\overline{\Psi}_R$ – complex value of rotor flux-linkage; L_μ – magnetization inductance; L_S – stator inductance; L_R – inductance of the rotor; z_P is the number of machine pole pairs.

To facilitate further transformations, we shall neglect the stator active resistance and its inductive scattering resistance. We shall also assume that the complex value of voltage is defined as $\overline{U}_R = U_R e^{j\delta}$ [4].

The scattering coefficient of the machine is determined by the expression:

$$\sigma = 1 - \frac{L_\mu^2}{L_S L_R} \quad (2)$$

Let us write down the expressions for the current coupling and flux-linkage of the machine windings taking into account (2):

$$\begin{cases} \overline{I}_S = \frac{1}{\sigma L_S} \overline{\Psi}_S - \frac{k_R}{\sigma L_S} \overline{\Psi}_R \\ \overline{I}_R = -\frac{k_S}{\sigma L_R} \overline{\Psi}_S - \frac{1}{\sigma L_R} \overline{\Psi}_R, \end{cases} \quad (3)$$

where k_S, k_R – are the stator and rotor coupling coefficients, respectively.

Substituting the expressions for currents (3) into system (1), we obtain:

$$\begin{cases} \overline{U}_S = \left(\frac{R_S}{\sigma L_S} + j\omega_0 \right) \overline{\Psi}_S - \frac{R_S k_R}{\sigma L_S} \overline{\Psi}_R, \\ \overline{U}_R = -\frac{R_R k_S}{\sigma L_R} \overline{\Psi}_S + \left(\frac{R_R}{\sigma L_R} + j\omega_0 s \right) \overline{\Psi}_R', \\ M = -\frac{3}{2} p_P \frac{L_\mu}{L_S L_R - L_\mu^2} \left[\overline{\Psi}_S \times \overline{\Psi}_R \right] \end{cases} \quad (4)$$

Let us consider system (4). Obviously, it is universal and can describe both an asynchronous motor (at $\overline{U}_R = 0$) and a dual-fed machine (at $\overline{U}_R \neq 0$). The product σL_R, called the equivalent machine inductance, is determined by the following relationship: $\sigma L_R = L_S$, where $\sigma L_R = L_S$ – the machine inductance in the short-circuit mode.

At $\overline{U}_R = 0$ and nominal conditions of inductance L_S corresponds reactive short circuit resistance X_S:

$$X_S = 2\pi f_C L_S = X_{S|fc} + X_{R|fc},$$

where f_C – nominal voltage frequency of the power mains; $X_{S|fc}$ – reactance of the stator at the nominal voltage of the power mains; $X_{R|fc}$ – rotor reactance at the nominal voltage of the power mains.

Let us divide both sides of the second equation of system (4) by ω_0:

$$\frac{\overline{U}_R}{\omega_0} = -\frac{R_R k_S}{\omega_0 \sigma L_R}\overline{\Psi}_S + \left(\frac{R_R}{\omega_0 \sigma L_R} + js\right)\overline{\Psi}_R.$$

With the assumption about the equivalent inductance and nominal operating conditions of the machine, the equality will be fulfilled:

$$\frac{R_R}{\omega_0 \sigma L_R} \approx \frac{R_{Rn}}{X_{Sn}} = s_{CRn}, \tag{5}$$

where R_{Rn} – nominal rotor active resistance; X_{Sn} – nominal short-circuit reactance of the machine; s_{CRn} is the nominal critical slip of the machine on its natural mechanical characteristic (under the condition $\overline{U}_R = 0$).

We shall neglect the stator active resistance, and write down the first two equations of system (4) taking into account (5) in the form:

$$\begin{cases} \overline{\Psi}_S = \dfrac{\overline{U}_S}{j\omega_0}, \\ \overline{\Psi}_R = \dfrac{\overline{U}_R}{\omega_0}\dfrac{1}{s_{CRn} + js} + \dfrac{\overline{U}_S k_S}{j\omega_0}\dfrac{s_{CRn}}{s_{CRn} + js}, \\ M = -\dfrac{3}{2} p_P \dfrac{L_\mu}{L_S L_R - L_\mu^2}\left[\overline{\Psi}_S \times \overline{\Psi}_R\right] \end{cases} \tag{6}$$

In the obtained system (6) there is an implicit connection between the moment developed by the machine and the rotor rotation speed (in the form of slip). This relationship is expressed through the machine flux-linkage. The direct computation of the moment according to (6) is difficult, since it requires a complete identification of the flux-linkages of machine windings (finding their current effective values and phases), which is a non-trivial task even with Hall sensors installed into the steel of the machine.

Turning to the method of generalized two-phase electrical machine and performing the necessary transformations [4], we obtain the following expression for the moment of the machine:

$$M = \frac{3}{2} z_P \frac{k_S^2 U_S^2}{\omega_0^2 \sigma L_R} \frac{s_{CRn} s}{s_{CRn}^2 + s^2}\left[1 - \frac{U_2^*}{s}\left(\cos\delta + \frac{s}{s_{CRn}}\sin\delta\right)\right]. \tag{7}$$

Under $\overline{U}_R = 0$ and other nominal conditions, the following relationship is satisfied:

$$M_{CRn} = \frac{3}{2} z_P \frac{k_S^2 U_S^2}{\omega_0^2 \sigma L_R}, \tag{8}$$

where M_{CRn} – nominal critical moment of the machine on its natural mechanical characteristic.

IOP Conf. Series: Earth and Environmental Science **84** (2017) 012030 doi:10.1088/1755-1315/84/1/012030

Substituting (8) in (7), we obtain an expression relating the moment developed by the machine, slip, phase and relative amplitude of the voltage in the rotor circuit:

$$M = \frac{2M_{CRn}}{s_{CRn}/s + s/s_{CRn}}\left[1 - \frac{U_R^*}{s}\left(\cos\delta + \frac{s}{s_{CRn}}\sin\delta\right)\right], \tag{9}$$

where U_R* is determined in accordance with the expressions:

$$U_{ROTH} = \frac{U_R}{U_S},$$

$$U_R^* = U_{ROTH}/k_S.$$

where U_{ROTH} – relative voltage across the rotor, U_R* – refined relative voltage across the rotor.

It should be noted that the critical slip and critical moment values in (9) correspond to their nominal values on the natural mechanical characteristic. The critical values of slip and moment on artificial characteristics when the additional voltage source is included into the rotor circuit are not equal to those of the natural characteristic and depend on the parameters of the additional voltage, as will be shown below.

It follows from expression (9) that a change in the amplitude and phase of the additional voltage in the rotor circuit makes it possible to adjust the moment developed by the motor and the speed of rotor rotation [10].

As it was mentioned above, in asynchronous mode the amplitude of the voltage applied to the rotor and its phase can be independently set, and its frequency is always equal to the current slip frequency (the frequency of the self-emf of the rotor). Thus, with the values of U_2^* and δ and substituting them into expression (9), it is possible to construct mechanical characteristics of the machine.

To determine the optimal parameters of voltage applied to the rotor (amplitudes and phases), we shall analyze expression (9). In accordance with (9), the moment developed by the machine is a function of three variables: slip, voltage amplitude across the rotor and its phase. Analysis of the function of three variables is difficult, therefore we shall consider the slip s as a parameter, and, accordingly, (9) as a function of two variables with the parameter:

$$M = f(U_R^*, \delta) = \frac{2M_{CR}}{s_{CRn}/s + s/s_{CR_n}}\left[1 - \frac{U_R^*}{s}\left(\cos\delta + \frac{s}{s_{CRn}}\sin\delta\right)\right]. \tag{10}$$

The analysis of (10) shows that this function is periodic in the variable δ with a period of 2π, which allows us to focus only on one period of its variation for this function. From considerations of physical realizability, it is enough to study the interval of phase variation $[-\pi \le \delta \le \pi]$.

In expression (10), we can distinguish two parts:

1) The part describing the natural mechanical characteristic of asynchronous motor with a phase rotor, to the rotor of which no additional voltage is applied $\dfrac{2M_{CR}}{s_{CR_n}/s + s/s_{CRn}}$;

2) The part describing the change in the mechanical characteristic of the machine when the voltage source is connected to the rotor $\left[1 - \dfrac{U_R^*}{s}\left(\cos\delta + \dfrac{s}{s_{CR_n}}\sin\delta\right)\right]$.

The part $\left[1-\frac{U_R^*}{s}\left(\cos\delta+\frac{s}{s_{CR_n}}\sin\delta\right)\right]$ includes two parameters of voltage across the rotor – amplitude (relative amplitude) and phase. Change in both parameters leads to a change of machine mechanical characteristic. Let us assume, as it was mentioned above, that for the voltage, directed counter to the self-emf of the rotor, U_R* is positive, and when codirected – negative. Then it follows from expression (10) that the amplitude increase of counter directed voltage across the rotor leads to a decrease in electromagnetic moment with the same slip, and the increase in the amplitude codirected voltage leads to its increase. It is easily explained by referring to the machine replacement scheme (Figure 1). Indeed, the inclusion into the rotor circuit of a voltage source counter directed to the rotor self-emf, will lead to a decrease in the rotor current in accordance with the second Kirchhoff law, and, consequently, to a decrease in the moment developed by the motor. Thus, the inclusion of the codirected voltage leads to an increase in the current of the rotor circuit and to the growth of the electromagnetic moment.

To find the values of the variables U_R^* and δ, for which the value (10) is maximal, it is necessary to determine the extremum points of function (10). Since (10) is a function of two variables and one parameter, the extremum points should be determined using partial derivatives with respect to the indicated variables. It suffices to consider the factor

$$k_y = 1-\frac{U_R^*}{s}\left(\cos\delta+\frac{s}{s_{CRn}}\sin\delta\right). \tag{11}$$

Let us find the values of phase δ for which (11) takes the maximum values. To do this, we find the partial derivative (11) with respect to δ, taking s and U_R^* as constants, and equating it to 0:

$$\begin{gathered}\frac{\partial k_y}{\partial\delta}=0,\\ -\frac{U_R^*}{s_{CRn}}\cos\delta+\frac{U_R^*}{s}\sin\delta=0.\end{gathered} \tag{12}$$

Solving equation (12) with respect to δ, we obtain a condition determining the maximum of (3.35) in phase:

$$\delta = arctg\left(\frac{s}{s_{CRn}}\right) \tag{13}$$

Finding the derivative (12) with respect to the voltage amplitude, we can come to the conclusion that the expression has no extremum with respect to the voltage amplitude:

$$\frac{\partial k_y}{\partial U_R^*}=0. \tag{14}$$

It follows from (11) – (14) that the maximum of the mechanical characteristic is determined by the value of voltage phase across the rotor, and its maximum value depends on the amplitude.

The phase surface of function (10) is shown in figure 2. As it can be seen from the constructed surface, the magnitude of electromagnetic moment of the machine is determined by the amplitude and phase of the additional voltage, while for various combinations of amplitude and phase the machine can develop both a positive and a negative moment, and, therefore, work both in the motoring and in the braking modes.

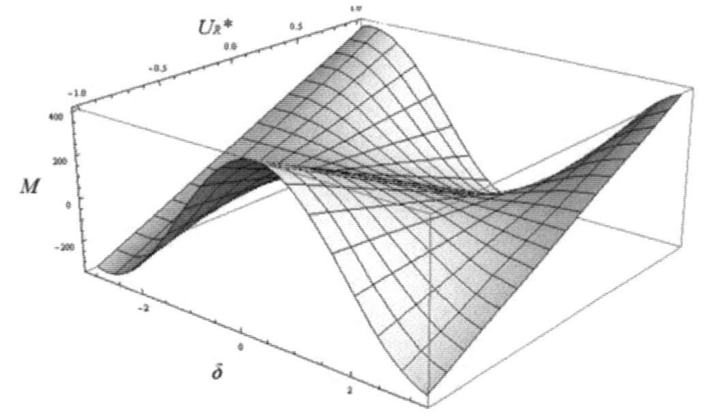

Figure 2. Phase surface of the DFM moment function.

The basic equation of drive motion (one of the forms of Lagrange equation following from D'Alembert principle) in the general case is written dowm as

$$\pm M \mp M_C = J\frac{d\omega}{dt}, \quad (15)$$

where M, Mc – electromagnetic moment of the machine and the moment of loading, respectively; $\pm M \mp M_C = M_{dyn}$ – dynamic moment; J – the moment of inertia of the working mechanism reduced to the machine rotor.

Depending on the signs of ω и M_{dyn}, the machine can operate in different quadrants of mechanical characteristic. For example, let us consider the operation of a mine hoisting installation [11, 12], the force diagram and speed diagram of which are given in figure 3.

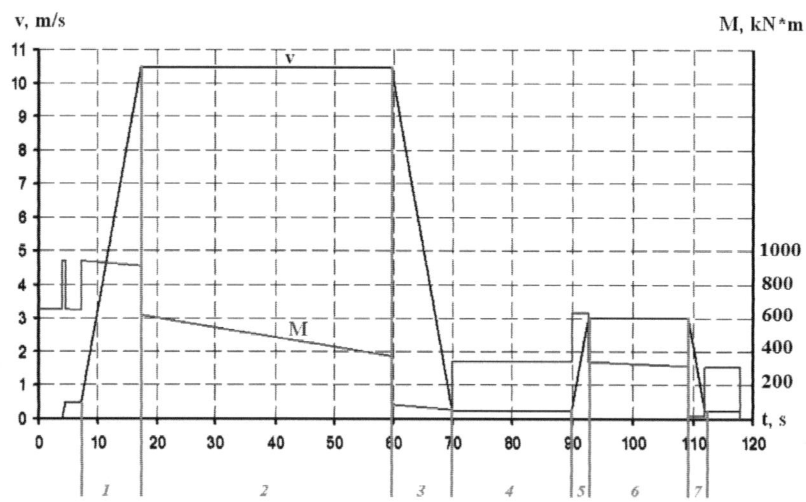

Figure 3. Typical tachogram and load diagram of the mine hoisting system.

Acceleration in sections 1, 5 is positive. In this case, expression (15) with regard to the sign of the speed takes the form:

$$\begin{cases} M_{MO} - M_C = J_\Sigma \dfrac{d\omega}{dt} > 0, \\ \omega > 0. \end{cases} \quad (16)$$

Acceleration in sections 3, 7 is negative, respectively, expression (15) takes the form:

$$\begin{cases} -M_{MO} - M_C = J_\Sigma \dfrac{d\omega}{dt} < 0, \\ \omega > 0. \end{cases} \quad (17)$$

In this case, the engine operates in the braking mode in these sections.

In sections 2, 4, 6 the acceleration is equal to 0, expression (17) takes the form:

$$\begin{cases} M_{MO} - M_C = 0, \\ \omega > 0. \end{cases} \quad (18)$$

In this case, the engine operates in the motoring mode in these sections.

It follows from expressions (15) – (18) that the electric drive must operate in different modes, to which correspond all coordinate quadrants of "speed-moment" plane.

The mechanical characteristics of experimental DFM based on the MTF-111 H6 engine are shown in figure 4.

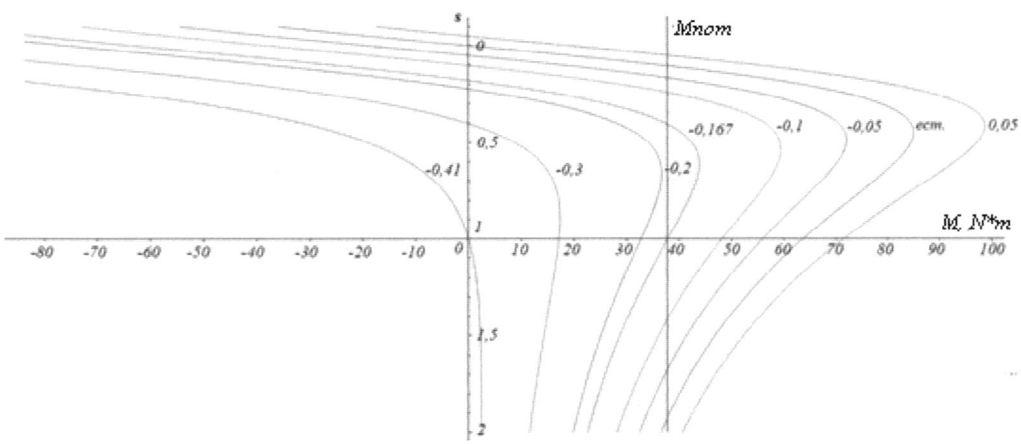

Figure 4. Family of DFM mechanical characteristics based on MTF-111-H6 engine under the condition of maximizing the moment by condition (13).

The numbers in the characteristics indicate the value of the relative voltage across the rotor, at which the corresponding characteristic is obtained. According to the characteristics shown in Figure 3, it is evident that when the moment is maximized by condition (13) the working sections of the machine mechanical characteristics are practically parallel and the rigidity in their limits is preserved. According to the given mechanical characteristics (I and II quadrants) of DFM in asynchronous mode, it is seen that the range of speed control down at the nominal load moment in the motoring mode is limited by the maximum value of the amplitude of additional voltage across the rotor. In general, the range of speed control down depends on the type of natural mechanical characteristic of the asynchronous motor with a phase rotor, on the basis of which DFM is constructed; its maximum estimation is 1:2 – 1:2.5 [10]. As it was established by the authors, the conclusions drawn for quadrants I and II are valid for quadrants III and IV. To organize the operation of machine in these quadrants, it is necessary to change the direction of rotation of the stator magnetic field.

In addition, as it can be seen from figure 4, it is possible to start the machine by changing the amplitude of the additional voltage across the machine rotor.

Thus, the obtained expression for mechanical characteristics allows us to identify the DFM during its operation it and determine the required operating mode. It makes possible to organize digital control of the electric drive using the information-control system [13].

4. Conclusions

The achieved in asynchronous mode the range of DFM speed control is quite satisfactory for mechanisms that do not require deep speed control (fans, pumps, compressors). However, this range is unacceptably small for the electric drive of mine hoisting installations. To extend the range of DFM speed control, it is advisable to use a combined mode of DFM operation, which is a combination of asynchronous and synchronous operating modes of the machine in different sections of the speed diagram.

The resulting expression for the DFM mechanical characteristic is convenient to use in electric drive control systems.

References

[1] Ostrovlyanchik V Yu and Popolzin I Yu 2016 *Vestnik KuzSTU* **2(114)** 75–84
[2] Meshcheryakov V N and Bezdenezhnykh D V 2010 *Bulletin of TulSU. Series: Engineering* **3(4)** 108–14
[3] Semykina I Yu and Nekhlebova S G 2011 *Bulletin of KuzSTU* **6** 40–4
[4] Onishchenko G B and Lokteva I L 1979 *Asynchronous Valve Cascades and Double-Power Motors* (Moscow: Energia) p 200
[5] Hopfensperger B, Atkinson D and Lakin R A 2002 *International Trans. on Electrical Energy Systems* **12(6)** 427–37
[6] Firago B I and Pavlyachik L B 2006 *Adjustable Electric Drives of Alternating Current* (Minsk: Tehnoperspektiva) p 363
[7] Ostrovlyanchik V Yu and Popolzin I Yu 2016 *Science-intensive Technologies for the Development and Use of Mineral Resources* **3** 303–9
[8] Klyuchev V I 2001 *Theory of Electric Drive* (Moscow: Energoatomizdat) p 704
[9] German-Galkin S G 2012 *Bulletin of USURT* **3 (15)** 24–34
[10] Lecocq D, Lataire P and Wymeersch W 1993 *5th European Conf. on Power Electronics and Applications* pp 419–423
[11] Bezhok V R, Dvornikov V I, Manets I G and Pristrom V A 2007 *Mine Lift* ed B A Gryadushchy and V A Korsun (Donetsk: Southeast Ltd) p 624
[12] Datskovskii L Kh et al 2010 *Russian Electrical Engineering* vol **81** pp 15–30
[13] Ostrovlyanchik V Yu and Popolzin I Yu 2015 *Bulletin of NSTU* **4 (61)** 7–20

Control system of mutually coupled switched reluctance motor drive of mining machines in generator mode

A S Ivanov, I Yu Kalanchin and E E Pugacheva

Siberian State Industrial University, 42 Kirova Street, Novokuznetsk, 654007, Russia

E-mail: sibsiuprk@gmail.com

Abstract. One of the first electric motors, based on the use of electromagnets, was a reluctance motor in the XIX century. Due to the complexities in the implementation of control system the development of switched reluctance electric machines was repeatedly initiated only in 1960 thanks to the development of computers and power electronic devices.

The main feature of these machines is the capacity to work both in engine mode and in generator mode. Thanks to a simple and reliable design in which there is no winding of the rotor, commutator, permanent magnets, a reactive gate-inductor electric drive operating in the engine mode is actively being introduced into various areas such as car industry, production of household appliances, wind power engineering, as well as responsible production processes in the oil and mining industries.

However, the existing shortcomings of switched reluctance electric machines, such as non-linear pulsations of electromagnetic moment, the presence of three or four phase supply system and sensor of rotor position prevent wide distribution of this kind of electric machines.

1. Introduction

Switched reluctance machines are one of the most promising types of electrical machines. Because of simple design and low cost they are more often used both in everyday life and in various industries. Noteworthy is the fact that such machines are capable of working both in engine mode and in generator mode. For each mode of operation own control system with a certain algorithm of operation is needed, which mainly depends on the operating angles.

Operating are the angles of rotor position relative to the stator, under which either the excitation of the stator coils occurs or their disconnection. To make the control system of the switched reluctance machine work more efficiently, it is necessary to determine with an accuracy at which angle value of the rotor position the induction profile has the maximum and minimum values.

The rotor and stator of a switched reluctance electrical machine are made of electrical steel, and only the stator poles have windings that are placed around the tooth, while the rotor is free of windings, magnets and a brush-collector device [1]. The windings of one stator phase are connected in series and located diametrically opposite each other. There are different configurations of valve inductor-reactive machines. They depend on the number of stator and rotor teeth, they can be 6/4, where 6 is the number of stator poles and 4 rotor poles, and configuration can be 8/6 and 10/8. The appearance of a switched reluctance machine 8/6 is shown in figure 1.

IOP Conf. Series: Earth and Environmental Science **84** (2017) 012031 doi:10.1088/1755-1315/84/1/012031

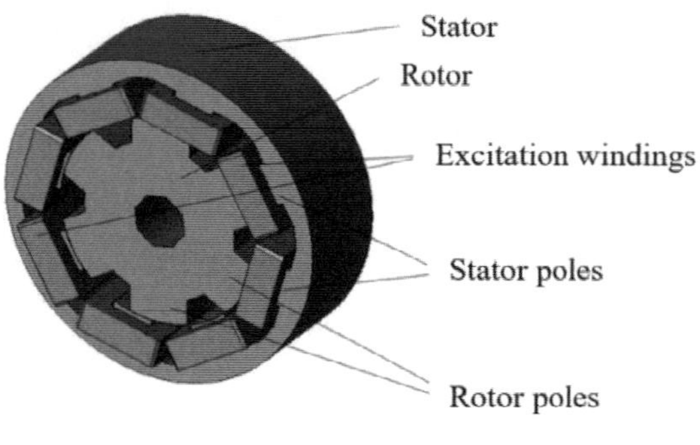

Figure 1. Appearance of switched reluctance machine 8/6.

The main advantages of a switched reluctance machine can be attributed to the simplicity of design, since the rotor consists only of electrical steel and the winding of phases are located only on the stator teeth, the absence of expensive rare-earth metals in the structure, operation at high speeds and temperatures and high reliability [2].

However, besides these advantages the switched reluctance machines have a number of serious disadvantages. A three-phase or four-phase power supply system is required depending on the configuration of the electrical machine, providing alternate excitation of the stator windings to create a rotating electromagnetic field. The disadvantages also include non-linear characteristics of these machines, high pulsations of electromagnetic moment and electromagnetic flow.

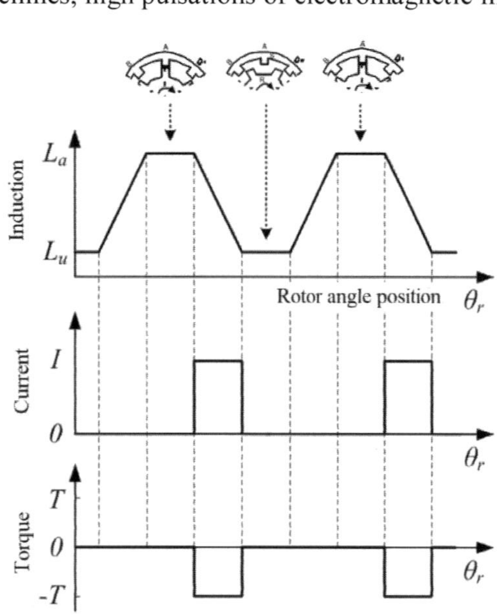

Figure 2. Dependence of induction profile magnitude on the position of rotor relative to the stator.

The principle of mutually coupled switched reluctance machine is based on the fact that ferromagnetic bodies in the external magnetic field are oriented so that the magnetic flux penetrating them is maximized, i.e. when the stator phase is excited, the nearest rotor poles tend to assume an aligned position with respect to the stator poles [3].

When the rotor poles move relative to the excited phases of the stator, the induction profile varies periodically from maximum to minimum. At the agreed position, when the stator pole aligned with the rotor pole, the electromagnetic induction assumes the maximum value, and if the poles are unaligned the induction tends to a minimum. The dependence of the magnitude of induction profile on the rotor position relative to the stator is shown in figure 2 [4].

When excitation of stator windings at the time of mismatching of rotor and stator poles a negative electromagnetic moment is created, which entails the generation of electrical energy in the network. However, a negative electromagnetic moment can also cause the stop of electrical machine; thus, an external source is required to maintain the necessary constant speed for different loads. The functional scheme for controlling the excitation of a switched reluctance generator is shown in figure 3 [5].

Performance of a mutually coupled switched reluctance machine in the generator mode depends on its control strategy, which consists in the correct switching of power electronic devices. It is necessary to determine carefully the excitation angles to ensure continuous operation of the generator and to maximize its efficiency. It should be noted that with increasing excitation current, the output power will increase comparable to the speed increase. Consequently, the value of the output power will depend both on the angular speed and on excitation voltage.

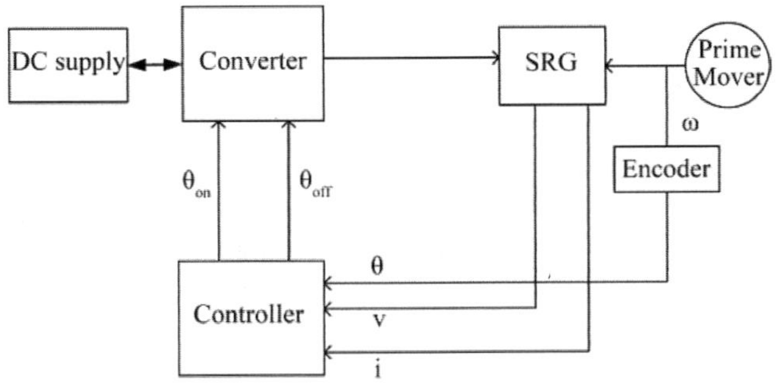

Figure 3. Control system of excitation of a mutually coupled switched reluctance generator.

However, because of the nonlinear characteristics of mutually coupled switched reluctance machines, it is difficult to determine the optimal control system [4]. The most effective way is to determine the optimal angles of the rotor poles relative to the stator poles for windings excitation, which allows the pulsations of the electromagnetic moment to be reduced and the energy efficiency of the switched reluctance generator to be increase. Currently, there are various combinations of excitation angles that can produce the maximum output power.

The control strategy, based on the determination of excitation angles, can be divided into three groups:

- fixed angles of switching on and switching off;
- fixed angles of switching on and varying angles of switching off (and vice versa);
- varying angles of both switching on and off.

Input values of the controller are current, voltage, as well as the angle of rotor position. The output signals from the programmable logic controller are transmitted to the predetermined angles of position rotor to the voltage converter for switching on and switching off the excitation circuit.

2. Methods of research

Rapidly developing information technologies alongside with digital computing allow complex mathematical problems to be solved, as well as physical and mathematical modeling to be performed. At present, numerical methods for modeling electromagnetic fields are widely used. These methods are based on spatial and spatial-temporal discretization. Their main advantage is sufficiently accurate results obtained taking into account the properties of materials and the geometric shape of the bodies [6]. The main requirement of the numerical method is the statement of boundary conditions of the calculated domains.

To study electromagnetic fields of mutually coupled switched reluctance generator FEMM program is used, since it is freely distributed and has a fairly convenient interface.

FEMM is a complex of programs designed to solve low-frequency electromagnetic problems in a two-dimensional plane. At the moment, this program is focused on solving linear and nonlinear magnetostatic problems, linear electrostatic problems, as well as static heating problems. FEMM includes three parts that make up this software:

- Interactive shell (FEMM.EXE). It consists of a 2D design system in which the investigated object is constructed, for each element of which certain material properties and boundary states are assigned. In 2D design environment, DXF files built in programs such as AutoCAD or KOMPAS 3D can be imported. The results of solutions can be reflected in the form of contours and density plots. The program also allows the field in its various points to be explored by the user, as well as the values of physical quantities of interest to be determined.
- The next component of the software (TRIANGLE.EXE) allows the solution area you to be discredited into the set of triangles required for the solution with the help of finite element method.
- The main part of this complex is solvers (FKERN.EXE) for magnetic; (BELASOLV.EXE) for electrostatic tasks. Each of the solvers uses a set of files that describes the problem

under investigation and produces a solution of partial differential equations preliminarily divided into domains (triangles) in order to obtain the necessary values within the solution domain of investigated field.

To study the electromagnetic field of a switched reluctance generator using CAD KOMPAS 3D, a model of mutually coupled switched reluctance machine with 8 stator poles and 6 rotor poles is created having the following main parameters: stator diameter 174 mm, stator pole height 23 mm, rotor pole height 16 mm and air gap 0.5 mm. Next, the model is imported into FEMM environment in DXF format, previously specifying the solution of magnetic problem. After that, assign materials for each element of the model, as well as determine the stator coils to which the current will be fed. The contours of the figures should be closed, otherwise the calculation will be unreliable [7].

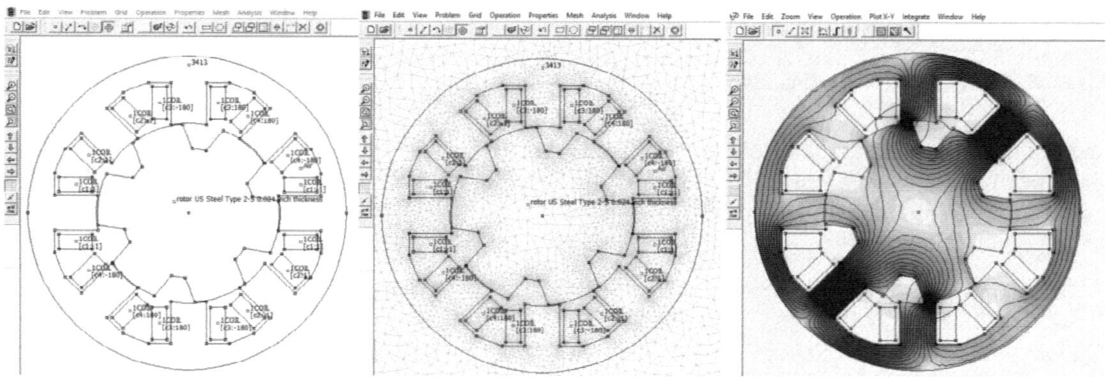

Figure 4. Screenshot of the FEMM interface.

Figure 1 shows FEMM interface, where the solution sequence of electromagnetic problem can be seen. First, the profile of the object under investigation is constructed, then the solution domain is discretized. After the solution is completed, magnetic flux lines are constructed, as well as the density of its distribution, sections with large density values have a darker color. Besides, it is possible to determine the values of flux linkage, magnetic flux, magnetic field induction and also the electromagnetic moment. To determine the working angles, it is necessary for different rotor positions relative to the stator, and also for different current values, to obtain the values of flux linkage for the constructed model [8].

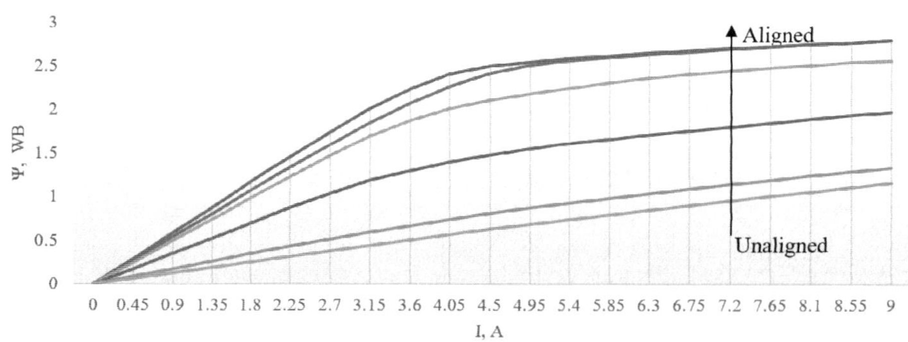

Figure 5. Magnetization curve.

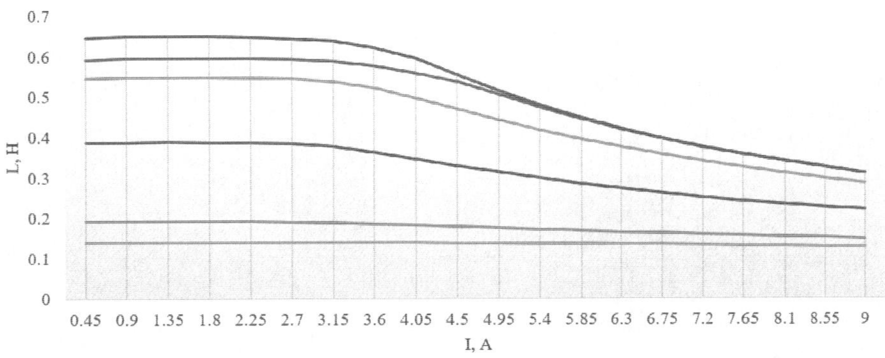

Figure 6. Profile of electromagnetic induction.

In figure 5 it can be seen that the flux linkage values depend on the rotor position relative to the stator. When the rotor pole and the stator pole are aligned the maximum induction value is observed (figure 6) [9]. When operating in the generator mode, the excitation of stator phase must be performed at the moment when the rotor pole moves past the stator tooth, i.e. at 6°-15° from the aligned rotor position. When the machine is operating in engine mode, excitation occurs at the minimum induction value, while the stator and rotor poles are in unaligned position, i.e. with rotor position 25°-36°.

Thus, the software for solution of electromagnetic problems using the finite element method allows us to determine the main characteristics of the examined machine at the modeling stage without creating a prototype. Due to the complex of programs Finite Element Methods Magnetics the sections of rotor position were determined in which the magnetic circuit has the maximum and minimum values of electromagnetic induction for determining the working angles of a mutually coupled switched reluctance electrical machine.

Figure 7. Appearance of a mutually coupled switched reluctance machine.

3. Results and discussion

On the basis of the results obtained, a mutually coupled switched reluctance electric machine was constructed on which various operating modes are currently being investigated. This electric machine has a power of 2.2 kW and 3 feeding phases. The appearance of the mutually coupled switched reluctance machine, rotor and stator are presented in figures 7 and 8, respectively. The investigated electric machine functions both in the engine mode and in generator mode.

Figure 9 shows the transient processes of electromagnetic torque, obtained as a result of computer simulation by the program complex Infolytica, at rotor speeds of 600, 750 and 800 rpm. The figure shows that the torque pulsations are minimal at a rotation speed of 750 rpm.

Also, using computer simulation the efficiency of the developed inductor-reactive electric machine was determined (figure 10). The maximum efficiency value of 0.86 is achieved at a rotor speed 1200 rpm.

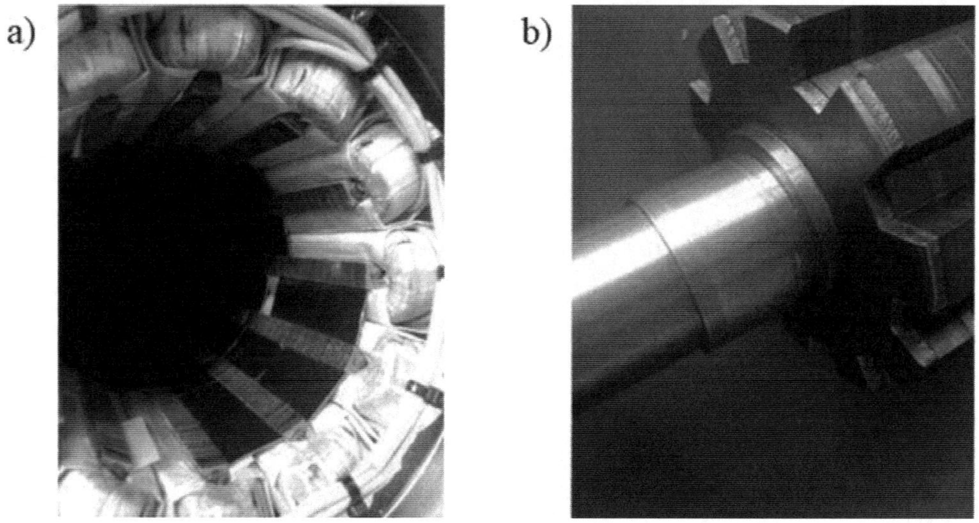

Figure 8. (a) appearance of the stator, (b) appearance of the rotor of mutually coupled switched reluctance machine.

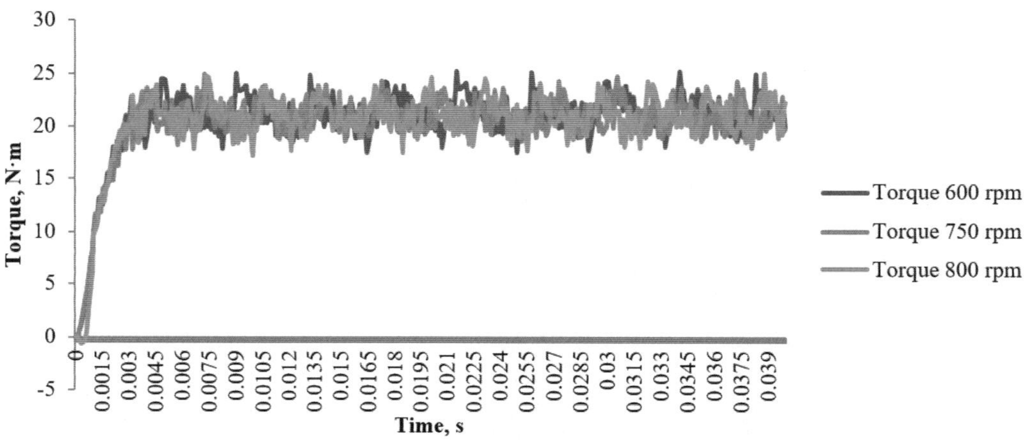

Figure 9. Graph of the torque of the examined mutually coupled switched reluctance machine.

Since the mutually coupled switched reluctance electric drive is structurally more simple and more technologically efficient for manufacturing, maintenance and repair, it has a lower cost price; has got increased energy efficiency and higher overload capacity in comparison with asynchronous and synchronous engines, constant power in a wide range of rotational speeds, which is crucial for responsible equipment in mining, oil and gas, processing industries.

One of the latest developments of the company Le Tourneau Technologies, Inc. is the world's largest loader included in the Guinness Book of Records, L-2350, 6.7 m high and 20.3 m long, weighs 266 tonnes and can lift 72 tons of cargo in a bucket with capacity of 40.5 m^3. The cost of one such loader is 7.6 million USD [10]. As the traction motor of motor-wheel a switched reluctance engine of 400 kW is used, which transmits the torque to the wheel through the planetary gear. Compared with the DC traction motor the service interval of the unit has increased from 500 to 20 000 h, only the bearings are replaced.

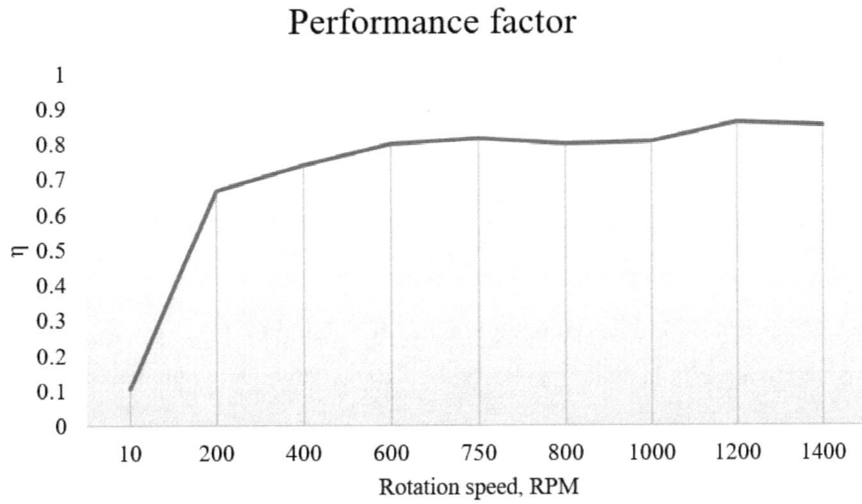

Figure 10. Graph of the performance factor of examined mutually coupled switched reluctance machine.

The manufacturer of construction and agricultural machinery American Deere&Company (Molin, Illinois, USA) in 2011 presented at the international exhibition ConExpo in Las Vegas a 944K diesel quarry loader with an engine 500 hp/367.5 kW with a hybrid electric transmission based on switched reluctance-technology [11].

4. Conclusions

Mutually coupled switched reluctance motors have several advantages over frequency-controlled asynchronous electric AC motors, machines with synchronous electromagnetic excitation and excitation from permanent magnets. At present, application of switched reluctance machines, operating in the generator mode, in the sphere of wind power engineering, as well as in the automotive industry are are carried out.

Mutually coupled switched reluctance machines can reduce the cost of electricity generation through wind farms, improve the energy efficiency of various production processes, and reduce maintenance costs due to its simple and reliable design [12]. The existing shortcomings of these electric machines hinder their wide spread and require prompt elimination.

References

[1] State Standard 27471-87 *Rotating Electric Machines. Terms and Definitions*
[2] Savvinov P V and Semenov A S 2013 *Modern High Technologies* **8-2** 342–3
[3] Semenov A S 2012 *Proc. IIth Intl. Conf. "Scientific Discussion: Issues of Technical Sciences"* (Moscow: Int. Center for Sci. and Ed.) pp 52–6
[4] Brandstetter P, Hrbac R, Polak M and Stepanec L 2006 *Advances in Electrical and Electronic Engineering* **5(1)** 68–71
[5] Anekunu A Y 2015 *Int. J. Adv. Eng. Research and Tech.* **3** 290–5
[6] Stepanov V M and Avdoshin V S 2014 *Bulletin of TulSU. Engineering* **8** 181–3
[7] Valetov E V 2011 *J. Sci. Publ. of Post-Graduate and Doctoral Students* **6 (60)** 75–8
[8] Galushko V N et al 2014 *Bulletin of Gomel State Technical University named after B Y Sukhoy* **58** 1–11
[9] Amissa Arifin et al 2012 *Energy and Power Engineering* **4** 447–58
[10] http://hontos.ru/tehnica/letorneau-l-2350-pogruzchik-ogromnyih-razmerov
[11] Arifin A and Al-Bahadly I 2011 *Smart Grid and Renewable Energy* **2** 27–36

IOP Conf. Series: Earth and Environmental Science **84** (2017) 012032 doi:10.1088/1755-1315/84/1/012032

Control system of roll crushes in emergency situations

A G Nikitin, K V Tagiltsev-Galeta and K A Chaynikov

Siberian State Industrial University, 42 Kirova Street, Novokuznetsk, 654007, Russia

E-mail: nikitin1601@yandex.ru

Abstract. The system of emergency control of a roll crusher is developed, which allows the hardness of the crushed material to be evaluated indirectly estimating the current strength of the unit, or directly by swing of the deflecting the crusher jaw, and the pieces with the increased hardness to be recrushed, increasing compression, reducing the motor rotation frequency. This system reduces the number of emergencies and provides less downtime for equipment in the event of an accident than crushers using conventional safety elements: toggle plates and restraining spring or hydraulic fuses. Besides, this system is extensible, i.e. it does not interfere with the inclusion of additional devices (for example, mechanical breakers such as toggle plates), sensors (for example, angle protractor to improve the accuracy of measurements). With the combined use of direct and indirect evaluation of the crusher condition the accuracy and efficiency of the crusher control increases improving the reliability of a crushing unit as a whole.

1. Introduction

Many industrial manufactures, including mining, process and use large quantities of bulk solid materials of various coarseness classes. In most cases the required coarseness is achieved by grinding larger pieces on crushers, including single-roll ones. Improvement of crushers productivity operating within the crushing and screening complexes is an important task.

The materials provided for crushing, as a rule, differ in significant variations in physical and mechanical properties, thus, the main requirement for controlling the crushing process is to obtain the specified size of the final product, to maximize the efficiency of electricity supplied to the crushers. Since the crushing and screening process is continuous, stopping of one element inevitably leads to a shutdown of the whole complex.

The accident-caused failure of a single-roll crusher can be caused by the non-crushable body in the crushing chamber. To prevent such accidents that lead to a long stop of crushers various safety devices are used. It is possible to use toggle plates consisting of two parts, which are interconnected by bolts.

Bolts are the main element of a safety device, as they are cut off when the non-crushable object is fed into the crushing chamber. However, this technical solution is not reliable enough, as the toggle plates often break without visible overloads, and not only when non-crushable bodies get into the crushing chamber. The imperfection of toggle plates caused the development of safety devices of non-destructive type [1], for example, spring or hydraulic ones.

For example, there is a crusher [2] in which the movable jaw is supported by the lower part on the toggle plate, connected with the back stop (movable traverse) interacting with the wedge-like locking mechanism. A hydraulic cylinder serves for a prompt change change in the size of the output gap. The disadvantage of the crusher is the complexity of the mechanism design providing fixation of the jaw and a large number of technological operations necessary to changing the dimensions of the output gap.

There is also a crusher [3] which includes a hydraulic system that is equipped with a block of hydraulic cylinders connected by a piston cavity with a throttle, hydraulic lock, a pressure switch and hydraulic relief valves, and also contains a control line connected via a time relay to a crusher drive and a hopper. Such safety devices are used in most crushing machines on the market [4, 5].

The functioning of the protective device on the jaw allows the dimensions of the crushing machine to be reduced and the size of the output gap to be changed; also the block of hydraulic cylinders is more reliable than a single hydraulic cylinder. However, this arrangement worsens the repairability of the crusher due to the position of the control unit above the hydraulic cylinders, and the use of the protective device with a hydraulic cylinder block significantly increases the cost of the crushing machine.

In addition, the block of hydraulic cylinders itself can cause an emergency situation, in the event that the hydraulic cylinders do not work simultaneously (this can lead to a skewing of the toggle plate and its failure even under small loads on the movable jaw).

Also to improve the reliability of crushers, control systems are used based on the analysis of main parameters of a electric drive [6], which are effective for determining the parameters of the state of an electric motor, but do not allow the overall state of the unit to be evaluated, including its mechanical parts.

2. Results of research and discussion

A system of combined automatic control over parameters of the crusher state alongside with the system of parametric control of electric drive has been developed to solve the problem of increasing the reliability and speed of crushers control.

The controlled object under consideration (in this case a roll crusher) is described by the following set of characteristics:

- current frequency;
- the state of the jaw breaker from the restraining spring;
- hardness of the crushing material;
- current strength;
- inclination angle of a jaw.

After describing the characteristics of the object it is necessary to select a control system in which the object will be operated.

The most reliable and easy in use object control system is a feedback system [7], that is, a closed system in which the controlled object acts on the control device, and the control device depending on the deviation of the controlled quantity from the prescribed value, in turn, acts on the governing body and, therefore, on the object so that the regime of the latter satisfies certain conditions. The presence of feedback underlies operation of any automated system working on the principle of inclination.

In a system with feedback the parameters are distributed as follows: input parameters (operated by the technological process) and output parameters (generated in the technological process). Input parameters are divided into input controlled ones, i.e. their values are currently known, and the input uncontrolled parameters with values that at the current moment cannot be directly estimated.

After describing the selected control system, it is necessary to perform decomposition, i.e. distribute the object characteristics dividing them into groups according to the structure of the chosen control system.

Based on the theory of automatic control and system analysis [7], the parameters characterizing the crusher can be distributed as follows:

- input parameters

input controlled (U): supply voltage, current frequency;

input disturbance (W): size of a piece, material hardness;

- output parameters (Y)

- current strength;
- size of the output piece;

In accordance with the above structure of parameters the crusher can be represented in the form of a control circuit with feedback as follows (figure 1):

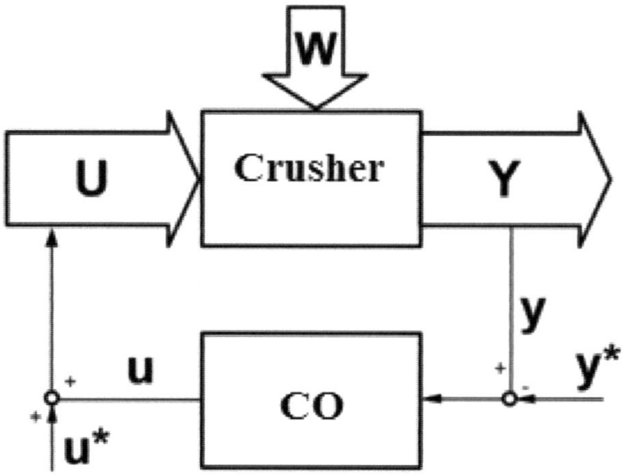

Figure 1. Crusher control ciruit.U – vector of input controlled values; W – vector of input disturbance values; Y – vector of output values; u – generated input value; u* - base values of the generated input value; y – measured output values; y* – basic values of output values; CO – controlling object.

The principle of operation of this circuit is as follows. The vector of input values U enters the crusher a which is known and is controlled. In this case, the frequency and voltage are included into this vector. Also a vector of disturbance parameters W enters into the crusher, which displays the properties of the crushed material that are not predetermined in advance – the size of the the fed pieces and its hardness. As a result of the materials processing by the crushing machine the input vectors U and W are converted into the output vector Y consisting of the current and the size of the fed piece. For direct measurement only the current signal is available which is the measured value y. This value is compared with the baseline y*, determined by the technologist or operator on the basis of the technology, personal experience, etc. If a significant (for example, above 5% of the base) deviation of the measured value is detected, the controlling object generates a change in the input value u (in this case the current frequency) or, in the event of a pre-emergency situation activates the crusher safety device.

The emergency control system for the crusher (figure 2) operates as follows. From the receiving hopper through the feeder *2*, the raw material enters the crusher. In the crusher the material is crushed by roller compression *13* and jaw *4*. The processed material along the guide *8* enters the conveyor 10 for further transportation.

If a material falls into the crusher with a strength exceeding the calculated one, the force developed by the roll increases.

The increase in force is characterized by an increase in the current in the roll drive *14*. The evaluation of the current signal is made according to the formula of the consumed electric power (P_{el}). Due to the rigid kinematic connection between these components the mechanical power (P_{mech}) produced by the electric motor can be considered equal to the power consumed by the crusher.

$$3 \cdot U_f \cdot I_f \cdot \cos\varphi = M \cdot \omega. \quad (1)$$

It is known [8] that the rotational speed of the motor shaft (ω) is proportional to the frequency of the electric current. When using frequency converters, the frequency of the current is a known quantity, thus, the rotational speed of the shaft is also known.

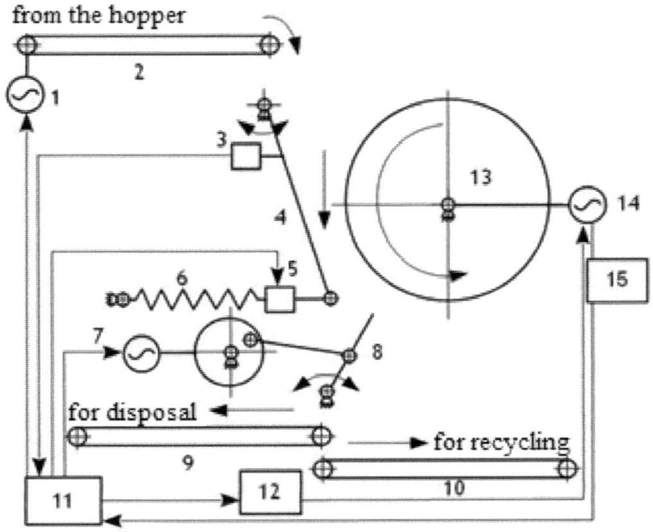

Figure 2. Schematic diagram of a crushing plant with a joint assessment of the state. 1 – feeder drive; 2 – feeder; 3 – sensor of the jaw position; 4 – jaw; 5 – fixing device; 6 – safety device; 7 – drive of the guide; 8 – the guide; 9 – conveyor "for disposal"; 10 – conveyor "for processing"; 11 – controller; 12 – frequency converter; 13 – roll; 14 – crusher drive; 15 – current sensor.

Assuming that the supply-line voltage remains unchanged and the angular velocity is known, the current of the motor load is functionally dependent on the torque on the shaft [9]:

$$I_f = f(M). \quad (2)$$

If the material is not crushed under the action of this force it is transmitted to the jaw. Due to the presence of the safety device 6 the jaw starts to deviate from the initial position which is fixed by the position sensor 3 [10].

The signal about the deviation is sent to the controller 11 which with the frequency converter *12* changes the current frequency at the drive of the crusher *14* to increase the crushing force, while stopping the drive of the feeder *1*. In this case it is possible to crash the material with increased hardness exceeding the calculated one.

In the event that the strength of the crushed material is greater than the values of the increased hardness, there is jamming, that is, a piece of material is fixed in the crusher fracture zone. When the piece is jammed the jaw deviates from its original position. When the inclination of the jaw is fixed the controller starts the drive the guide *8*. After changing the position of the guide *8* a signal is sent to the fixing device *5* (figure 3) which results in a weakening of the piece clamping, and it falls on the conveyor to the disposal *9*.

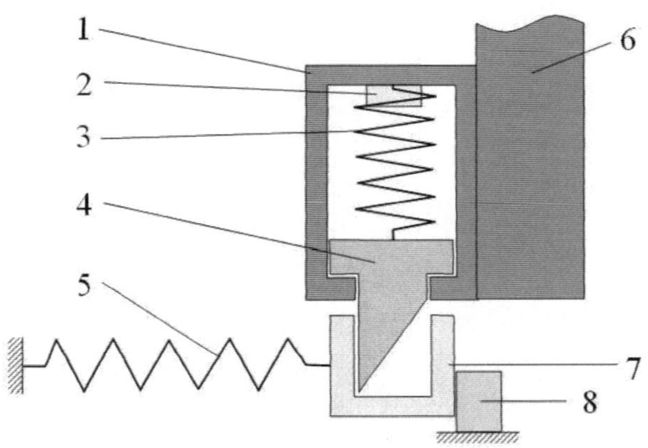

Figure 3. Schematic diagram of the fixing device. 1 – body of the fixing device; 2 – electromagnet; 3 – spring; 4 – retainer; 5 – spring; 6 – jaw; 7 – groove; 8 – stop.

The fixing device works as follows (figure 3): when the non-crushed body gets into the fracture zone, the jaw deviates from its original position, which is fixed by the angle sensor directing the signal to the controller, which generates a signal aimed at turning on the electromagnet.

When a signal from the controller is received the electromagnet *2* is activated, which drives the retainer 4 breaking the kinematic connection between the jaw and the safety device, which reduces the clamping of the non-crushed body allowing it the fracture chamber to be left when the roll rotates.

After removing the non-crushed body from the crushing chamber the electromagnet is disconnected and the retainer under the action of the spring *3* returns to its original position. In this case the torque that develops under the weight of the jaw returns it to its original position.

After removing the non-crushed body from the working chamber the jaw returns to its working position and the crusher continues to operate in the nominal mode.

3. Conclusions

A method for joint evaluation of the mechanical and electrical parameters of a crushing unit is developed, which makes it possible to use the combined object control by joint analysis of the measured signals in order to recognize characteristic informative signals describing the pre-emergency or emergency state of an object.

References

[1] Klushantsev B V, Kosarev A I, Muizemnek Yu A 1990 *Crushers. Construction, Calculation of Operation Features* (Moscow: Mechanical Engineering) p 320
[2] Pavlov N A et al 2009 *Grinding by Crushes with Reciprocating Motion of Working Elements* Patent of the Russian Federation 2345837 Bul 7 p 8
[3] Kulikov A A and Degtyarenko V A 1995 Patent of the Russian Federation 2043788 V02S1/04
[4] Sandvik – Roller Crushers www.mining.sandvik.com/en/products/equipment/crushing-and-screening/hybrid-crushers
[5] Maekawa-kogyosho – Roller Crushers www.maekawa-kogyosho.com/product/single/index.html
[6] Petukhov V S 2005 *News of Electrical Eng.* **1** 54–61
[7] Tsypkin Ya Z 1977 *Fundamentals of the Theory of Automatic Systems* (Moscow: Nauka) p 560
[8] Ilyinsky N F et al 2008 *Electric Drive: Energy and Resource Saving* (Moscow: Academiya) p 208
[9] Makarov A V and Tagiltsev-Galeta K V2014 *Ferrous Metallurgy* **6**
[10] Nikitin A G and Tagiltsev-Galeta K V 2014 *Ferrous Metallurgy* **8** 34–6

IOP Conf. Series: Earth and Environmental Science **84** (2017) 012033 doi:10.1088/1755-1315/84/1/012033

Diagnosis of the rock crushing modes to increase the efficiency of one-roll crusher operation

A G Nikitin, S A Laktionov and K S Medvedeva

Siberian State Industrial University, 42 Kirova Street, Novokuznetsk, 654007, Russia

Abstract. In the paper the hypotheses of destruction of brittle materials are given, as well as the analysis of energy costs for crushing. The methods of crushing are considered and a comparative analysis of their energy capacity and quality of the product is carried out. The parameters of the crushing process are determined from the point of view of their productivity and, accordingly, their energy efficiency. It is shown that the process of crushing in the one-roll crushers is due to the frictional forces acting between the surfaces of the roll and a piece of the crushed material, and also between the piece and the immovable jaw. The comparison of initial sizes of the crushed pieces in the crushes with the vertical and inclined surfaces of immovable jaw, which shows that with the increase in the bend angle of the immovable jaw the degree of crushing increases. Based on the analysis of forces acting on the crushed material in the one-roll crusher with the inclined surface of the jaw, the calculation of maximal possible value of the bend angle of the immovable jaw is performed.

1. Introduction

Many industrial plants process and use in large amounts the bulk solid materials of various coarseness grades. In most cases the required size is achieved by grinding lumps on crushers, including one-roll crusher. The need for processing a large volume of materials results in the need for fundamentally new approaches, first of all, to the problem of destructive effect of the machine on the material and to the realization of a new method, which is achieved by solving design problems.

The process of fracture of brittle rocks by impact and the process of destruction by compression differ from each other and require different energy inputs necessary for the destruction of the material (compression failure requires 1.5 times more energy compared to impact destruction [1]). However, impact crushers have a serious drawback associated with the quality of the finished product; only 25-30% of the finished product is obtained with a required range of fineness after processing in the impact crushers [2].

2. Methods of research

Under a brittle piece we shall understand the material that is crushed without noticeable plastic deformations. The complete cycle of crushing of individual pieces to the required size consists of successive independent acts of crushing the pieces of smaller size. Determination of the necessary work of a single act of destruction allows energy costs for the entire crushing process to the required size to be established [3].

The amount of energy needed to crush the material to a certain size depends on many factors: size, shape, relative position of the pieces, hardness, brittleness, homogeneity of the initial material, its humidity, type and condition of the working surfaces of the machine, etc. Therefore, it is possible to establish an analytical relationship between the energy consumption for crushing, physico-mechanical

IOP Conf. Series: Earth and Environmental Science **84** (2017) 012033 doi:10.1088/1755-1315/84/1/012033

properties of the material to be crushed and the features of the process only in the most general form [4].

In 1867 Prof. P.V. Rittinger for the first time hypothesized that the work for crushing or grinding the initial material is directly proportional to the newly formed surface. Subsequently this hypothesis was called the first law of fragmentation. On the basis of experimental work on grinding it was concluded that Rittinger's hypothesis is valid in a certain interval of values of the specific surface area of the crushing products from 250 to 1000 mm^2/mm^3. However, when the values of the specific surface of the products do not correspond to this interval the proportionality of the increase in the specific surface and the energy consumption for grinding are disturbed [5].

In work [6] the process of destruction is divided into two stages: deformation and destruction. The results of the destruction of glass balls with sizes from 3 to 15 mm made it possible to state that the total energy required for destruction can be determined by the useful work of destruction multiplied by the proportionality coefficient. This hypothesis is called the second law of grinding or "volumetric".

F.H. Tartaron [7] believes that the "volumetric" hypothesis characterizes the work that should be spent on overcoming the connection between atoms and ions of the crystal lattice, and destruction should occur up to particles with micron sizes. In turn, R.T. Hookey [8] believes that the "volumetric" hypothesis is better to apply for coarse grinding, and other hypotheses can be applied for a narrow range of sizes of the crushed pieces. Statements of F.H. Tartaron and R.T. Hooks were not experimentally confirmed.

I. Brach [9] made an attempt in practical calculations to take into account the work of crushing depending on the sizes of the crushed rock by introducing the notion of specific work, and recognized that it is impossible to correctly determine the work of crushing taking only elastic deformation into account. Having proposed a calculated dependence I. Brach did not formulate the physical essence of the hypothesis.

Majima Hirosi and Oka Ikkitosi [10] on the basis of the theory of elasticity and the results of experiments showed that the destruction of an individual piece under the action of compressive loads occurs when the maximum permissible tensile stresses are exceeded. They found that the work of deformation of some ideally rigid body in the process of grinding is directly proportional to the particle size in a variable degree.

Since plastic deformations are practically absent in brittle rocks, their destruction at the point of contact occurs immediately after the elastic deformation and the compressive stresses reach the limiting values. The hypotheses proposed to date do not give accurate results in the theoretical determination of energy consumed by crushing. Numerous amendments and additions to the main hypotheses of crushing, as well as new hypotheses did not make the methods of calculating energy for crushing more reliable. This, in many respects, is due to the peculiarities of the initial material, most often anisotropic, the physico-mechanical characteristic of which varies considerably even within a single deposit, and also due to the very process of crushing which depends on many random circumstances.

The main indicators of the process of crushing are the degree and efficiency of crushing [5]. The degree of crushing is estimated by the ratio of the size of the crushed piece to the size of the gap between the roll and the fixed jaw, which depends on the value of the angle of nip, and the greater the angle of capture, the greater the degree of crushing. The efficiency of crushing is determined by the mass of the crushed material obtained by consuming a unit of electricity and depends mainly on the hardness of the crushed material.

With all the variety of crushing processes and types of loads used for crushing materials (crushing, splitting, breaking, etc.), normal stresses arise in the crushing body, the ultimate strength of which is almost twice as high as the tensile strength under the action of tangential stresses. Consequently, it becomes necessary to develop such a process of crushing in which tangential stresses arise in the material to be crushed, so that fracture occurs due to the shift.

In a two-roll crusher compressive forces act on the crushing material and, accordingly, normal stresses arise. In order to reduce the energy consumption for material crushing in a two-roll crusher it is necessary to create such conditions when not only normal stresses but also tangential stresses act on the crushed material. This crushing process can provide a one-roll crusher. In the one-roll crushers the crushing process is provided by the presence of frictional forces acting between the roll surfaces and a piece of a material, as well as between the piece and the immovable jaw. If we carry out a force analysis of the crushed material in a single-roll crusher it will be seen that not only compression forces on the crushed piece occur during the operation of the one-roll crusher but also the internal torque which results in shearing stresses [11]. The disadvantage of one-roll crushers is consists in their low degree of crushing.

3. Results and discussion

In order to increase the degree of crushing of brittle materials in a one-roll crusher the design of a one-roll crusher [12] has been developed at Siberian State Industrial University, in which the fixed jaw has a bend (figure 1).

As it can be seen from figure 1 when using the fixed jaw with a bend, the size of the nipped piece is larger than when using the vertical fixed cheek with the unchanged nipping conditions, which are determined by the nip angle, the coefficient of friction between the roll and the crushed piece and the friction coefficient between the fixed cheek and the crushed piece, due to which the degree of crushing increases.

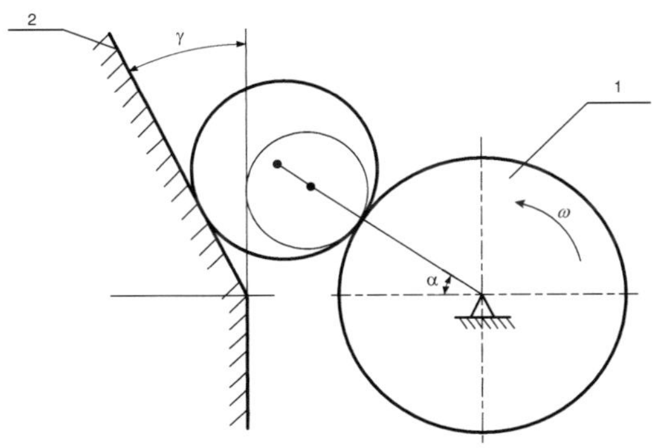

Figure 1. Comparison of the crushed pieces radii. 1 – driven roll, 2 – fixed jaw.

To ensure the maximum degree of crushing it is necessary to determine the maximum permissible bend angle of the fixed jaw.

The crushed piece is in contact with the roll and the fixed jaw at some points A and B (figure 2). The force of gravity determines the forces of interaction $\vec{N}_1$ and $\vec{N}_2$ between the contacting surfaces. These forces are directed perpendicular to the tangent to the roll circumference and perpendicular to the jaw surface. The angle between the horizontal line and the force line $\vec{N}_1$ is the nip angle α. The bend angle of the surface of the fixed jaw γ.

When the roll rotates between it and the crushed piece, a frictional force $\vec{F}_1 = f_1\vec{N}_1$ arises directed along the tangent in the opposite direction of rotation of the roll preventing its rotation. Then the reactive frictional force, equal in magnitude to the force of friction $\vec{F}_1$ and directed at point A perpendicular to the line of action of the normal force in the direction of roll rotation, will act on the fractional piece. The projection of the resultant two forces $\vec{R} = \vec{N}_1 + \vec{F}_1$ determines at point A the compression force acting on the crushed piece. Also the reaction of this resultant force creates at the point of contact *B* of the crushed piece with a fixed jaw having a bend, the force of normal pressure $\vec{N}_2$ (figure 2).

When the roller begins to rotate, it pulls the crushed piece into the crushing gap, while the normal pressure forces increase manyfold due to the compression force and thereby increase the frictional force that drags the piece into the crushing zone. As a result the forces of normal pressure, that arise when the engine is running, exceed manyfold the forces of normal pressure that arise under the action of gravity, that is why gravity is not taken into account because of its insignificance in comparison with the forces arising in the process of crushing the material.

The condition for nipping the crushed piece into the gap will be satisfied if the vertical component of the resultant of all forces participating in the crushing process is directed downward. These forces are determined by the nip angle α, the bend angle γ of the fixed jaw with a bend and the friction coefficients f_1 and f_2 between the crushed piece and the roll, the crushed piece and the jaw, respectively.

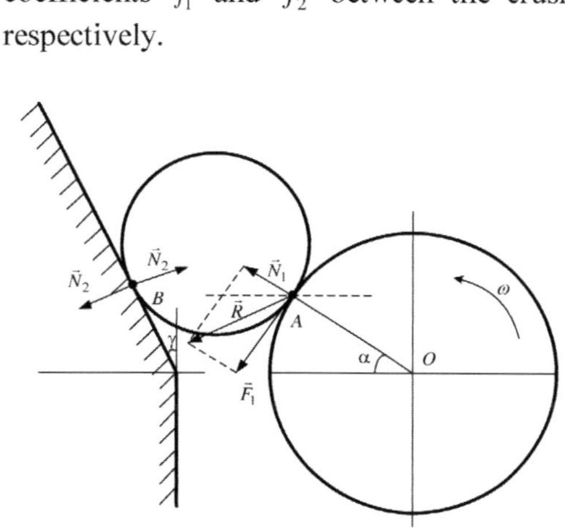

Figure 2. Distribution of forces in a one-roll crusher with a jaw having a bend.

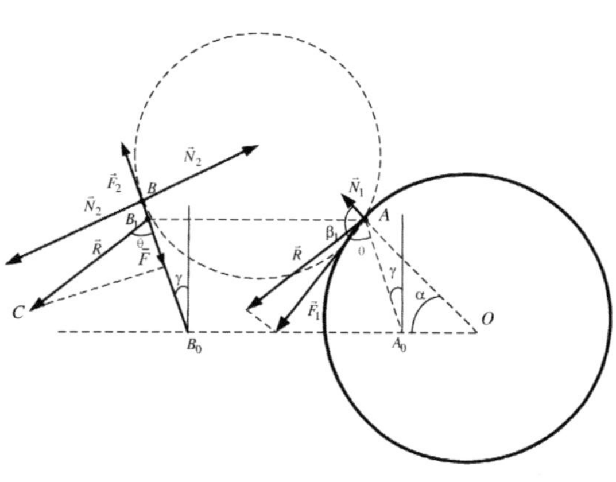

Figure 3. Scheme of the forces acting on a crushed piece in the crusher with a jaw having a bend.

The condition for nipping the crushed piece into the gap will be satisfied if the vertical component of the resultant of all forces participating in the crushing process is directed downward. These forces are determined by the nip angle α, the bend angle γ of the fixed jaw with a bend and the friction coefficients f_1 and f_2 between the crushed piece and the roll, the crushed piece and the jaw, respectively.

The equal force $\vec{R}$ at the point of contact A of the crushed piece and roll relative to the horizontal makes the angle $\beta_1 - \alpha$, where α is the nip angle, and β_1 is the friction angle $(tg\beta_1 = f_1)$ between the roll and the crushed piece. It is obvious that to capture the piece, it is necessary for this angle to be positive, so that the vertical component of the resultant R to be directed downward, i.e. the condition $\alpha < \beta_1$ must be satisfied. This condition is the determining factor for the crusher with an inclined jaw surface, while the component $\vec{F}$ of the resultant force R will be directed downward along the section of the jaw at the contact point B, therefore, the angle θ must be sharp (figure 3).

In the parallelogram $B_0B_1AA_0$, the angle $B_1B_0A_0$ equals to angle B_1AA_0, from which it follows that $\gamma + \frac{\pi}{2} = \theta + \beta_1 - \alpha$, and, consequently, $\theta = \gamma + \frac{\pi}{2} - \beta_1 + \alpha$. On the plane of the inclined section the resultant $\vec{R}$ determines the force $\vec{F}$ acting on the crushed piece and directed downward along the plane of the jaw. The magnitude of this force is determined by equality $F_2 = R\cos\theta$.

In this plane it is counteracted by the frictional force $F_2 = f_2N_2 = f_2R\sin\theta$, where f_2 is the coefficient of friction between the piece and the stationary jaw. As long as the force F is greater than the friction force F_2, the crushed piece will be drawn into the crushing gap. To carry out the crushing process in the one-roll crusher, in which the fixed jaw has a bend, it is necessary that $F > F_2$ or $R\cos\theta > f_2R\sin\theta$. Thus, $tg\theta < \frac{1}{f_2}$.

Then solving $tg\left(\gamma + \frac{\pi}{2} - \beta_1 + \alpha\right) < \frac{1}{f_2} \Rightarrow \gamma + \frac{\pi}{2} - \beta_1 + \alpha < arctg\frac{1}{f_2}$ we find out that the value of the inflection angle of the jaw equals $\gamma_{max} = arctg\frac{1}{f_2} + arctgf_1 - \frac{\pi}{2} - \alpha$.

From the analysis of the equation obtained, it is seen that the inflection angle depends on the values of friction coefficients between the roll and the crushed piece and also between the fixed jaw and the crushed piece. And the greater the coefficient of friction between the roll and the crushed piece is, the smaller the friction coefficient between the fixed jaw and the crushed piece is, the greater the angle of inflection can be set.

4. Conclusions

Based on the analysis of the forces acting on the crushed piece in a one-roll crusher with an inclined jaw surface, the maximum possible bend angle of the fixed jaw is calculated, which ensures the maximum degree of crushing and increases the efficiency of the crusher.

References

[1] Lagunova Yu 1996 *Izv. Vuzov. Gorniy Zhurnal* **10-11** 121–4
[2] Grubachic V 2006 *Cement and its Application* **2** 56 –9
[3] Rodin R A 1987 *Izv. Vuzov. Gorniy Zhurnal* **6** 84–9
[4] Klushantsev B V, Kosarev A I and Muyzemnek Yu A 1990 *Crushers. Construction, Calculation, Operation Features* (Moscow: Mechanical Engineering) p 320
[5] Rodin R A 1989 *Izv. Vuzov. Gorniy Zhurnal* **4** 71–8
[6] Buss B and Shubert H 1971 *Zur Bruchfunction bei der Einselkornd Ruckzerkleinerung, Zerkleingen, Zerkleinerun in Canns*
[7] Tartaron F X 1964 *Foundation of General Theory Comminution American Inst. of Mining Metallurgical and Petroleum Engineers Trans.* **vol 229**
[8] Hukki R T 1954 *Trans. AIME* **199** 728 –30
[9] Brach I 1965 *L. Techn. Koohschkle. Otto Nongueriche* ***1-2***
[10] Hirosi M and Ikkitosi A 1971 *Canadian Metallurgical Quaterly* **vol 10** 1–4
[11] Nikitin A G and Sakharov D F 2010 *Ferrous metallurgy* **10** 41–2
[12] Nikitin A G, Laktionov S A, Bazhenov I A and Medvedeva K S 2016 *Roll Crusher* Patent of the Russian Federation No. 2603923 Bul. 34

Optimal resources allocation for independent and joint test of it service releases

V V Zimin, T V Kiseleva and E V Maslova

Siberian State Industrial University, 42 Kirova Street, Novokuznetsk, 654007, Russia

E-mail: kis@siu.sibsiu.ru

Abstract. The problem of resources allocation for testing IT service releases, which are built into the operational environment for reducing information risks, is considered. Testing can be either independent or system (joint). The problem was formulated, for which the method of network (dichotomous) programming was used allowing the solution of the optimization problem to be considerably simplified. The scheme for changing the current baseline of IT environment when embedding releases into it is introduced, which is necessary to prevent the destruction of the operating environment. The scheme of the network representation of cost functions and incidents probability function, which are structurally similar, is shown. Specific examples of this problem solution for four releases are given, in which the specific values of incidents probability, costs and quality of testing were assigned.

1. Introduction

IT service is a complex of interacting IT assets, the purpose of which is to produce value for consumers, determined by usefulness, availability, capacity, continuity and security of service, or a combination of assets.

The life cycle of any IT service consists of stages of strategy, design, implementation and operation. At the stage of implementation into the operational environment, the IT service is exposed to the risk of full or partial loss, in addition, the destruction of the operational environment itself is possible [1 - 4]. To reduce the probability of such risk, it is proposed to break the implemented IT service into releases, integrate them into a specially created test environment and conduct pre-test. Releases can be built in and tested sequentially one by one and such test is called independent. If the releases are tested in parallel, then it is a system test. The higher the quality of a conducted test, the less likely the occurrence of any incident during the operation of the IT service.

2. Methods

To create a test environment and conduct the test itself, it is necessary to provide the organization with additional resources, and the important task is to minimize the costs required for IT service releases implementation into the operational environment.

Below in figure 1 there is a diagram of the change of the current base states of operation environment into new states as a result of embedding IT service releases into the environment, that update the technological assets (release A_1), application assets (release A_2), service portfolio assets (release A_3) and business assets (release A_4).

Let us denote the probability of occurrence of IT incidents in the operational environment when embedding the appropriate release after testing by $P(A_i), i=\overline{1.4}$. Test quality K is assessed on a three-point scale: 1 –"poor", which corresponds to a large risk of IT incidents, 2 –"satisfactory", which corresponds to an average risk, 3 –"good", which corresponds to low risk.

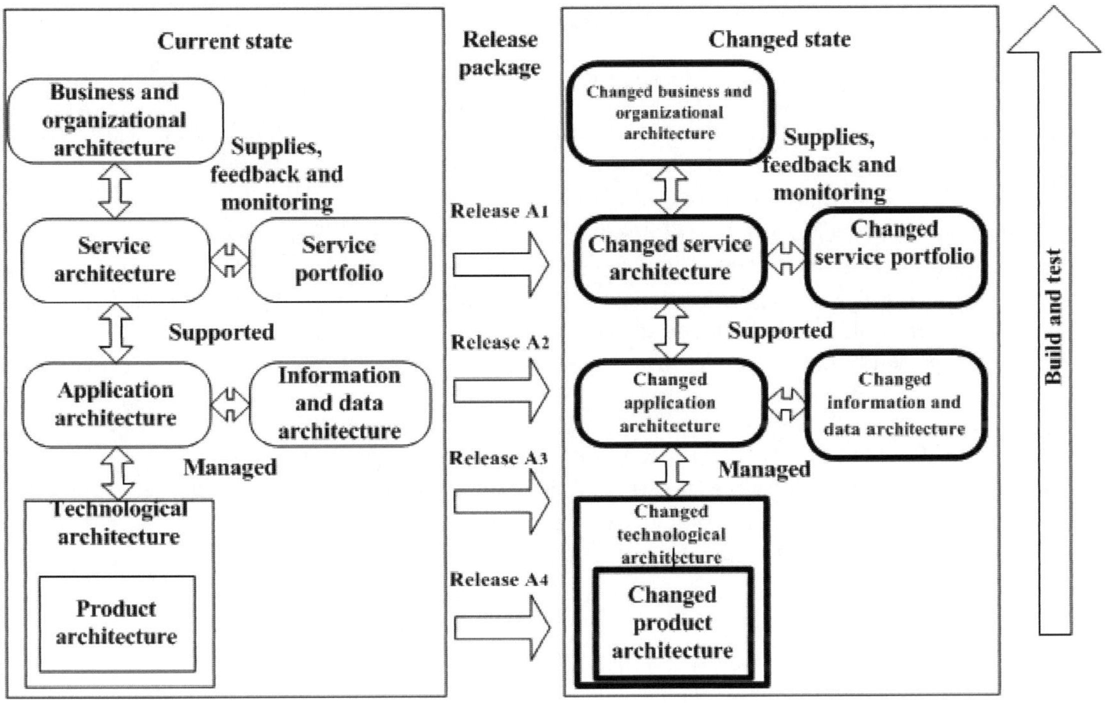

Figure 1. Changes of the current states of IT environment into new base states.

Let us denote the costs for release testing by $z(A_i), i=\overline{1.4}$. The cost functions from the release of test quality $z_K(A_i), K=\overline{1.3}, i=\overline{1.4}$ are specified, it is shown in table 1.

Table 1. Cost functions depending on the release of test quality.

K(Ai)	3	2	1
$z_K(A_i)$	$z_3(A_i)$	$z_2(A_i)$	$z_1(A_i)$

Let us state the problem of optimal resources allocation for independent and system test of IT services releases.

Given:

1. Operating environment, the composition of which is shown in figure 1.
2. Package of releases A_i = 1, 2, 3, 4.
3. Quality of releases testing K (A_i).
4. Incidents probability $P(A_i), i=\overline{1.4}$.
5. Costs of releases testing: $z_K(A_i), K=\overline{1.3}, i=\overline{1,n}$.
6. Limitation: $K(A_1 \cup A_2 \cup A_3 \cup A_4) \geq K^*$ for independent test of IT service and $K(A_i \mid A_1, A_2, ..., A_{i-1}) \geq K^*$ for system test of IT services.
7. Criterion: total costs for releases testing: $\sum_{i=1}^{4} z_K(A_i)$.

Required:

To optimize the resources allocation for testing, follow restrictions and minimization of the criterion, i.e. $\sum_{i=1}^{4} z_K(A_i) \to \min$.

Otherwise, it is required to determine such minimum costs $z_K(A_i), i=\overline{1,n}$, which provide quality $K(A_1 \cup A_2 \cup A_3 \cup A_4) \geq K^*$ for independent test of IT service and $K(A_i|A_1,A_2,...,A_{i-1}) \geq K^*$ for system test of IT services, i.e. not lower than the given level K^*.

To simplify the procedure of problems solution, it is proposed to use the method of network programming [5 - 7]. Complex functions are represented in the form of a composition of simpler ones, and the function representation itself is convenient to use in the form of a network (network representation), where the inputs are function variables, the outputs are the functions themselves, the intermediate vertices – functions included in the composition. Two functions are structurally similar if their network representations coincide, which is a prerequisite for using the network programming method.

Consider the solution of the problem of optimal allocation of resources first for the case of system test of releases of IT services when they are introduced into the operational environment [8].

Figure 2 shows the network representation of costs functions for a test and the incidents probability during testing. Obviously, that they coincide, i. e. these functions are structurally similar, therefore, it is possible to use the network programming method.

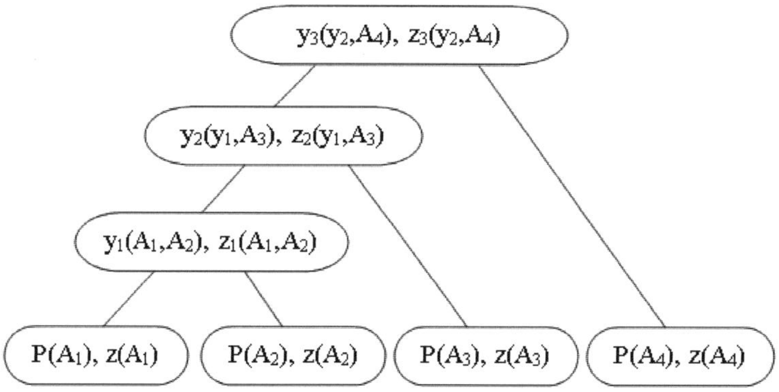

Figure 2. Network representation of functions P (A_i) and z (A_i).

The implemented IT service S consists in the general case of n releases A_i, i.e. $S=(A_1, A_2, \ldots A_n)$. The incidents probability $P(S)=P(A_1, A_2, \ldots A_n)$ during the service implementation is related to the number of service links between each other and with the environment. According to the theorem of multiplication of probabilities, the probability of incidents occurrence will be calculated by the formula:

$$P(A_1, A_2, ..., A_n) = P(A_1)P(A_2 \mid A_1)......P(A_n \mid A_1, A_2, ..., A_{n-1}). \quad (1)$$

In figure 3, for example, a graph is given that describes the relations between four releases with the components of the environment itself, and the numbers on the graph branches are the numbers of releases between each other and the environment.

IOP Conf. Series: Earth and Environmental Science **84** (2017) 012034 doi:10.1088/1755-1315/84/1/012034

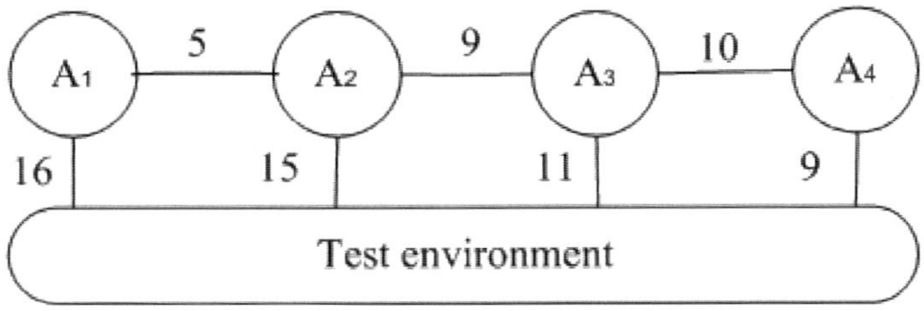

Figure 3. Mutual relations of services and their relations with the operational environment.

Table 2 shows the numerical values of mutual relations of services and relations with the operational environment.

Table 2. Numerical values of service relations

	A_1	A_2	A_3	A_4
A_1	16	5	0	0
A_2	5	15	9	0
A_3	0	9	11	10
A_4	0	0	10	9

Suppose that:

$$P(A_i \mid A_1, A_2, ..., A_{i-1}) = \frac{A_{ii} + A_{i(i-1)}}{\sum_{i=1}^{n}(A_{ii} + A_{i(i-1)})}, \quad (2)$$

whereA_{10}=0.

Then:

$$P(A_1, A_2, ..., A_n) = \prod_{i=1}^{n}\left(\frac{A_{ii} + A_{i(i-1)}}{\sum_{i=1}^{n}(A_{ii} + A_{i(i-1)})}\right). \quad (3)$$

The interval (0 ÷ P (A_i | A_1, A_2, ..., A_{i-1})) is divided into three equal subintervals, after which the test quality is evaluated depending on the probability value of a test. The costs functions $z_K(A_i|A_1,A_2,...,A_{i-1}), K = \overline{1.3}, i = \overline{1,n}$ from the quality of releases testing are known.

Then there are following relations for the costs function:

$$z_1 = \sum_{i=1}^{2} z(A_i);\ z_2 = z_1 + z(A_3);\ z_3 = z_2 + z(A_4). \quad (4)$$

For probabilities the following relations are valid:

$$y_1 = P(A_1);\ y_2 = P(A_1)P(A_2 \mid A_1); ...;\ y_{n-1} = P(A_1)P(A_2 \mid A_1)...P(A_n \mid A_1, ..., A_{n-1}). \quad (5)$$

To solve the initial problem by the method of network programming it is necessary to solve successively (n-1) problems:

The task of the first level:

$$z_1 = \sum_{i=1}^{2} z(A_i) \to \min;\ K(y_1) = K(P(A_1)P(A_2 \mid A_1)) \geq K^*. \tag{6}$$

The task of the second level:

$$z_2 = z_1 + z(A_3) \to \min;\ K(y_2) = K(P(A_1)P(A_2 \mid A_1)P(A_3 \mid A_1, A_2)) \geq K^*. \tag{7}$$

...

The task (n-1), which corresponds to the output from the network, is solved the last:

$$z_{n-1} = z_{n-2} + z(A_n);\ K(y_{n-1}) = K(P(A_1)P(A_2 \mid A_1)...P(A_n \mid A_1,..., A_{n-1})) \geq K^*. \tag{8}$$

The solution of (n-1) problem is the solution of the original problem.

In case with testing four releases, the links of which are shown in Fig. 3 and in Table 2, according to the formula 2 we obtain [9]:

$$\begin{gathered} P(A_1) = \frac{16}{75} = 0.21;\ P(A_2|A_1) = \frac{20}{75} = 0.7; \\ P(A_3|A_1,A_2) = \frac{20}{75} = 0.27;\ P(A_4|A_1,A_2,A_3) = \frac{19}{75} = 0.25. \end{gathered} \tag{9}$$

Suppose the costs for testing releases A_1 and A_2, depending on the testing quality, are described by functions $z_K(A_i)$, their values are specified and given in tables 3 and 4. The calculated intervals $P(A_i)$ are divided into three equal subintervals.

Table 3. Values of costs $z_K(A_1)$ for different intervals of probabilities and test quality indicators.

$K(P(A_1))$	3	2	1
$P(A_1)$	$(0 \div 0.07)$	$(0.07 \div 0.14)$	$(0.14 \div 0.21)$
$z_K(A_1)$	15	12	8

Table 4. Cost values $z_K(A_2|A_1)$ for different intervals of probabilities and test quality indicators.

$K(P(A_2\|A_1))$	3	2	1
$P(A_2\|A_1)$	$(0 \div 0.09)$	$(0.09 \div 0.18)$	$(0.18 \div 0.27)$
$z_K(A_2\|A_1)$	21	17	9

Multiplying the mean values of each obtained intervals $P(A_1)$ and $P(A_2|A_1)$ with the corresponding quality values, we obtain the probability values y_1. The costs required for a test are calculated by summing the costs $z_K(A_1)$ and $z_K(A_2|A_1)$. These results are shown in table 5.

Table 5. Values of probabilities y_1 and costs $z(y_1)$ depending on the quality of releases testing.

$K(P(A_1))$	3	3	3	2	2	2	1	1	1
$K(P(A_2\|A_1))$	3	2	1	3	2	1	3	2	1
y_1	0.0016	0.0047	0.079	0.0047	0.0141	0.0236	0.0079	0.0236	0.0394

$z(y_1)$	36	32	24	33	29	21	29	25	17

Then the interval (0 ÷ max (y_1)) is divided into three equal subintervals, the inclusion of each obtained probability value y into these intervals is checked. From each interval such probability value is chosen, at which the costs will be minimal, thus, we obtain the solution of the problem of the first level, given in table 6.

Table 6. Final values of probabilities and costs for testing releases (A_1) and ($A_2|A_1$).

$K(y_1)$	3	2	1
y_1	0.0079	0.0141	0.0394
$z_K(y_1)$	24	29	17

Next, we calculate the probability of incidents occurrence and costs for testing releases A_1 and ($A_2|A_1$) with release of A_3. y_1 and $z(y_1)$ are computed at the previous stage, values of $P(A_3|A_2, A_1)$ and z_K ($A_3|A_2, A_1$) are shown in table 7.

Table 7. Cost values $z_K(A_3|\ A_2,A_1)$ for different intervals of probabilities and quality indicators of the test.

$K(P(A_3|\ A_2,A_1))$	3	2	1
$P(A_3|\ A_2,A_1)$	(0 ÷ 0.09)	(0.09 ÷ 0.18)	(0.18 ÷ 0.27)
$z_K(A_3|\ A_2,A_1)$	22	19	11

The calculation of the values of probability y_2 and costs $z(y_2)$ is carried out as before, that is, the second level problem is solved, its result is given in tables 8 and 9.

Table 8. Values of costs $z(y_2)$ and the probability y_2 for different quality of the test.

$K(y_1)$	3	3	3	2	2	2	1	1	1
$K(P(A_3|A_2,A_1))$	3	2	1	3	2	1	3	2	1
y_2	0.0004	0.0011	0.0018	0.0006	0.0019	0.0032	0.0014	0.0041	0.0068
$z(y_2)$	46	43	35	51	48	40	39	36	28

Table 9. Final values of incidents probability for release ($A_3|A_2,A_1$) testing and the costs required for that.

$K(y_2)$	3	2	1
y_2	0.0014	0.0041	0.0068
$z_K(y_2)$	39	36	28

The probability values and costs for testing release A_4 release are calculated in the same way. The probabilities and costs for testing A_4release are described by functions $P(A_4|A_3,A_2,A_1)$ and $z_K(A_4|A_3,A_2,A_1)$, their values are given in table 10, they are also specified in advance.

By analogy with the above calculations, we calculate the probability and costs required for system testing of releases with A_4 release. All results obtained are given in tables 11 and 12.

Table 10. Value of costs $z_K(A_4|A_3,A_2,A_1)$.

$P(A_4|A_3,A_2,A_1)$	(0 ÷ 0.08)	(0.08 ÷ 0.16)	(0.16 ÷ 0.25)
$K(P(A_4|A_3,A_2,A_1))$	3	2	1
$z_K(A_4|A_3,A_2,A_1)$	15	10	8

Table 11. The probability of incidents occurrence y_3 and costs $z(y_3)$.

$K(y_2)$	3	3	3	2	2	2	1	1	1
$K(P(A_4\|A_3,A_2,A_1))$	3	2	1	3	2	1	3	2	1
y_3	0.00006	0.00020	0.00029	0.00016	0.00049	0.00086	0.00027	0.00081	0.001394
$z_K(y_3)$	54	49	47	51	46	44	43	38	36

Table 12. Total values of incidents probability and costs for release testing $(A_4|A_3,A_2,A_1)$.

$K(y_3)$	3	2	1
y_3	0.00027	0.00086	0.001394
$z_K(y_3)$	43	44	36

Therefore, the minimum costs for testing IT services with the best evaluation of testing quality corresponding to 3 ("good") are 43 units of resources, and the probability of incidents during IT services implementation is negligible and equals to 0.00027.

In the case of independent testing of IT service releases, the probability of IT incidents is calculated differently [9, 10].

Network representations of cost functions and occurrence probabilities for this type of a test are similar to the network representations discussed above (figure 2). The costs in this case are also calculated by the formula (4).

The probability of the sum of two joint events A and B is determined by the formula:

$$P(A+B)=P(A)+P(B)-P(AB). \quad (10)$$

Then probability y_1 of IT incidents is calculated by the formulas:

$$y_1(A_1,A_2)=P(A_1+A_2)=P(A_1)+P(A_2)-P(A_1\cap A_2), \quad (11)$$

$$y_2=(y_1,A_3)=y_1+P(A_3)-y_1P(A_3), \quad (12)$$

$$y_3=(y_2,A_4)=P(A_1+A_2+A_3+A_4)=y_2+P(A_4)-y_2P(A_4). \quad (13)$$

To solve the problem of optimal resources allocation for releases testing, as in the previous task, it should be divided into several sub-tasks. First, the values $y_1(A_1, A_2)$ and $z_1(A_1, A_2)$ are calculated. Wherein

$$z_1=\sum_{i=1}^{2} z(A_i)\rightarrow \min;\ K(y_1)=K(A_1+A_2)\geq K^*. \quad (14)$$

Further, using the obtained values, $y_2(y_1, A_3)$, $z_2(y_1, A_3)$ are calculated under the following conditions:

$$z_2=z_1+z(A_3)\rightarrow \min\ ;\ K(y_2)=K(A_1+A_2+A_3)\geq K^*. \quad (15)$$

After that, the values $y_3(y_2, A_4)$; $z_3(y_2, A_4)$ are calculated in a similar way under the following conditions:

$$z_3=z_2+z(A_4)\rightarrow \min\ ;\ K(y_3)=K(A_1+A_2+A_3+A_4)\geq K^*. \quad (16)$$

This is the solution of the original problem.

3. Results and discussion

Let us explain the above stated ideas on the example of the four releases from the previous example. As the initial values of the incidents probabilities, we use values $P(A_i)$, calculated from formula (2). Similarly, the obtained intervals $(0 \div P(A_i))$ are divided into three equal subintervals. The values of the cost functions $z(A_i)$ are given. The quality of testing, depending on the values of incidents probabilities, is evaluated in the same way as in the previous example. The specified values of costs, probabilities and quality of testing during implementation of A_1 and A_2 releases are given in tables 13 and 14. Depending on the quality of A_1 and A_2 releases testing and the incidents probabilities during implementation of these releases, probabilities $y_1$are calculated (according to formula 11)) and costs $z(y_1)$ (according to formula (4)),besides, probabilities $P(A_i)$ of occurrence of IT incidents in the course of release implementation in each interval are taken as maximum, the final calculations are shown in table 15.

Table 13. Cost values of $z_K(A_1)$ for different intervals of probabilities and indicators of testing quality.

$K(P(A_1))$	3	2	1
$P(A_1)$	$(0 \div 0.07)$	$(0.07 \div 0.14)$	$(0.14 \div 0.21)$
$z_K(A_1)$	15	12	8

Table 14. Cost values $z_K(A_2)$ for different probability intervals and indicators of testing quality.

$K(P(A_2))$	3	2	1
$P(A_2)$	$(0 \div 0.09)$	$(0.09 \div 0.18)$	$(0.18 \div 0.27)$
$z_K(A_2)$	21	17	9

Table 15. Values of incidents probability during releases testing A_1, A_2 and costs required for that.

$K(P(A_1))$	3	3	3	2	2	2	1	1	1
$K(P(A_2))$	3	2	1	3	2	1	3	2	1
y_1	0.154	0.237	0.321	0.217	0.295	0.372	0.281	0.352	0.423
$z(y_1)$	36	32	24	33	29	21	29	25	17

Thus, an interval of probabilities $y_1 = (0.154 \div 0.423)$ was obtained, which in turn is divided into three equal sub-intervals, and the quality of testingA_1 and A_2 releases is reduced to a three-point scale. For each sub-interval, the probability value y_1 is chosen, at which the costs of testing are minimal. If there are several such probability values, then the minimal value of y_1 is chosen. The final solution of the first level problem is given in table 16.

Table 16. Total values of $y_1$and $z(y_1)$.

K(y1)	3	2	1
y1	0.237	0.321	0.423
z(y1)	32	24	17

Next, the task of the second level is solved, the values of costs, probabilities and quality of a test during the implementation of A_3 release are specified and given in table 17. The values of incidents probabilities y_2 during A_3 release implementation and costs of $z(y_2)$, necessary to ensure the appropriate quality, are determined. The results of these calculations are presented in table 18.

Table 17. Cost values $z_K(A_3)$ for different intervals of probabilities and indicators of the test quality.

$K(P(A_3))$	3	2	1
$P(A_3)$	$(0 \div 0.09)$	$(0.09 \div 0.18)$	$(0.18 \div 0.27)$
$z_K(A_3)$	22	19	11

Table 18. Values of probabilities y_2 and costs $z(y_2)$.

$K(y_1)$	3	3	3	2	2	2	1	1	1
$K(P(A_3))$	3	2	1	3	2	1	3	2	1
y_2	0.306	0.375	0.443	0.382	0.443	0.504	0.475	0.527	0.579
$z(y_2)$	54	51	43	46	43	35	39	36	28

The defined interval $y_2=(0.306 \div 0.579)$ is divided into three equal sub-intervals, in each of them such probability value is chosen, so that the value of cost function would be minimal. The solution obtained is given in table 19.

In the example above, the third-level task is the last one, it is solved with the values of probabilities y_2 and costs $z(y_2)$, as well as the pre-set cost values z_4 for A_4 release testing and estimated incidents probabilities $P(A_4)$. The initial data for the solution of the problem of the third level are given in table 20, the intermediate results – in table 21.

Table 19. Total values y_2 and $z(y_2)$.

$K(y_2)$	3	2	1
y_2	0.382	0.443	0.579
$z(y_2)$	46	43	28

Table 20. Cost values $z_K(A_4)$.

$P(A_4)$	$(0 \div 0.08)$	$(0.08 \div 0.16)$	$(0.16 \div 0.25)$
$K(P(A_4))$	3	2	1
$z_K(A_4)$	15	10	8

Table 21. Values of probabilities y_3 and costs $z(y_3)$.

$K(y_2)$	3	3	3	2	2	2	1	1	1
$K(P(A_4))$	3	2	1	3	2	1	3	2	1
y_3	0.438	0.493	0.549	0.493	0.543	0.594	0.617	0.655	0.693
$z(y_3)$	61	56	54	58	53	51	43	38	36

Similar to the previous steps, the obtained interval $y_3=(y_{min} \div y_{max})$ is divided into three sub-intervals, from each such value $y_3$is selected, at which costs would be minimal. This is the final result presented in table 22.

Table 22. Final values y_3 and $z(y_3)$.

$K(y_3)$	3	2	1
y_3	0.493	0.594	0.693
$z(y_3)$	56	51	36

4. Conclusion

Thus, the minimal costs for testing of an IT release with a "good" grade are 56 units of resources. At the same time, the probability of occurrence of IT incidents when implementing IT services is 0.493.

When comparing the results obtained in solving the optimal resources allocation for independent and system testing, it was found that performance of a system testing is more efficient, since with its best quality and lower costs required for this, the probability of occurrence of IT incidents is significantly lower than in the case with independent testing.

References

[1] Zimin V V, Kulakov S M, Purgina M V and Koinov R S 2013 *Steel in Translation* **43(12)** 793–7
[2] Zimin V V, Kulakov S M and Purgina M V 2013 *Steel in Translation* **43(4)** 176–9
[3] Zimin V V, Kulakov S M and Zimin A V 2011 *Steel in Translation* **41(2)** 109–16
[4] Zimin V V and Kulakov S M 2010 *Steel in Translation* **40(6)** 539–48
[5] Burkov V N and Burkova I V 2014 *Automation and Remote Control* **75(3)** 470–80
[6] Burkov V N and Burkova I V 2012 *Automation and Remote Control* **73(7)** 1242–55
[7] Burkova I V 2009 *Automation and Remote Control* **70(10)** 1606–12
[8] Kiseleva T V and Maslova E V 2016 *Proc. Int. Conf. on Technologies for the Development of Information Systems TRIS-2016* (Taganrog: SFU) pp 40–4
[9] Avdeev V P, Burkov V N and Kiseleva T V 2001 *Automation and Remote Control* **62(10)** 1645–50
[10] Zimin V V, Kiseleva TV and Maslova E V 2016 *Proc. IVth Int. Conf. on Modeling and Science-intensive Technologies in Technical and Socio-economic Systems* (Novokuznetsk: SibSIU) pp 220–5

IOP Conf. Series: Earth and Environmental Science **84** (2017) 012035 doi:10.1088/1755-1315/84/1/012035

Automated information and control complex of hydro-gas endogenous mine processes

K S Davkaev[1], M V Lyakhovets[2], T M Gulevich[2] and K A Zolin[2]

[1] Synergo Soft Systems LLC, 91a Stroiteley Ave, Novokuznetsk, 654005, Russia
[2] Siberian State Industrial University, 42 Kirova Street, Novokuznetsk, 654007, Russia

E-mail: lyakhovets_mv@siu.sibsiu.ru

Abstract. The automated information and control complex designed to prevent accidents, related to aerological situation in the underground workings, accounting of the received and handed over individual devices, transmission and display of measurement data, and the formation of preemptive solutions is considered. Examples for the automated workplace of an airgas control operator by individual means are given. The statistical characteristics of field data characterizing the aerological situation in the mine are obtained. The conducted studies of statistical characteristics confirm the feasibility of creating a subsystem of controlled gas distribution with an adaptive arrangement of points for gas control. The adaptive (multivariant) algorithm for processing measuring information of continuous multidimensional quantities and influencing factors has been developed.

1. Introduction

From the point of view of the safety in mining operations, the airgas control (AGC) of the mine atmosphere is the most important issue. At present, it is regulated by the Provision on airgas control in coal mines [1], which establishes the procedure for organization of continuous automatic control over the parameters of mine atmosphere, dust content and air consumption in mine workings; detection of underground fires and the initial stages of their occurrence according to the parameters of the mine atmosphere; monitoring and control of the units and equipment operation to maintain a safe airgas regime.

In this issue, an important factor is the continuous monitoring of the concentration of three gases – methane, carbon monoxide and oxygen using individual portable gas analyzers that are handed over to miners before their going underground [2, 3]. The registration and processing of harmful gases measurements, time and place of measurement with the help of individual gas analyzers allows the following goals of ensuring safety during minerals extraction to be achieved:

- reduction of the explosion hazard of mine atmosphere (lowering of flammability and explosion hazard) in the mine workings when the methane concentration exceeds the allowed standards (or increases) and the alarm of gas analyzers belonging to three or more miners goes off;
- detection of underground fires and the initial stages of their occurrence according to the parameters of the mine atmosphere;
- provision of information for analysis and elimination of the causes of endogenous and exogenous fires;
- control of individual binding of the gas analyzer to each miner;

- provision of information on controlled parameters to specialists working at the mine who perform operational management of mining operations and ensure the mining operations safety;
- storage of information and possibility of its use for the development of comprehensive measures ensuring industrial safety at the mine for the purpose of current (operational) detection of natural and technogenic hazards that affect directly or indirectly the state of mine atmosphere;
- provision of remote control by indirect data.

2. Automated information and control complex

The developed automated system is designed to prevent emergencies, incidents and violations of safety rules associated with violations of aerological environment in the underground workings and leading to life and health risks of personnel. In addition, the tasks of integrated automation of handing over (receiving) and accounting of individual portable devices (cap lamps, radio stations, gas analyzers), as well as data transmission from gas analyzers and display of the controlled gases concentration in the mine atmosphere, were solved.

The typical structure of information means of the mine consists of underground and above-ground complexes separated by an intrinsic safety unit. In the example under consideration, the underground complex was implemented on the basis of mine information complex “Talnakh”, intended for construction of mining and underground radio communication systems and automation for various purposes in the underground part of mines [4, 5]. Current information from individual gas analyzers comes through “Talnakh” complex to the central server, where it is processed and stored on the database servers of the system under consideration.

The structure of the developed system includes the following subsystems:

- subsystem for accounting individual gas analyzers;
- subsystem for controlling the handed over and handed in individual portable devices (gas analyzers, radio stations, cap lamps);
- subsystem for displaying the content of oxygen, methane, carbon monoxide by individual means of continuous monitoring;
- subsystem for data exchange about mine workers.

With the help of the gas analyzer subsystem the user is able to keep a record of the gas analyzers of the organization. The following functions are realized: automated input of information about gas analyzers (sensor and radio tag) into the system; control over the number of gas analyzers (total, handed over, under repair); control over the timing of gas analyzers calibration.

The subsystem that controls the handed in and handed over individual portable devices is designed to record the facts of handing in and handing over the individual portable gas analyzers, radio stations and cap lamps using the device itself (equipped with automatic identification tools) and passes (contactless cards) of access control systems, automated verification and/or setting technological parameters, prohibition of device handing over, etc.

The subsystem for displaying the oxygen, methane, and carbon monoxide content by the individual continuous monitoring devices is designed to display the results of monitoring methane, carbon monoxide and oxygen by individual automatic portable gas analyzers, which every underground worker, when he is in the underground workings and hazardous aboveground facilities, should have. The subsystem provides automatic recording of threshold values for methane, oxygen, carbon monoxide in the database with time recording and indication of the location of the miner from the database “Gases”. Figure 1 shows the automated workplace of an air gas control operator.

The subsystem for exchange of data about mine workers imports from the organization’s access control system information about the mine employees and contractors performing underground works.

The developed system has ample opportunities for interaction with existing automated systems and databases at the enterprise, in particular, from the mine information complex “Talnakh”, access control system “KODOS” and the “Gases” database. Data streams during data import/export operations are schematically shown in figure 2.

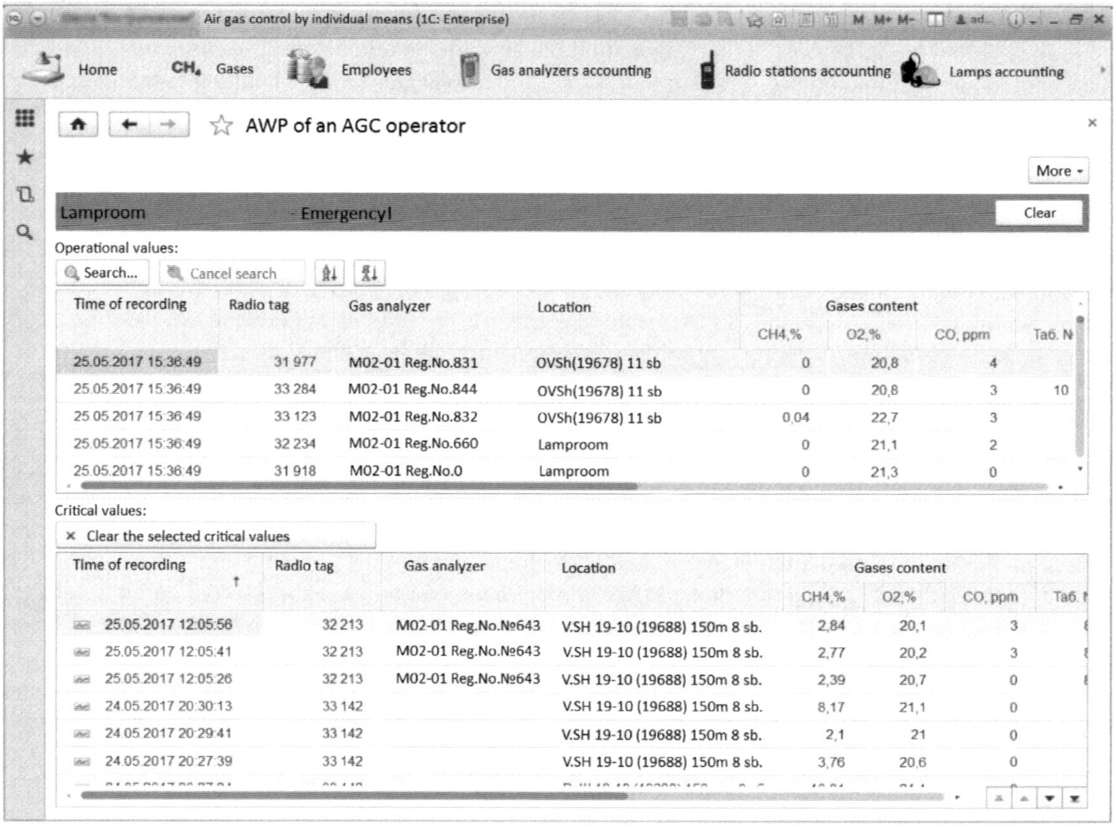

Figure 1. Automated work place of an operator of airgas control by individual means.

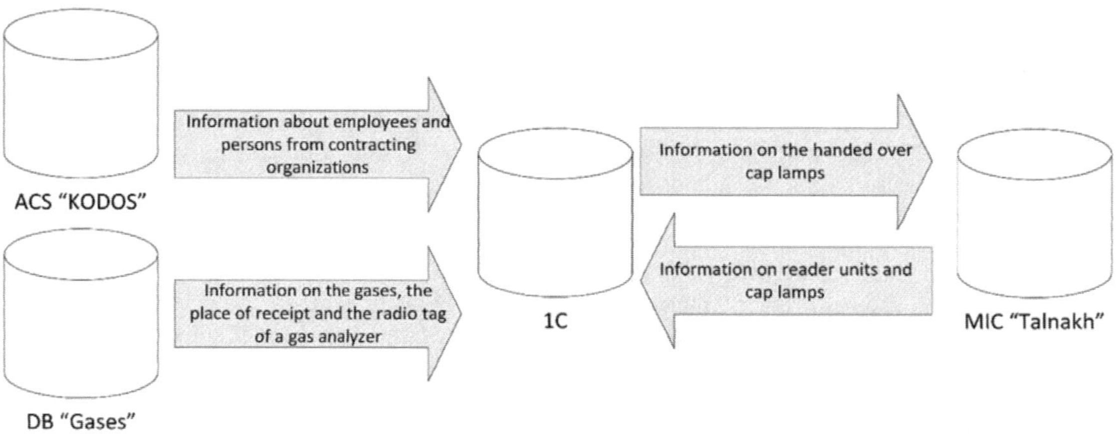

Figure 2. Data streams.

The system has been successfully implemented at several mines in Kemerovo region, but the work continues on improving the current information control system, in terms of processing the information received from gas analyzers. A lot of works are devoted to the problem of improving airgas control methods, for example [6 - 9]. Nevertheless, in real conditions of distribution and non-stationarity of the airgas parameters of mine processes, sudden increases in local gas concentrations and, as a consequence, exceedance of the maximum permissible concentrations of explosive gas mixtures (in particular, methane), as well as various disturbances in the flows of air, causes an additional danger of explosions and fires.

Therefore, it is especially urgent to ensure the safety and reliability of the mines by creating a modern automated technology for obtaining measurement information about a number of air gas parameters of the mine atmosphere and the ventilation regime, its joint processing, in order to determine timely the characteristic changes in the informational parameters of the state of controlled objects.

An adaptive automated information and control complex is designed to prevent emergencies, incidents and violations of safety rules related to aerological situation in the underground workings, as well as to transmit and display measurement data and take preemptive actions. A distinctive feature of this complex is the integration into the basic stationary system of a mobile information system based on the indications of individual gas analyzers handed over to miners. And also a distinctive feature is the diagnostic subsystem built into the existing production control and regulation system based on the operational multi-variant joint processing of the measuring information and development of advanced solutions. The simplified functional structure of the proposed complex is shown in figure 3.

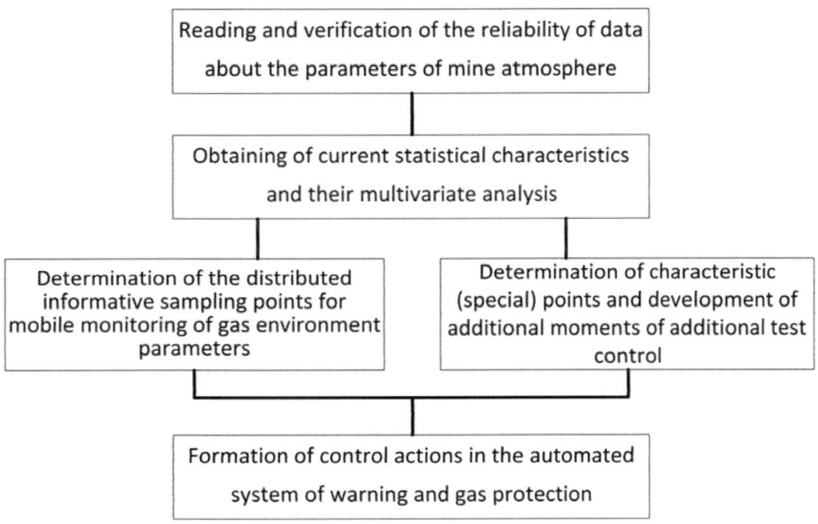

Figure 3. Functional structure of the complex.

In connection with the set tasks, there are increased requirements to the mathematical apparatus and, accordingly, to the algorithmic support of the described complex.

3. Multivariate algorithm for processing the measurement information

An adaptive (multivariant) algorithm for processing the measurement information of continuous multidimensional quantities and influencing factors is developed. Its multivariant algorithmic structure is presented in figure 4.

The synthesis of these algorithms is based on the procedures of a robust smoothing of non-stationary series of primary and calculated data. The need for a robust estimation of structural and statistical characteristics of airgas control, for example, in the form of a polynomial description for a slowly varying vector component of correlation moments measurement of quantities coupling in it, conditioned by the presence of coarse control interferences and sharp individual changes in the course of operation in the workings.

IOP Conf. Series: Earth and Environmental Science **84** (2017) 012035 doi:10.1088/1755-1315/84/1/012035

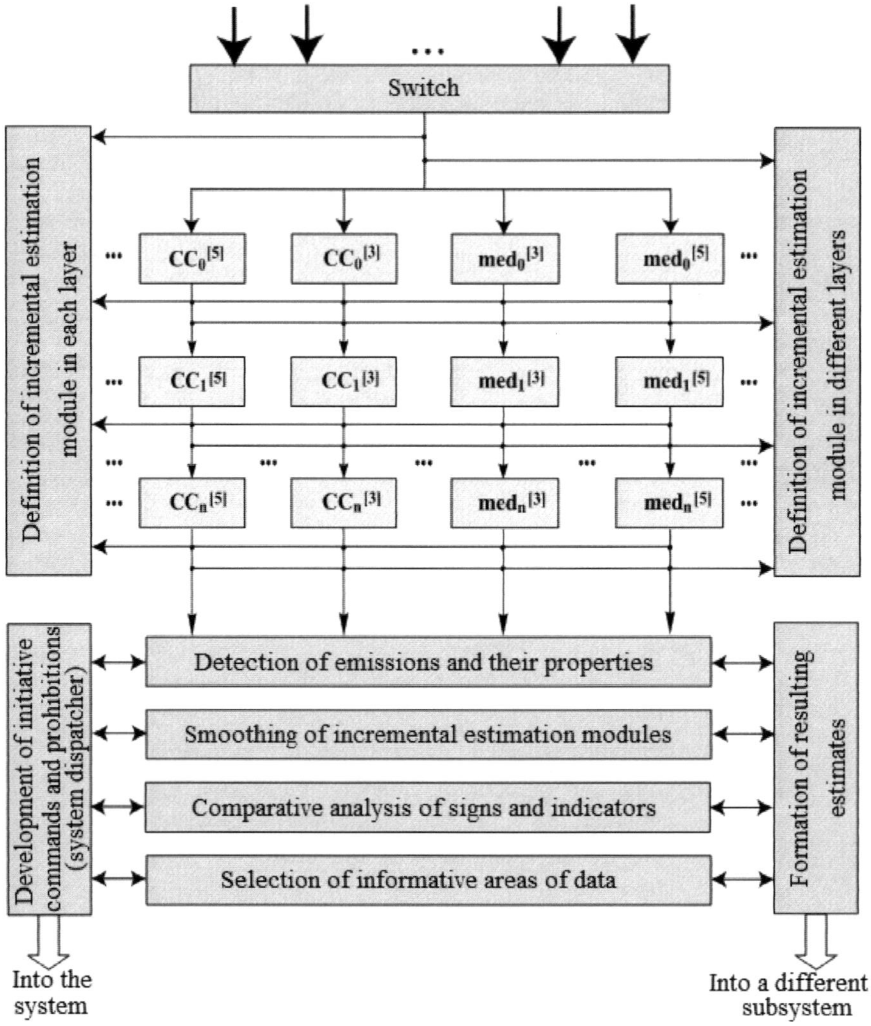

Figure 4. Multivariant algorithmic structure.

Data processing based on conventional averaging results leads in such conditions to the overestimated range of estimations variation compared to significant deviations in the real object itself, which is confirmed by numerous results of histograms and correlation moment matrices plotted with the use of the recorded data on gas distribution and other parameters of endogenous mine processes (figure 5).

Comparison of the robust and conventional procedures allows the former to be preferred, because they lead to more reproducible statistical characteristics at different time intervals of mine operation, including using simple median and relay-exponential smoothing algorithms. At the same time, there are cases of insufficient speed performance of both conventional and the described robust procedures in the operative tracking of significant changes in the trends of endogenous processes. Therefore, additional developments were made to develop adaptive robust algorithms of smoothing multidimensional data series with a joint analysis of deviations of actual and smoothed values for all interrelated variables.

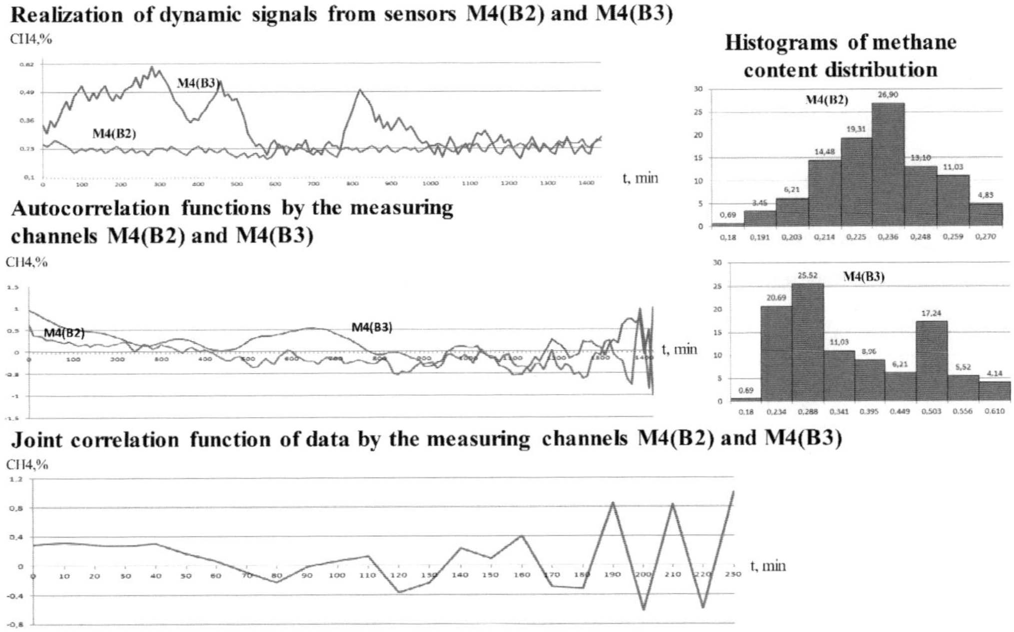

Figure 5. Statistical characteristics.

A multidimensional version of the algorithm, implemented in [10] and presented in a simplified form in figure 4, can be considered as one of the most adapted algorithms for a wide application in automated information systems. Such analysis is carried out in order to timely identify significant changes that are consistent in the changes in most primary taken into account and calculated data.

The obtained results of the trend estimation (non-stationary mathematical expectation), auto- and intercorrelation moments in the joint processing of data for all monitored gas control points proves the satisfactory reliability and speed performance of the proposed procedures.

4. Conclusions

The dynamic estimates of the correlation coefficients between the methane measurement values in the fixed points, found with the help of a multidimensional multivariant algorithm, have a pronounced non-stationarity. At different time intervals there is a "drifting" dispersion of information, which confirms the expediency of creating a subsystem of controlled gas distribution monitoring with an adaptive arrangement of gas control points. The version of a control action is developed, according to which the distance between adjacent points is subsequently adjusted depending on the difference between the current cross-correlation coefficients and their average value over all pairs of control points. Together with this, the increments of smoothed variables in the adjacent time intervals of the real object should be used. In case of exceedance of the established threshold values by the instability indicators, the control solutions are developed, including special test control measurements of the gas concentration.

The command control of gas distribution contributes to a more timely and reliable assessment of the actual state of endogenous mine processes, increasing the representativeness and speed in collecting and analyzing multidimensional measurement information, as well as expanding the functional potential of human-machine system for development of effective solutions and eliminating emergency situations in operation of the mine complex.

Acknowledgements

The research is supported by the grant of the Russian Foundation for Basic Research, Project No. 15-07-02231.

References

[1] Order of Rostekhnadzor from 01.12.2011. No. 678 “On Approval of the Regulation on Airgas Control in Coal Mines”
[2] Medvedev V N and Osipov V M 2014 *Technical supervision* **4(89)** 96–8
[3] Medvedev V N and Osipov V M 2013 *Safety of work in industry* **7** 36–8
[4] Company “Information Industry” 2008 *Coal* **9(989)** 24–5
[5] Barabanschikova S V and Simanovsky Yu A 2005 *Mining* **6(64)** 26–7
[6] Pat. 2526033 Russian Fed. No. 2013114143/03; Appl. 29.03.2013; Publ. 20.08.2013, Bul. No. 23 (part II) p 6
[7] Orlov A S and Shvets S N 2016 *Proc. Conf. on Biotechnical, Medical and Ecological Systems and Complexes. Biomedical Systems-2016* (Ryazan: Russian State Archives) pp 132–6
[8] Azbel M D 2012 *Mining Information and Analytical Bulletin* (Sc. And Tech. J.) 8264–70
[9] Babenko A G and Vilgelm A V 2016 *Proc. Conf. on Occupational Safety and Efficiency of Mining Enterprises with Underground Development* (Ekaterinburg: USMU) pp 49–55
[10] The Multivariate Filter: Certif. about the State Reg. of a Computer Program No. 2017613343; Appl.30.11.2016; Registered 15.03.2017

IOP Conf. Series: Earth and Environmental Science **84** (2017) 012036 doi:10.1088/1755-1315/84/1/012036

Social-economic processes of fuel and raw materials region: diversification, mathematical modeling and regularities

A V Novichikhin, V N Fryanov, T V Petrova, L D Pavlova and M V Temlyantsev

Siberian State Industrial University, 42 Kirova Street, Novokuznetsk, 654007, Russia

E-mail: novitchihin@bk.ru

Abstract. The paper proposes the structure of production diversification in the fuel and raw material region. A distinctive feature of diversification is the provision of regional reproduction of natural resources – reserves of mineral deposits, as well as fuel and energy deposits – traditionally consumed energy products. The algorithm for determining the regularities of socio-economic processes in the region is developed and the results of its concretization are presented. The difference of the proposed algorithm is the joint accounting of quantitative and qualitative information, the emergence properties of social and economic systems under favorable and the following problem situations: crisis, resource scarcity, uncertainty of the prediction of changes in external conditions, changes in legislative bases. It is suggested to implement identification of possible negative processes at the initial stage and when eliminating them step by step: the formation of a hypothesis about the existence of regularity, cognitive modeling, formulation of statements. Based on the results of cognitive modeling of development scenarios for fuel and raw material regions, a set of statements that form the methodological basis for correcting hypotheses about regularities is justified. Four regularities of social and economic processes of the fuel and raw materials region are revealed.

1. Introduction

In modern conditions of economy management the most important strategic goal of development of fuel and raw material regions (FRMR) is the release of finished products of high technological conversion, which is confirmed by the “Program for the development of the Russian coal industry for the period until 2030” [1].

In this regard, the actual applied task of managing social and economic systems (SES) of FRMR is the development of effective options for technological conversion and diversification of enterprises of fuel and energy complex to ensure sustainable development. The mathematical statement of the problem of the SES FRMR following the trajectory of sustainable development is presented in [2]. When forming options for technological conversion and diversification of enterprises, it is proposed to consider the following directions of structural transformations that are regional clusters [3]: society, natural resources potential, ecology, economy and poles of growth.

2. Methods of research

To implement and improve the regional processes of reproduction of natural, human, financial, information and other resources, methods of identification of social and economic processes of FRMR were used. Identification of the significance of their influence on SES development indicators and integration of production processes, including technological conversions of various energy products, was carried out using methods of cognitive simulation.

3. Results and discussion

One of the promising areas for improving regional reproduction processes is the diversification of enterprises of fuel and energy complex and related industries [4 - 6].

figure 1 shows the structure of diversification of SES FRMR enterprises. A distinctive feature of diversification of FRMR production from other regions is the provision of regional reproduction of natural resources – reserves of mineral deposits, as well as fuel and energy deposits – traditionally consumed energy products. It should be noted that the current conditions of SES FRMR functioning result in higher planning requirements [7 - 11], which consists in the development of strategies, programs of operation and development, including the periodic formation of short-, medium- and long-term forecasts and the flexible modification of policies and programs, as well as their updating.

In this regard, in order to develop effective sets of technological conversions and diversification of enterprises in the fuel and energy complex, it is important to study the regularities of social and economic processes of FRMR.

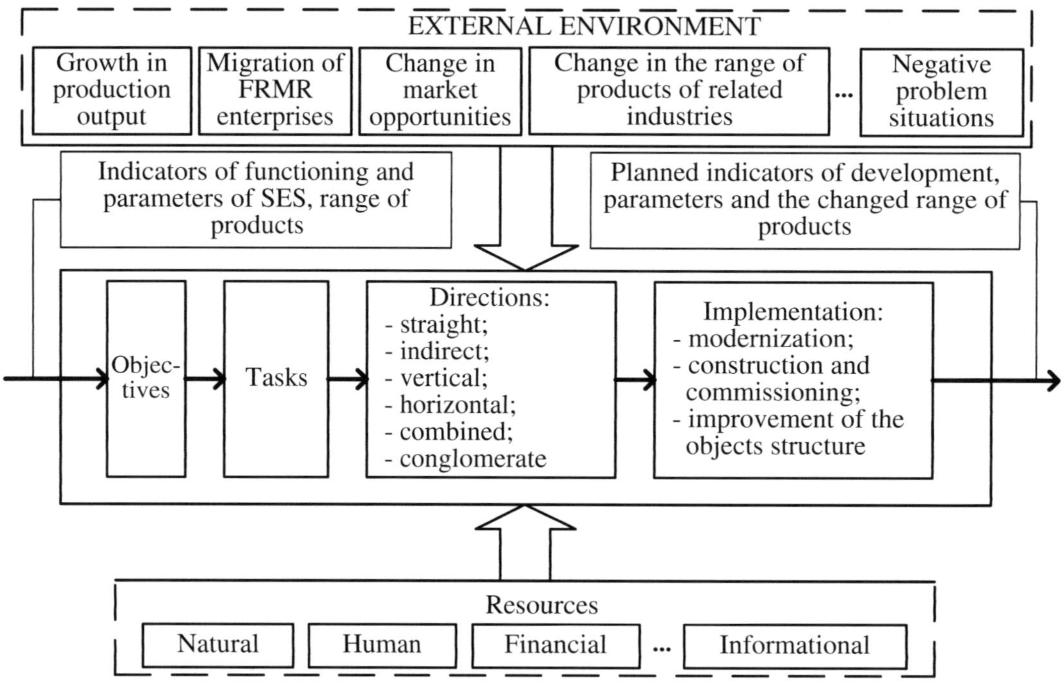

Figure 1. Structure of the production diversification in FRMR.

The algorithm for revealing the regularities of social and economic processes of FRMR (figure 2) is based on complexes of cognitive and imitation models [12 - 14] corresponding to regional clusters (society, natural resources potential, ecology, economy and poles of growth).

The difference between the proposed algorithm is the joint accounting of quantitative and qualitative information, emergence properties of SES in favorable and the following problem situations: crisis, resource scarcity, uncertainty of forecast of changes in external conditions, changes in the legislative basis of SES functioning.

Accounting of qualitative and quantitative information during the formation of decisions allows possible negative processes at the initial stage of their manifestation to be identified and problematic situations to be effectively eliminated. At stage 1 "Formation of hypothesis about regularity", a hypothesis is made about the existence of regularity that may or may not be present in the database. The hypothesis about regularity characterizes favorable or problem situations in FRMR. In the absence of a hypothesis or difficulty in its formation (only at the first cycle of regularity identification), this stage is skipped and diagnostics of the directions of SES FRMR development are performed on the basis of stage 2 "Cognitive modeling". In the absence of a hypothesis about the regularity, at stage 3 "Formulation of statements" its formation is performed. Correction of the hypothesis about the regularity is carried out when analyzing the results of modeling at a qualitative and quantitative level – stages 2 and 5. At stage 7, "Formulation of regularities" when confirming the hypothesis about regularity its formulation is carried out taking into account the results of cognitive and simulating modeling, describing the studied social and economic process of FRMR (figure 2).

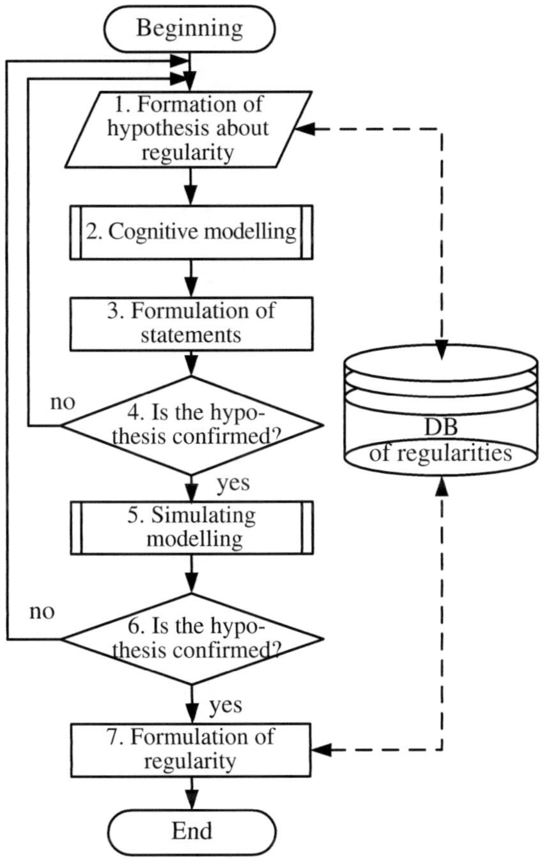

Figure 2. Algorithm for revealing the regularities of social and economic processes of FRMR.

Statements about the socio-economic processes of FRMR are obtained on the basis of analysis of the results of cognitive modeling of 275 different priority development scenarios at a qualitative level. The statements about socio-economic processes (stage 3) form the basis for correcting hypotheses about the regularities and consist of several provisions:

Statement 1:

1.1 The determining factor in ensuring a sustainable improvement in the life quality of FRMR population is the creation of new jobs.

1.2 The priority of improving the factor "state of labor market" provides a less socio-economic effect, compared with a combination of factors – improvement of the state of labor market, incomes and minimum of subsistence of population in FRMR.

Statement 2:

2.1 Strengthening or maintaining the natural resources potential of FRMR at a given normalized level is provided by a combination of three factors: the state of industrial reserves; investments in the reproduction of resources and legislative norms.

2.2 The state of natural resources potential depends on the effectiveness of problem solution connected with agreement of the interests of SES FRMR agents and the state.

Statement 3:

3.1 The ecological environment of FRMR depends on the type and production structure of SES agents and investment opportunities.

3.2 The load on the environment corresponding to one of the conditions of sustainable development depends on the balance of production volumes and emissions, except for non-waste production.

3.3 Diversification of FRMR production, focused on products of high technological conversions, in some cases also can ensure the stabilization or preservation of environmental conditions at a given normalized level.

Statement 4:

4.1 In crisis and post-crisis conditions, efficient operation of SES FRMR agents is provided through plans, programs and strategies coordinating their joint activities, developed and implemented by the center (regional and federal authorities).

4.2 The effectiveness of SES FRMR management in these conditions does not imply the suppression of the activity of SES agents, but is provided by joint integrated directed actions to overcome crisis problem situations and reduce potential joint losses.

Statement 5:

5.1 The level of production development of SES FRMR depends on investment opportunities, the state of main productive assets, structure and production volumes and causes the increase or maintenance of the gross regional product at the achieved level as one of the conditions for sustainable development.

5.2 Effective management of the level of production development, in accordance with the proposed approach, aimed at modifying the production structure of FRMR, can ensure the sustainable functioning of SES, even in crisis conditions.

Statement 6:

6.1 The level of development of the growth poles depends on the conditions of functioning created by the authorities on the basis of legislative norms (restrictions), on the state of the production structure and the level of diversification, which integrally determines its contribution to the economy.

6.2 The change in the production structure of growth poles is possible if favorable conditions for the functioning and action effectiveness of the authorities are created, and there is an effectiveness in interactions between them.

6.3 The location and "migration" of growth poles is substantially limited by territorial concentration and the markets of domestic resources.

The analysis of the results of simulation modeling of development scenarios (stage 5) is performed for the following types: pessimistic; moderate; rational; optimistic; dominant. 72 scenarios for technological conversions, consisting of 14 pessimistic, 15 moderate, 14 rational, 15 optimistic and 14 dominant variants, respectively, and 388 complex scenarios of five types (100, 64, 64, 80, 80, respectively) were modeled in the course of research.

At stage 7 of the algorithm for revealing the regularities of social and economic processes of FRMR (figure 2), the following confirmed regularities are formulated:

Regularity 1. A large amount of investments does not guarantee high quality of life for the FRMR population, due to the wavy character of its economy development – the presence of short waves with a period of 6-7 years and pronounced turbulence in raw materials markets. The hypothesis about this regularity is formulated on the basis of the results analysis of scenarios cognitive modeling – rational, optimistic and dominant types by the maps G_1 "Society" (Statement 1) and G_4 "Economy" (Statement 5) at stage 3. The hypothesis is confirmed by the simulation results of scenarios modeling for the SES FRMR development of optimistic and dominant type.

Regularity 2. Orientation to high technological conversions can provide a change in the production structure of FRMR and help to overcome its mono-specialized specialization alongside with the current negative trends in the economy: dependence on gross regional product, pronounced raw material orientation and high competition in transnational markets. The hypothesis about this regularity was formulated at stage 1 and confirmed by cognitive modeling of five types of scenarios on the basis of maps G_2 “Natural Resources Potential” (Statement 2), G_4 “Economy” (Statements 4 and 5), and G_5 “Poles of Growth” (Statement 6), as well as by the results of simulation.

Regularity 3. The distribution and migration of “growth poles” is substantially limited by the level of territorial concentration of enterprises in the regional economic space and environmental safety. Formation of the hypothesis about this regularity was carried out at stage 1 and confirmed by the results of cognitive modeling of scenarios of moderate, rational, optimistic and dominant types on the basis of maps G_2 “Natural Resources Potential” (Statement 2), G_3 “Environment” (Statement 3) and G_5 “Poles of Growth” (Statement 6). Confirmation is also the results of simulation modeling of scenarios –rational, optimistic and dominant types of SES FRMR development.

Regularity 4. The level of diversification of the FRMR subjects stimulates the emergence of new “poles of growth”, depends on the state of the market and is weakly dependent on existing legislative norms. The hypothesis about this regularity was formed at stage 1 and confirmed by cognitive modeling on the basis of map G_5 card “Poles of Growth” (Statement 6) and scenarios simulation modeling of all types of SES FRMR development.

4. Conclusion

Thus, the use of the revealed regularities of social and economic processes taking place in FRMR ensures a reduction in iterative formation of variants for technological conversion and diversification of enterprises operating within the fuel and energy complex, and the development of effective development strategies in favorable and problem situations of functioning. The created algorithm provides identification of regularities which is especially important in the unfavorable operating conditions and the need for new knowledge about current and prospective socio-economic processes.

References

[1] Program of the development of the Russian coal industry for the period until 2030: Order of the Government of the Russian Federation No. 1099-r from June 21, 2014 *Collected Legislation of the Russian Federation* 2014 No. **27** Art. 3786

[2] Novichikhin A V and Fryanov V N 2014 *Economy and Management of Control Systems* **3.1** 165–72

[3] Novichikhin A V 2015 *Economy and Management of Control Syst* **2.3(16)** 375–82

[4] Novichikhin A V 2014 *Problems of Management Theory and Practice* **8** 39–45

[5] Pristupa Yu D, Fryanov V N and Pavlova L D 2016 *IOP Conference Series: Earth and Environmental Science* 2016 **45** 012014

[6] Bishop P and Gripaios P 2007 *Urban Studies* **9(44)** 1739–57

[7] Burkov V N, Novikov D A and Shchepkin A V 2008 *Mechanisms of Management of Ecological and Economic Systems* (Moscow: Fizmatlit) ed S N Vasilyev p 244

[8] Shniper N I 1991 *Region: Economic Methods of Management* (Novosibirsk: Nauka) p 320

[9] Prangishvili I V 2000 *System Approach and System-wide Regularities* (Moscow: SINTEG) p 528

[10] De Haan M 2004 *Accounting for Goods and for Bads: Measuring Environmental Pressure in a National Accounts Framework* (Voorburg: Statistics Netherlands) p 216

[11] Ericsson M and Noras P 2005 *Minerals and Energy – Raw Materials Report* **1(20)** pp 29-39

[12] Novichikhin A V and Fryanov V N 2014 *Economics and Management of Control Systems* **2** 72–83

[13] Tatarinova O A 2011 *GIAB. Single issue* **6** 122–7

[14] Gammerschmidt A A 2015 *Kuzbass Fair* **6** 12–4

Analysis of waste coal from the enterprises of Kemerovo region as raw materials for production of ceramic materials

A Yu Stolboushkin, D V Akst, O A Fomina, A I Ivanov and V A Syromyasov

Siberian State Industrial University, 42 Kirova Street, Novokuznetsk, 654007, Russia

E-mail: daniel_axt@mail.ru

Abstract. The analysis of waste coal from mining enterprises of Kemerovo region as raw materials for production of building ceramics is given. The results of studies of material, chemical and mineralogical compositions of waste coal from Abashevskaya processing plant (Novokuznetsk) are presented. It was established that the chemical composition of waste coal refers to aluminosilicate raw materials with a high content of alumina and coloring oxides, the residual carbon content in the wastes is 12-25 %. According to the granulometric composition the waste coal is basically a sandy-dusty fraction with a small amount of clay particles (1-3 %). Additional grinding of coal waste and the introduction of a clay additive in an amount of up to 30 % are recommended. The results of the study of the mineral composition of waste coal are presented. Clay minerals are represented in the descending order by hydromuscovite, montmorillonite and kaolinite, minerals-impurities consist of quartz, feldspar fine-dispersed carbonates. The results of the investigation of ceramic-technological properties of waste coal, which belong to the group of moderately plastic low-melting raw materials, are given. As a result of a comprehensive study it was been established that with chemical, granulometric and mineralogical compositions waste coal with the reduced residual carbon can be used in the production of ceramic bricks.

1. Introduction

As a result of the work of many enterprises from mining, energy and metallurgical industries of Kemerovo region there is an ecological disturbance. A huge amount of industrial wastes have been accumulated today [1 - 3]. In the south of Kuzbass in Novokuznetsk administrative area the annual output of waste coal from Myskovskaya and Kuznetsk coal processing plants is about one million tonnes, from Abashevskaya plant – more than 400 thousand tonnes. There are problems with the placement of slurries from coal-enrichment wastes at Shchedrukhinskaya plant [5].

Simultaneously with the "industrial pollution" of Kemerovo Region there is a shortage of natural resources for the construction industry [6]. Therefore, for the first half of the 21st century the creation of new "green" technologies with efficient utilization of heavy industry waste is important for the production of building materials [7, 8]. A positive example is the successful operation of a unique factory for the production of ceramic bricks from 100 % wastes of coal-enrichment at Abashevskaya factory in Novokuznetsk [9].

The authors carried out research into the manufacture of bricks from carbon waste with a different content of residual carbon and proposed a scheme for obtaining additional fuel for burning ceramic products [9, 10]. The technological idea of the authors is to organize deep processing of waste coal directly at a brick factory. At the same time the secondary enrichment of wastes is carried out to produce energy coal for the plant operation.

Waste coal is characterized by an unstable composition and high content of coal. This can cause problems in the manufacture of bricks and lead to the destruction of the material during firing. Consequently, an important stage in the development of wall ceramics technology from technogenic raw materials is a complete study of the composition and properties of coal waste.

The aim of the work was to conduct a comprehensive study of waste coal as a potential technogenic raw material for the production of ceramic bricks.

2. Methods and materials

The coal wastes from Abashevskaya plant, which were obtained by processing coal from four coal mines in Kemerovo region: Erunakovo coal mine, Uskovo coal mine, Esaulskaya coal mine, Osinnikovskaya coal mine, were analyzed as raw materials. For the study of carbon waste methods of physical and chemical analysis were used: X-ray fluorescence; X-ray diffraction; IR absorption spectroscopy; complex thermal analysis and laser granulometry. The numbering of waste coal samples is given in table 1.

Table 1. Samples of waste coal from Abashevskaya plant.

Sample	Enterprise	Initial particle size, mm
1	Erunakovo coal mine	5-100
2	Uskovo coal mine	5-100
3	Esaulovka coal mine	5-50
4	Osinniki coal mine	5-150

Waste coal in its initial form is a large-lump material (table 1, figure 1).

Figure 1. Initial waste coal.

X-ray fluorescence (XRF) was done for four samples of waste coal using spectrometer Shimadzu XRF 1800. The chemical composition is given in table 2.

Table 2. Chemical composition of waste coal samples.

Sample	Mass fraction of the components (%)									
	SiO_2	*TiO_2*	*Al_2O_3*	*Fe_2O_3*	*MnO_2*	*MgO*	*CaO*	*K_2O*	*Na_2O*	*LOI**
1	51.30	0.76	14.72	2.91	0.06	1.44	1.26	2.80	0.81	23.31
2	58.49	0.76	15.07	3.07	0.11	1.73	3.28	3.07	1.30	12.99
3	55.68	0.75	15.28	3.23	0.07	1.69	2.01	3.40	1.00	16.63
4	56.87	0.78	15.17	3.98	0.13	1.96	4.78	2.94	1.36	12.01

*LOI – loss on ignition

The chemical composition of waste coal is related to aluminosilicate raw materials with a high content of alumina (14.72-15.28 wt.%), The carbon content is 11.2-22.1 wt. %. The amount of alkali and alkaline earth metal oxides in the samples is 6.3-11.0 wt. %. According to the content of coloring oxides in the calcined state waste coal is referred to a group of raw materials with a high content of coloring oxides.

In the diagram of clays for industrial applications, depending on their chemical composition [11], the samples belong to raw materials suitable for the production of building ceramics. The limitation is a high content of residual carbon (more than 10 % by weight) for all samples which implies the need for their secondary enrichment. The work of the brickyard on the raw materials from waste coal and the authors' research showed that the content of carbonaceous particles in the composition of the batch should not exceed 5-6 wt. %, which ensures their complete burn-out and the absence of a black core during firing of ceramic masses [10].

3. Results and discussion

The granulometric composition of the four waste coal samples after two-stage grinding in the jaw crusher and laboratory runners is provided in table 3 and in figure 2.

Table 3. Granulometric composition of waste coal samples.

Sample	Fraction content in %, particle size in mm				
	>1.0	1.0-0.5	0.5-0.1	0.1-0.05	<0.05
1	12.8	39.52	30.96	10.91	5.81
2	0.0	7.79	54.76	21.2	16.25
3	0.31	17.08	52.1	17.73	12.88
4	5.07	30.66	38.52	14.89	10.86

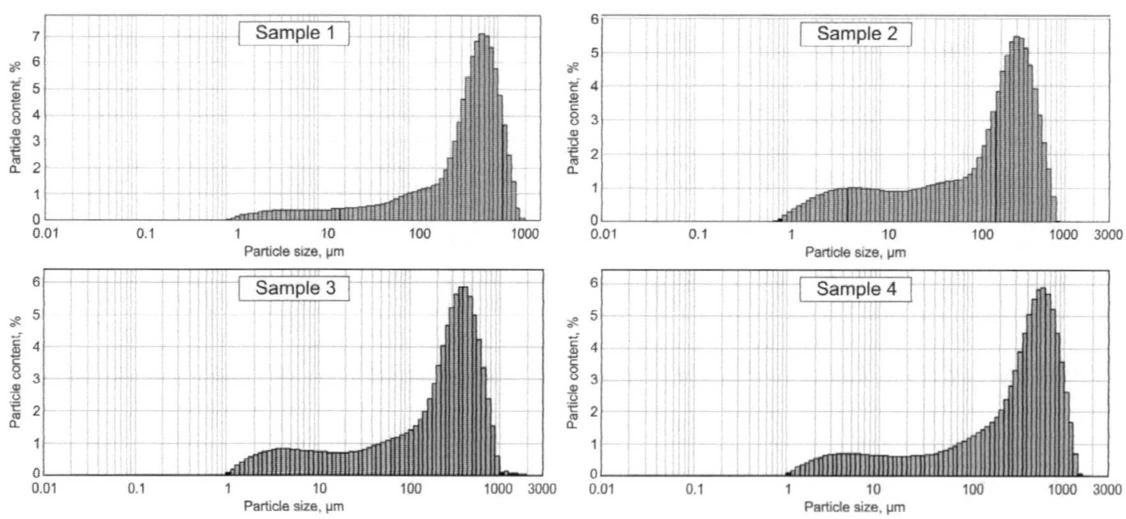

Figure 2. Histograms of the particles content depending on the size of wastes.

Studies of the granulometric composition of waste coal showed that the samples have almost the same particle size distribution. Carbon waste consists of a sandy-dusty fraction and a small amount of particles smaller than 1 μm (1-3 %). According to the diagram of clays for industrial applications depending on the particle size distribution [12] the material is not suitable for the production of ceramic bricks and needs to be adjusted for the fractional composition. The authors recommend an additional grinding of coal waste in a rod-type mill and the introduction of a clay additive in the amount up to 30 %.

The mineral composition of the waste coal was studied by X-ray diffraction (XRD) methods; and complex thermal analysis. Thermograms of coal waste are given in figure 3. Thermal analysis of four samples of waste coal showed the same endothermic effect in the temperature range of 88.3-100.1 °C. The thermal effect is associated with the removal of adsorbed water, while the mass loss is 1.27-2.86 % (figure 3). The DTA curve shows exothermic effects at temperatures of 367.4-370.7 °C and 449.5-460.7 °C which correspond to the processes of pyrolysis and carbon burn-out. The total weight loss in these reactions is 7.72-9.49 %.

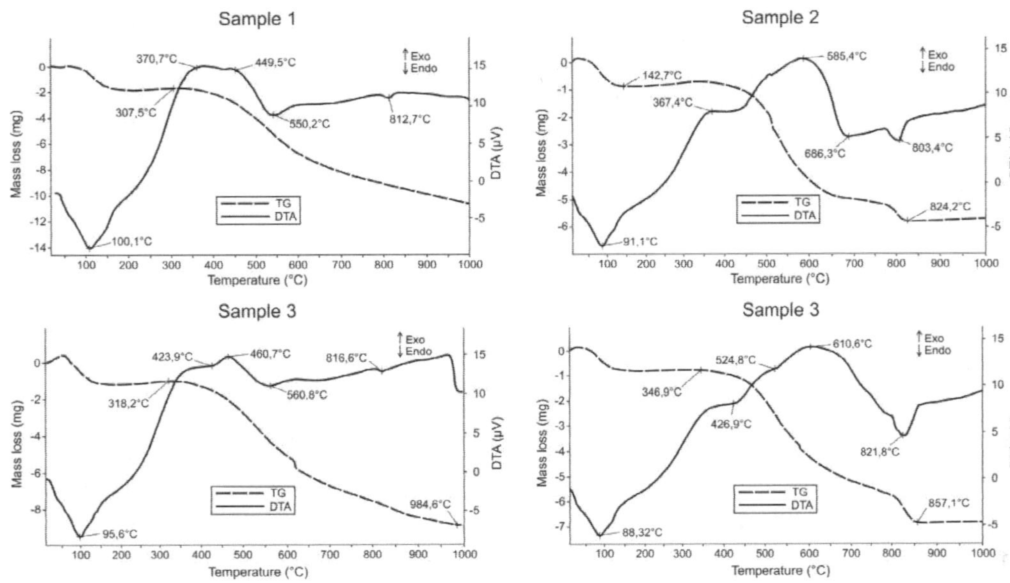

Figure 3. TG and DTA curves of the waste coal samples.

The second endothermic effect in samples 1, 3 at a temperature of 550.2-560.8 °C is associated with the removal of interlayer water and the destruction of the crystalline lattice of layered silicates. The absence of this effect in samples 2 and 4 can be explained by the simultaneous occurrence of an exothermic oxidation reaction in the temperature range 500-650 °C. The oxidative reaction is caused by a shift in the temperature peak of the coal burn-out reaction and is explained by a higher content of clay minerals (table 2). The reaction of removal of OH groups from the lattice of hydromuscovite slows down the process of carbon burn-out. In this regard the oxidation of carbon occurs at temperatures 100-150 °C higher (figure 3).

The endothermic reaction of decarbonization of waste coal samples occurs at a temperature of 803.4-821.8 °C, and the mass loss in the samples is 1.25-3.16 %. At a temperature of 1000 °C the total mass loss in waste coal samples is 12.01-23.31 % and depends on the amount of carbonaceous particles (table 2).

IOP Conf. Series: Earth and Environmental Science **84** (2017) 012037 doi:10.1088/1755-1315/84/1/012037

X-ray diffraction of four waste coal samples was done using Shimadzu LabX XRD-6000 spectrometer. The XRD results are given in figure 4 and table 4.

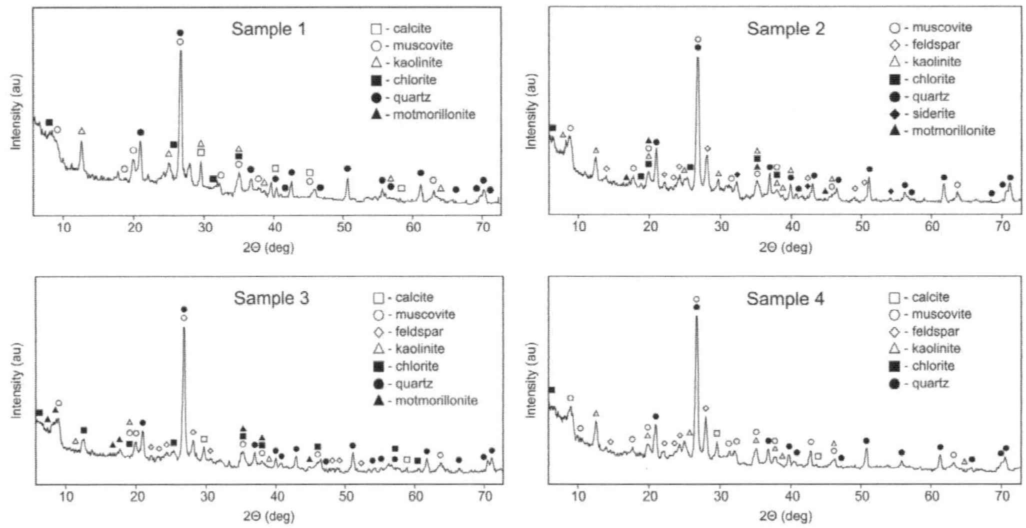

Figure 4. XRD analysis of the average waste coal samples.

Table 4. Mineralogical composition of wastes coal.

Sample	Name of minerals		
	Clay	Non-clay	Impurities
1	Hydromuskovit, kaolinite, montmorillonite	Quartz, calcium feldspar	Calcite, chlorite
2	Hydromuskovit, kaolinite, montmorillonite	Quartz, calcium feldspar	Siderite, calcite, chlorite
3	Hydromusovit, montmorillonite, kaolinite	Quartz, calcium feldspar	Hematite, calcite, chlorite
4	Hydromuskovite, kaolinite	Quartz, calcium feldspar	Calcite

According to XRD analysis the samples of waste coal have the following mineralogical composition: quartz (d/n = 0.425, 0.334, 0.223, 0.128), kaolinite (d/n = 0.710, 0.357; 0.234); hydromuscovite (d/n = 0.447, 0.239, 0.150); calcium feldspar (d/n = 0.636, 0.319, 0.183) and montmorillonite (d/n = 1.390, 0.448, 0.167). Calcites are identified as impurities (d/n = 0.303, 0.209, 0.160), chlorite (d/n = 1.380, 0.353, 0.280), siderite (d/n = 0.279, 0.172) and hematite (d/n = 0.271, 0.169).

The results obtained are confirmed by thermal analysis of four samples. Minerals of hydromica group are predominant from layered silicates. The impurities are predominantly quartz-feldspar in nature with an insignificant amount of dispersed carbonates.

In the study of ceramic-technological properties the ductility of finely grinded waste was determined (table 5). Definitions were carried out by two methods: according to GOST 21216.1-93 “Clay raw materials. Method for plasticity determination” and Pfefferkorn method.

Table 5. Plasticity of finely grinded waste coal*

Sample	Humidity, % On the fluidity boundary	Humidity, % At the rolling limit	Plasticity index	Classification according to GOST 9169-75
1	11.8 / 11.2	22.8 / 22.4	11.0 / 11.2	moderately plastic
2	11.8 / 11.1	19.7 / 18.6	7.9 / 7.5	moderately plastic
3	12.4 / 11.9	22.6 / 21.8	10.2 / 9.9	moderately plastic
4	11.0 / 10.3	17.5 / 16.7	6.5 / 6.4	low-plastic

*Values according to GOST 21216.1-93 (numerator), values by Pfefferkorn method (denominator)

To determine the ceramic-technological properties of waste coal ceramic samples-cylinders were prepared by the method of semi-dry pressing. From the grinded waste coal with moisture of 8-9 % raw samples were pressed at a pressure of 15-17 MPa. The dried samples were fired at a temperature of 1000 °C.

The physical and mechanical properties of ceramic samples from waste coal are presented in table 6.

Table 6. Physical and mechanical properties of ceramic samples from waste coal.

Sample	Air shrinkage, %	Fire shrinkage, %	Average density, kg/m^3	Compressive strength, MPa	Water absorption, %	Strength-density ratio
1	1.0	0.6	1693	6.2	16.1	3.7
2	1.6	0.9	2053	32.4	10.7	15.8
3	1.2	0.7	1764	8.1	15.5	4.6
4	1.5	0.9	1984	21.7	10.5	10.9

The comparative analysis showed a wide dispersion in the physical and mechanical properties of ceramic samples depending on a number of waste coal sample. The compressive strength of waste coal samples 1, 3 is 6-8 MPa, for samples 2, 4 is 21-32 MPa. The difference in strength is more than 3 times due to the high content of carbonaceous particles in samples 1, 3 (figure 2).

During firing waste coal samples 1, 3 carbon does not burn out which leads to low strength. During testing a black core is observed inside the samples. Incomplete burn-out of C disrupts the sintering of ceramic material from waste coal. For samples 1, 3 the water absorption was increased by 1.5 times with an average density of 1700-1750 kg/m^3.

4. Conclusions

The conducted researches of four samples of coal waste from Abashevskaya plant established:

- according to the chemical composition the samples refer to aluminosilicate raw materials with high alumina content, coloring oxides, the residual carbon content in the waste is 12-25 %;
- according to the granulometric composition the samples have mainly a sandy-dusty fraction with a small amount of clay particles (1-3 %);
- mineral composition of waste coal is represented by clay minerals: hydromuscovite, montmorillonite and kaolinite. Minerals-impurities consist of quartz, feldspar, fine-dispersed carbonates. In a small amount there are chlorite, siderite and hematite;
- according to the ceramic-technological properties the samples belong to the group of moderately plastic low-melting raw materials.

As a result of a comprehensive study of the waste coal from mining companies of Kemerovo region it is established that waste coal can be used in the production of ceramic bricks with the corrected composition of batch.

Acknowledgements
The results of the research were obtained within the framework of the state assignment of the Ministry of Education and Science of the Russian Federation, the code of the project No. 7.7285.2017 / 8.9 “Fundamental research in the field of building ceramic composites with a matrix structure based on technogenic and natural raw materials”

References
[1] Petrov I V et al 2013 *Proc. Int. Conf. on Ecology. Nature Management. Economy* (Moscow: Rolix) pp 43–56
[2] Rakhimov R Z et al 2009 *Building Materials* **12** 8–11
[3] Chernyshov E M 2010 *Proc. XV Int. Sci. Conf. on Achievements and Problems of Materials Science and Modernization of the Construction Industry* (Kazan: Kaz-SASU) 1 8–9
[4] Ivanov A *The Enrichment Factory harmed the Environment for Millions* http://nk-tv.com/128049.html
[5] Kikava O Sh et al 1997 *Environment and Industry of Russia* **12** 23–8
[6] Rasskazov V F et al 2009 *Glass and Ceramics* **1** 5–9
[7] Shpirt M Ya 1986 Wasteless *Technology. Utilization of Wastes of Extraction and Processing of Solid Fossil Fuels* (Moscow: Nedra) p 255
[8] Burmistrov V P et al 1989 *Building Materials* **8** 18–9
[9] Storozhenko G I et al 2013 *Building Materials* **4** 57–61
[10] Stolboushkin A Yu and Storozhenko G I 2011 *Building Materials* **4** 43–6
[11] Avgustinik A I 1975 *Ceramics* (Leningrad: Stroyizdat) p 592
[12] Stolboushkin A Yu, Karpacheva A A and Ivanov A I 2011 *Wall Ceramic Products on the Basis of Waste Coal and Iron-containing Additives* (Novokuznetsk: Inter-Kuzbass) p 156

IOP Conf. Series: Earth and Environmental Science **84** (2017) 012038 doi:10.1088/1755-1315/84/1/012038

Development of decision- making mechanism in engineering design of phased coal mines technical upgrade

V Yu Kulak[1], T V Petrova[2] and A V Novichikhin[2]

[1]Promugleproyekt CJSC, 4 Nevskogo Street, Novokuznetsk, 654027, Russia
[2]Siberian State Industrial University, 42 Kirova Street, Novokuznetsk, 654007, Russia

E-mail: ptrvt@mail.ru

Abstract. The approach to a choice of a new mine design and technical upgrade of operating coal mines is substantiated. The choice of the option is made in the following way: the elements of the mine technological system are defined, for each element of the system two levels of costs are allocated – capital and operational; a graph of alternative options of the system is formed by matrix enumeration taking into account the possibility of simultaneous application of different elements, up to 10 000 scenarios are formed; capital and operating costs of options are estimated in the form of coefficients as the cost-to-cost ratio in the base variant, which has already been implemented and the costs of which are already known; ranking of the options at the level of costs and the definition of the 10 preferred are performed. It is established that the application of partial enumeration allows the costs relative to the base variant to be reduced by 10 %; the main constraint of costs reduction is the need to comply with all conditions that ensure industrial safety.

1. Introduction

Technical and economic indicators of coal mines of Kuzbass are less stable in comparison with the corresponding performance of coal mines. The implementation of the underground coal mining technology requires, in comparison with the open development, with other equal external influences, a higher qualification of the personnel, significant investments in maintaining the design level of production and compensation of the probability of occurrence of hazardous industrial situations.

The main factors hampering the development of underground coal mining technology, including the reduction of costs of underground coal mining, can be divided into two groups: external, which are manifested by the impact of coal market, and internal natural and technogenic conditions [1 - 6].

The first group of factors should include the volatility of prices and volumes of coal sales in the coal market and the low competitiveness of coal products in comparison with oil and natural gas.

Internal complicating factors are:

- reduction in the volume of geological exploration and depletion of in-situ coal reserves favorable for the application of modern high-intensive technologies of underground coal mining;
- restriction of investments for the construction and technical re-equipment of coal mines with a full range of buildings, structures, machinery and equipment on the surface and underground;
- spatial variability of mining, geological and technological characteristics and parameters of the massif within the field and mining allotment, including those not identified during geological exploration and at the design stage;

- "human factor", the influence of which is especially pronounced in underground conditions as a result of improper actions of employees performing technological processes and operations, use of technical means, implementation of incompetent control actions and other cases of abnormal behavior of personnel. The influence of "human factor" has been repeatedly pointed out in the investigation protocols of many accidents in coal mines, for example, [7, 8].

2. Methods of research and results

The operative step-by-step implementation of the project solutions is proposed to be carried out according to the following scheme: analysis of a specific market, mining-theological and mining-technical situation, the formation of an alternative and selection of a rational scenario for the technical re-equipment of a mine under construction or operation.

The general idea behind the choice of design solutions is as follows. After the development of the project documentation and its partial implementation, the author's audit and analysis of the specific situation and indicators of the enterprise under construction or the launch complex is carried out. As a tool for choosing the options for design solutions, the partial matrix enumeration method [9 - 12] is applied. If the design and actual indicators do not coincide, or if there is a need to improve the design parameters in the event of changes in internal or external conditions, scenarios for improving indicators and introducing new technological and organizational solutions in the real situation are developed.

The graph of technological system of an operating mine is shown in figure 1.

In the matrix (table 1), for each element of the mining technical system of the mine (STSM), two levels of costs are identified: capital for the construction or reconstruction of facilities and operational. Alternative options for each element of STSM are formed by experts. One of the options, as a rule, already implemented and the costs for which are known, is taken as the base one with indication of real costs. The costs of other alternatives are estimated as coefficients, as the cost ratio in the base case. In a matrix enumeration in accordance with the data from table 1, an additive convolution of coefficients for all elements is carried out. By comparing the coefficients for each element with respect to the costs of the basic options for the STSM element, the costs are accumulated [13 - 15].

Table 1. Elements of mine technical system of a mine for matrix enumeration of alternative variants.

Level number on the graph	Name of STSM elements and cost elements	Numbers on the graph and the names of variants of STSM elements, costs					
		1	2	3	4	5	6
	Group of MD complexity	1st group, simple MD	2nd group, middle complexity MD	3rd group, complex MD			
1	Costs for license purchase, rub/t cost-to-base ratio	0/1.2	90 (base)/1	0/0.8			
	Costs for operational supplementary exploration, rub/t / cost-to-base ratio	0/0.8	75 (base)/1	0/1.25			
	Technological complex on the surface	constant	modular	temporary			
2	Construction costs, rub/t / cost-to-base ratio	360 (base) /1	0/0.6	0/0.3			
	Operating costs, rub/t / cost-to-base ratio	60 (base)/1	0/0.9	0/1.2			
3	Method and scheme of mine opening	main vertical shafts	main inclined shafts	central-related vertical shafts	central-related vertical and inclined shafts	central and boundary with inclined shafts	boundary with inclined

							shafts
	Construction costs, rub/t / cost-to-base ratio	0/1.3	150 (base)/1	/1.4	0/1.3	0/1.2	0/1.1
	Operating costs, rub/t / cost-to-base ratio	0/1.2	135 (base)/1	0/1.5	0/1.2	0/1.1	0/1.1
4	Method and scheme of mine development	panel in-seam	panel in-stone	horizon	block	one- horizon	multi-horizon
	Construction costs, rub/t / cost-to-base ratio	75 (base)/1	0/1.3	0/2.3	0/2.4	0/1.2	0/2.4
	Operating costs, rub/t / cost-to-base ratio	60 (base)/1	0/0.9	0/1.4	0/1.3	0/1.4	0/2.0
5	Method of mine ventilation	blowing	vacuum	combined			
	Construction costs, rub/t / cost-to-base ratio	60 (base)/1	0/1.2	0/1.6			
	Operating costs, rub/t / cost-to-base ratio	75 (base)/1	0/1	0/1.5			
6	Development system	langwall with caving along strike	langwall with caving to the dip	pillar-and-room method			
	Costs for preparation of working area, rub/t / cost-to-base ratio	120 (base)/1	0/1.1	0/0.8			
	Operating costs, rub/t / cost-to-base ratio	600 (base)/1	0/1.2	0/1.4			
7	Degassing	pre-degassing during mining	degassing by wells in the rock cavity and from the ground	degassing over rock cavity from the parallel working	degassing of overworked and underworked seams	degassing by long wells of directed drilling	in-situ degassing
	Construction costs, rub/t / cost-to-base ratio	0/1.2	0/1.4	0/1.1	0/0.5	0/0.8	150 (base)/1
	Operating costs, rub/t / cost-to-base ratio	0/1.1	0/0.8	0/1.1	0/0.3	0/0.9	90 (base)/1
8	Industrial safety	MSS	MSS + nitrogen supply units	MSS + conveyor positioning system	MSS + monitoring of gas-dynamic processes		
	Construction costs, rub/t / cost-to-base ratio	240 (base)/1	0/1.2	0/1.1	0/1.3		
	Operating costs, rub/t / cost-to-base ratio	150 (base)/1	0/1.1	0/1.2	0/1.2		
9	Environmental safety	treatment facilities	treatment facilities + monitoring of the Earth's surface displacement	treatment facilities + methane utilization	treatment facilities + dust collecting plants	treatment facilities + mine reclamation	
	Construction costs, rub/t / cost-to-base ratio	0/0.8	0/0.8	0/1.2	90 (base)/1	0/0.8	
	Operating costs, rub/t / cost-to-base ratio	0/0.7	0/1.1	0/0.7	150 (base)/1	0/1.4	

To select a rational design scenario using the data from table 1 by means of partial matrix enumeration up to 10 000 scenarios the ranking according to the level of production costs of 10 preferred variants is carried out, from which the decision maker selects the rational variant.

As it follows from table 2 and the graphs in figure 2 optimization of variants of technical re-equipment for the conditions shown in figure 2, allows the costs in relation to the base variant to be reduced by 8.5 %, and in relation to the most costly variant – by 47.6 %.

Table 2 shows fragments of total costs for one of the scenarios of technical re-equipment of "Erunakovskaya VIII" mine: the top line in each cell is a rational design variant; the middle line is the base variant, the bottom line is the most expensive variant. The histogram of costs distribution for these options is shown in figure 2.

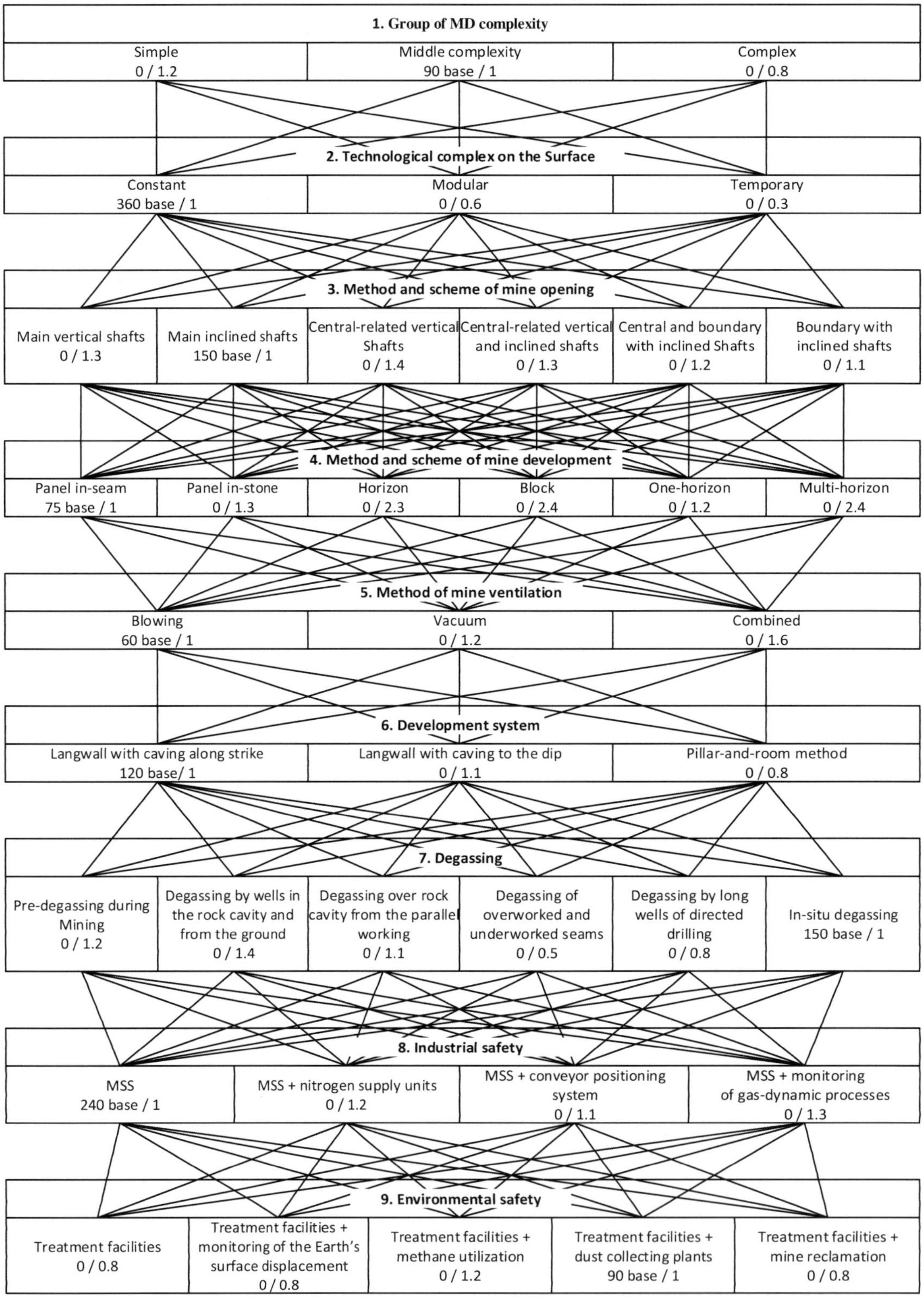

*MD – mineral deposit; **MSS – multifunctioned safety system

Figure 1. Graph of alternative variants of the mine technological system.

Table 2. Costs of one of the scenarios of coal mine technical re-equipment.

Level number and name of STSM elements	Number of STSM element vertex on the graph	Sum of costs by STSM elements, rub/t		
		capital	operating	total
1. Group of MPI complexity	2	90.00	75.00	165.00
	2	90.00	75.00	165.00
	2	90.00	75.00	165.00
2. Technological complex on the surface	3	198.00	147.00	345.00
	1	450.00	135.00	585.00
	1	450.00	135.00	585.00
3. Method and scheme of deposit opening	6	363.00	295.50	658.50
	2	600.00	270.00	870.00
	3	660.00	337.50	997.50
4. Method and scheme of mine development	1	438.00	355.50	793.50
	1	675.00	330.00	1005.00
	4	840.00	415.50	1255.50
5. Method of mine ventilation	1	498.00	430.50	928.50
	1	735.00	405.00	1140.00
	3	936.00	528.00	1464.00
Development system	1	618.00	1030.50	1648.50
	1	855.00	1005.00	1860.00
	3	1032.00	1368.00	2400.00
7. Degassing	3	783.00	1129.50	1912.50
	6	1005.00	1095.00	2100.00
	6	1182.00	1458.00	2640.00
8. Industrial Safety	1	1123.00	1279.50	2330.50
	1	1245.00	1245.00	2490.00
	4	1494.00	1638.00	3132.00
9. Environmental Safety	3	1131.00	1384.50	2515.50
	4	1335.00	1395.00	2730.00
	5	1566.00	1848.00	3414.00

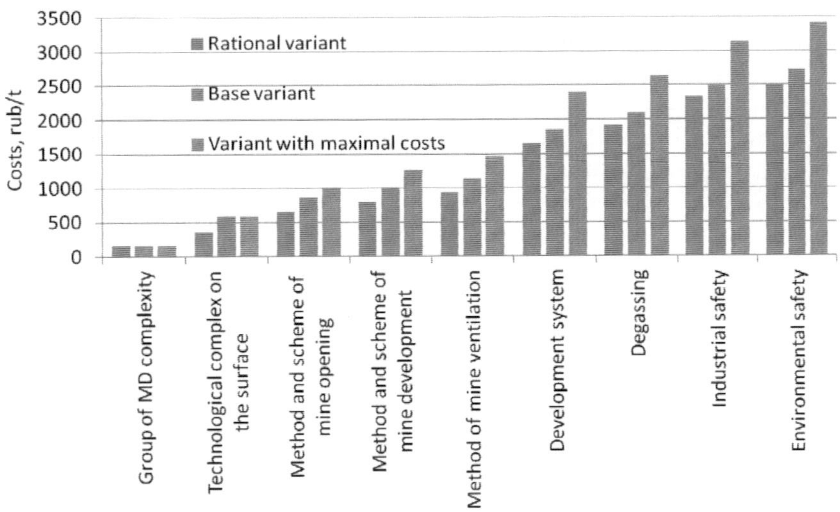

Figure 2. Costs distribution by elements of mining-engineering system of a mine.

3. Conclusions

A mechanism was developed for selecting project solutions of technical re-equipment of the existing mine, including a partial enumeration of alternative variants and selection of a rational scenario for the technical re-equipment of the mine in operation.

Based on the results of a search for new progressive design technological and technical solutions, it was established that there are reserves for exceeding the actual production as compared with the designed one. However, the excess of actual production compared to the design project is a violation

IOP Conf. Series: Earth and Environmental Science **84** (2017) 012038 doi:10.1088/1755-1315/84/1/012038

of the requirements of Article 8, paragraph 2 of the Federal Law "On industrial safety of hazardous production facilities".

Taking into account the methodological possibilities of the phased design system, the correction of project documentation was performed. The implementation of new design solutions in this documentation ensured almost complete compliance of the project and actual volumes of coal mining, while meeting the requirements of the current regulatory documents.

References

[1] Yankovsky A O 2015 *Coal* **3** 9–11

[2] Tarazanov I G 2016 *Coal* **3** 58–73

[3] Agafonov V V 2008 *Development of Scientific and Methodological Support for the Formation of a Strategy for Sustainable Development of Mining Systems of Coal Mines: the author's abstract. Dr. thes.* (Moscow: MSMU) p 47

[4] Melnik V V and Shulyatyeva L I 2010 (Moscow: MSMU) **8** 229–33

[5] Kulak V Yu and Petrova T V 2017 *Bulletin of the TSU* **23 (3)** 12–21

[6] Auty R 1993 *Sustaining Development in Mineral Economies: The resource curse thesi* (London: Routledge) p 273

[7] Podobrazhin S N 2007 *Industrial Safety in Industry* **5** 70–6

[8] Reason J 2015 *Man: Threat or Security Guarantor* (Novokuznetsk: KemSU) p 151

[9] Silov V B 1995 *Making Strategic decisions in an uncertain environment* (Moscow: INPRO-RES) p 228

[10] Kazanin O I and Romashkevich A A 2012 *Notes of the Mining Institute* **198** 104–7

[11] Kulba V V, Kononov D A et al 2002 *Scenario Analysis of Behavior Dynamics of Socio-Economic Systems* (Moscow: Institute for Control Sciences, RAS) p 122

[12] Abramova A N and Avdeeva Z K 2008 *Management Problems* **3** 85–7

[13] Pristupa Yu D, Fryanov V N and Pavlova L D 2016 *IOP Conference Series: Earth and Environmental Science* **45** 012014

[14] Fryanov V N and Pavlova L D 2016 *IOP Conference Series: Earth and Environmental Science* **45** 012002

[15] Fryanov V N and Pavlova L D 2017 *IOP Conference Series: Earth and Environmental Science* **53** 012028

IOP Conf. Series: Earth and Environmental Science **84** (2017) 012039 doi:10.1088/1755-1315/84/1/012039

Use of the method for addressing the challenges of resources procurement management at a mining enterprise

T V Petrova, S V Strekalov and A V Novichikhin

Siberian State Industrial University, 42 Kirova Street, Novokuznetsk, 654007, Russia

E-mail: ptrvt@mail.ru

Abstract. The article is devoted to the analysis of possible application of the total cost of ownership method for the purchase of resources at a mining enterprise. The description of the total cost of ownership method and experience of using this method in other spheres is provided. The article identifies the essential components needed to calculate the total cost of ownership of a resource. Particular attention is paid to the ratio of the price of the purchased resource and the total cost of ownership. To justify the relevance of application of this method at a mining enterprise for resources purchase, the technical and economic conditions of mining enterprises have been analyzed, which are quite specific and force to introduce certain adjustments to the application of the considered method and opens up new possibilities for its use. Specific spheres for application of this method at a mining enterprise are determined. The main result of the study is the proposed practical recommendations for the introduction and extension of the practice of using the method when a mining enterprise purchases resources.

1. Introduction

In the system of procurement of material resources at enterprises, various competitive procedures are currently widely used, the essence of which is to identify proposals with a minimal price, quality of the acquired item of fixed assets, auxiliary materials and acceptable terms of delivery. However, today as an alternative to a price-based approach the method oftotal cost of ownership (TCO) is increasingly being used (also the term "concept" is common in the literature).It is assumed that the purchase price is only the "tip of the iceberg", a visible part of the costs that the company incurs in connection with purchases. Orientation to the price only may, in the end, entail higher aggregate costs [1].

An exact, unambiguously understood definition of the concept of "total cost of ownership –TCO" does not exist. In some sources under TCO a methodology designed to determine the cost of information systems (and not only) are understood, which are calculated at all stages of the life cycle of the system, in others – the amount of direct and indirect costs incurred by the owner of the system over the lifetime [2].

It is believed that the term TCO was introduced in 1987 by the American consulting corporation Gartner, Inc. – the world leader in the field of independent information technology analytics, development problems of their strategic planning, development and effective application [3].

There are also alternative views on the authorship of the concept "total cost of ownership". For example, in [4] it is stated that the very concept of TCO arose much earlier: The administration of the American Railway Engineering Association (1929) refers to the total cost of ownership in its financial records.

The essence of TCO analysis is the study and evaluation of not only the price paid for the resource itself, but also the costs arising from the acquisition, use, and utilization of the purchased resource. In the field of procurement, the price is only a part of the total costs incurred by the enterprise. In addition to the price, it is necessary, at least, to take into account the cost of acquisition, operation, maintenance of the purchased resource, its service life, costs associated with its disposal.

Due to the application of this method the enterprise has the tool to assess its purchases not from the point of view of short-term benefits (acquisition at the minimum price), but from the point of view of long-term benefit (acquiring a resource with a minimum total cost of ownership).

The method of determining the total cost of ownership is widely used to carry out capital investments, acquisition of durable objects, auxiliary materials. The greatest spread of TCO is observed in the field of IT technologies – determination of the total cost of information systems, software products, which is confirmed by the publications on this topic.

For example, in [2] it is said about the relevance of the use of TCO method for the field of automation: it is necessary to know the possible cost of TCO, since this allows the structure and level of costs to be determined in advance, as well as to analyzed and correlated with the real capabilities of the potential user.

It is also noted in [3] that "the concept of TCO is applicable to any asset of an enterprise".

More often than not, companies spend most of their efforts on lowering purchase prices, and ignore spending in other areas – and they may be comparable to procurement prices, or even an order of magnitude higher. Only having a clear idea of the total costs, the company, while making important decisions, will be sure that reducing costs in one area will not lead to growth in another. Moreover, companies that do not estimate total costs do not use most of the potential opportunities to cut costs (not related to purchase prices) [1].

In the total cost of ownership quite a lot of components can be included, as an example, the following can be specified:

- cost of delivery;
- cost of recycling or income from resale, depending on the condition of the resource;
- internal costs of the enterprise, for example, for personnel training, for re-engineering.

From the supplier's point of view, the use of TCO allows the profitability of the services or products offered to be proved and partnerships with its customers to to be developed. Therefore, it can be argued that the need to determine TCO is relevant not only to the resource purchaser, but also to the supplier.

2. Results and discussion

The world leaders in the production of equipment for mining industry are focusing their attention on reducing TCO for their consumers, which underlines the urgency of applying this method. For example, CUMMINS Inc. announced a number of initiatives aimed at reducing TCO engines. "Faced with challenging economic conditions, our customers are keenly focused on conserving cash and optimizing equipment operation. Over 60 percent of a miner's TCO is tied to fuel, parts and service, and we have developed solutions which can be applied in a wide range of mining applications" [7].

In the mining industry TCO is used for the analysis of ownership cost of fixed assets. For example, in [8] it is noted that "in SUEK OJSC the choice of vehicles for transportation of rock mass is based, inter alia, on the analysis of total cost of ownership taking into account the individual mining and geological conditions of the enterprise and the specifics of the supplier's services. Calculations show that in conditions of SUEKOJSC it is more efficient to use less reliable but much less expensive equipment. For example, the cost of ownership of BelAZ-131 dump truck is 26% lower than the similar one of Caterpillar".

The TCO method is useful when purchasing durable materials from a mining enterprise. Conveyor belts, ventilation pipes, anchor fastening elements – these are examples of such materials. Purchase of such material and technical resources in most cases is now carried out through competitive procedures based on the minimum purchase costs.

In mining industries the use of TCO method is possible, both at open-cast mining enterprises and at enterprises with underground mining. In this regard, it can be noted that in general in 2015 in the structure of the cost of 1 tonne of coal the material costs made 44.8%, in 2014 – 42.3% [9]. Therefore, the task of optimizing the material and technical supply, including in the field of the criteria used for selecting competitive procedures, is very relevant.

At open-cast mining enterprises TCO method can be used for the purchase of fuels and lubricants. As an example, lubricants for quarry engines can be considered. Shell estimates [10] that 47% of mining companies do not believe that better lubricants can reduce maintenance costs, but 56% of companies recognize that unplanned shutdowns over the past three years have been associated with the wrong choice of lubricants.

For the correct identification and analysis of TCO it is necessary to develop a sequence of operations to be carried out. For example, in [11] the algorithm given in figure 1is offered.

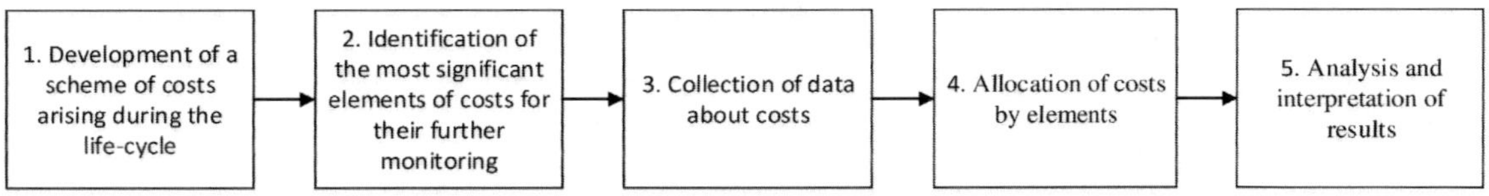

Figure 1. Process of TCO identification and analysis.

To determine the formation of costs and identify the most significant elements of costs it is necessary to study the process of receipt, use, disposal of the resource which is depicted in figure 2.

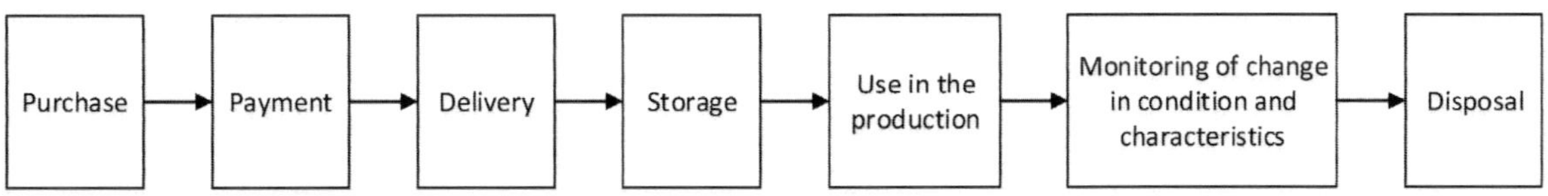

Figure 2. Life cycle of the purchased resource within the enterprise.

Thus, it is possible to form the list of costs on the basis of which TCO will be assessed. As an example, the application of TCO in the field of lubricants for mining equipment can be considered. Properly selected oils can have a significant impact on the service life of components, maintenance costs and unplanned downtimes, so they can help reduce costs more than the price of the lubricating oil itself.

The effective lubrication in the engine plays an important role in the protection of expensive equipment, since it can minimize downtime due to frequent oil changes, maintenance or even failure of individual equipment components.

In this case, the evaluation based on the following components can be performed:

- costs for selection of the required oils;
- costs for the purchase of oils, preparation of documents for the resource purchase (based on the retro information on the quality of the document circulation of the supplier in a temporal aspect and correctness);
- directly the cost of oils;
- costs for delivery and storage;
- costs for disposal;
- labor intensity and labor remuneration for oil replacement;
- cost of equipment downtimes during the oil replacement;
- cost of materials used for oils replacement;

IOP Conf. Series: Earth and Environmental Science **84** (2017) 012039 doi:10.1088/1755-1315/84/1/012039

- costs associated with the loss of quality by oil over time (increase in the fuel consumption).

As noted in [12] the main issue in the TCO estimation remains the problem of identifying and quantifying the most important components of the total cost of ownership with respect to a particular object. And this is due, first of all, to the fact that the number of indicators that might be taken into account when estimating TCO can reach several dozens.

Of course, not all these costs can and should be appropriately identified and evaluated, so the most significant of them must be used.

An analysis of the possibilities of applying TCO method has already been carried out with regard to the purchase of conveyor belt by V.V. Bushuyev [13]. The author suggests usage of the division of costs for the direct purchase of conveyor belt and operating costs. Operating costs include:

- costs for preventive maintenance;
- electricity costs;
- costs for installation and dismantling of conveyor belts;
- costs for elimination of failures consequences.

The author concludes that the total cost of ownership is a modern method for assessing the effectiveness of acquisition, ownership and operation of equipment and components and can be used to conduct competitive procedures.

However, the potential of TCO method is not limited to a conveyor belt, as noted in the table 1.

Table 1. Groups of auxiliary materials for which the use of TCO method is advisable.

Mine 1			Mine 2		
Auxiliary materials	Amount, thousand rubles	%	Auxiliary materials	Amount, thousand rubles	%
Pipes, metalware, isolation valves	12 315	5.7	Pipes, metalware, isolation valves	11 270	4.3
Roof bolts	17 822	8.3	Roof bolts	14 495	5.6
Conveyor belt	15 049	7	Conveyor belt	22 371	8.7
Wood materials	4 188	2	Wood materials	4 766	1.8
Spare parts	29 555	13.7	Spare parts	44 309	17
Metal and-metal products	10 259	4.8	Metal and-metal products	17 791	6.9
Steel support	44 910	21	Steel support	34 516	13.3
Electro-cable products	32 397	15	Electro-cable products	29 485	11.4
Fuels and lubricants	5 422	2.5	Fuels and lubricants	5 366	2
Other	43 381	20	Other	74 033	29
Total	215 298	100	Total	258 402	100

For underground mining it is possible to use TCO for the materials indicated in the table which reflects their share of costsin the composition of material costs for 2013.

Nevertheless, TCO method is critically evaluated by the author of the paper [14], which indicates that the existing methods differ in the composition and classification of the accounted cost items, methods of quantitative assessment of values of each cost element, and a number of other features. In addition, for all known approaches to TCO calculation, the following drawbacks are typical:

- existing methods do not give unambiguous recommendations on how to allocate a subset of the determining costs for specific conditions and objects;
- there are no well-founded procedures for obtaining quantitative estimates for individual cost items;
- the results of calculations for similar objects obtained by different methods for TCOestimation vary significantly;

IOP Conf. Series: Earth and Environmental Science **84** (2017) 012039 doi:10.1088/1755-1315/84/1/012039

- when using expert assessments, there are no clear and well-grounded procedures for conducting a group examination, procedures for coordinating expert opinions, which make it possible to improve the accuracy of calculations;
- does not take into account the dynamics and random nature of individual items of costs.

In addition, procurement managers do not have comprehensive information and knowledge to cover in full the characteristics and properties of all purchased auxiliary materials.

3. Conclusions

In connection with the above analysis, for the wide spread of TCO method enterprises need to solve the following tasks:

- generate a list of significant costs for each group of auxiliary materials, fixed assets purchased by coal-mining enterprises;
- create a universal algorithm for determining TCO supplementing it with specific requirements for each group of auxiliary materials;
- based on the algorithm to develop a calculation forms for rapid and accurate determination of TCO.

Creation of aworking algorithm and maintenance of the information base in an actual state will considerably simplify and increase the effectiveness of not only the activities of procurement structures and procurement procedures of the enterprise, but will also increase the efficiency of the entire enterprise.

References

[1] Chalabyan A 2003 *Bulletin of McKinsey* 4 http://vestnikmckinsey.ru/
[2] Hubaev GN and Rodina OV 2011 *Models, Methods and Software Tools for Estimating the Total Cost of Ownership of Long-lived Objects* (on the Example ofSoftware Systems) (Rostov-n/D: RSEU) p 336
[3] Karpychev V Yu 2015 *Economic Analysis: Theory and Practice* **8(407)** 25–37
[4] *TotalCostofOwnership (TCO)* https://www.techopedia.com/definition/159/total-cost-of-ownership-tco
[5] Polyansky AM 2013 *Issues of Territorial Development* **9(9)** p 2
[6] Total Cost of Ownreship 2016 NIGP Business Council: White Paperp 40
[7] *Cummins Announces Latest Total Cost of Ownership Initiatives at Minexpo* 2016 Equip. J.http://www.equipmentjournal.com
[8] Yasyuchenya SV 2012 *Mining Industry* **6** p 23
[9] TarazanovIG 2016 *Coal* **3** 58–72
[10] *Unlocking Efficiency in Mining How Lubricants Can Help Increase Productivity Reduce Total Cost of Ownership* http://www.shell.ae/
[11] MatthewsaK 2012 A Tool for Lubricant Management in the Mining Industry Thesis for Bach. Degree, UniversityofPretoria, SouthAfricap 149
[12] Motherland O In 2010 *UECS* **24** 318 –26
[13] Bushuev V V 2013 *Mining Industry* **1** p 32
[14] Rodina O V 2011 *UECS* **31** p 26

IOP Conf. Series: Earth and Environmental Science **84** (2017) 012040 doi:10.1088/1755-1315/84/1/012040

Basic tasks for improving spectral-acoustic forecasting of dynamic phenomena in coal mines

A V Shadrin and A A Kontrimas

Institute of Coal of the Federal Research Centre of Coal and Coal Chemistry, Siberian Branch of the Russian Academy of Sciences, 10 Leningradsky Prospect, Kemerovo, 650065, Russia

E-mail: avsh-357@mail.ru

Abstract. A number of tasks for improving the spectral-acoustic method for forecasting dynamic phenomena and controlling stress condition in coalmines is considered. They are: considering the influence of a gas factor on the danger indicator, dependence of a relative pressure coefficient on the distance between the source and the receiver of the probing acoustic signal, correct selection of operating frequencies, the importance of developing the techniques for defining the critical value of the outburst danger index The influence of the rock mass stress condition ahead of the preliminary opening face on the relative pressure coefficient defined for installing the sound receiver in the wall of the opening behind the opening face is also justified in the article.

1. Introduction

Spectral-acoustic method for forecasting dynamic phenomena manifestation starts to be very popular with the coal mines operating on the former USSR territory. It is widely used in Donbass for forecasting high rock pressure zones where while entering, some rock bumps, sudden coal and gas outbursts and other similar phenomena can manifest themselves in case the danger mitigating measures are not performed. At present, the equipment supporting this method is used in nearly all coal mines of "SUEK-KUZBASS" JSC. However, the method itself has a number of drawbacks and the authors introduce several issues which being solved can improve the method and subsequently increase the adequacy of the forecast.

The work aim: to analyze basic drawbacks of a spectral-acoustic method for forecasting dynamic phenomena manifestation and to propose possible ways of their elimination.

About naming the method

The analyzed forecasting method carries different names. Historically, its first name was the forecasting method based on amplitude-frequency response characteristics of a seam. The name did not come into use, as it would seem to be too long. In various sources, one can find the reference to the acoustic method. This name also failed to be used as there were plenty of acoustic methods for controlling rock mass and the name did not bear its true nature. In current regulatory documents this method is referred as "....based on the characteristics of artificial acoustic signal". As we see it, this name is overly long and is not correct in revealing the nature of the forecasting method.

We offer to name it as a method of spectral-acoustic analysis as it fully reveals the idea of this forecasting method and we believe it will be widely used both in scientific literature and in regulatory documents. This supposition is based on the idea that the spectral-acoustic method for controlling rock mass stress condition is in researching the frequency content of acoustic vibrations. The source of the vibration is the operating mining machinery effecting on the rock mass by mining combine or plough-cutting tools, drilling rig, a coal pick hummer etc. Any impact device that generates broadband acoustic signal can be the source of the vibration. The acoustic vibrations should get through the controlling rock mass area starting from the cutting unit of the operating mining machinery that produced it to the receiver installed in the wall of the opening. (It is obvious that the source of the signal is rather natural and the definition "artificial" is incorrect.)

2. The idea of the method

Primarily, this method was developed by S. Mirer's researching team. It was applied for the forecasting of coal and gas sudden outbursts [1]. This method was based on experimentally identified regularity which revealed that as soon as the opening face approaches to the outburst danger zone the amplitude of high frequency components of the noise spectrum of the operating mining machinery grew stronger than the amplitude of low frequency components. So this relation was identified as an outburst danger index. Therewith the authors of the method supposed that thus defined outburst index took into account all basic outburst factors: rock mass pressure, inter-bedding gas pressure, coal strength etc. and such outburst danger criteria based on the results obtained by analyzing the noise of the mining machinery operation on several Donbass mines was true for all world coal-mining fields.

However, the case record of Kuznetsk, Vorkuta and Karaganda coal-mining fields proved that it was not true. The analysis of acoustic wave propagation revealed that the growth of the relation of the amplitude of high frequency components of the noise spectrum to low frequency components while retaining the structure of a coal seam and the strength of its layers was caused by the increase of an average stress around the controlling rock mass zone. It can be explained as in a solid body in a first approximation the acoustic vibration decay coefficient α is linearly proportional to vibration frequency f and inversely proportional to the current average stress σ_c [2]:

$$\alpha \sim \frac{f}{\sigma_c}. \quad (1)$$

Herewith the non-associated gas pressure quantity in interstices, coal macro- and micro- cracks does not influence on the acoustic vibration decay coefficient. Thus, the danger indicator for dynamic phenomena manifestation in the form of the relation of high frequency and low frequency components of the noise is also called relative stress coefficient.

3. The tasks that may help in improving spectral-acoustic forecast of dynamic phenomena manifestation

Practical application of the spectral-acoustic method for forecasting gas-dynamic phenomena (GDP) in Donbass mines proved to be of high integrity. As it seems, in Donbass, (with deep deposits of coal seams and their low thickness) sudden outbursts start from the local rock burst and then transform into a sudden outburst. Here the stress condition factor has a greater influence on the rock mass instability than the gas factor [3]. Taking into consideration this peculiarity of a spectral acoustic method it started to be successfully applied in Kuzbass mines for forecasting dynamic phenomena manifestation and basically for rock bursts [4].

Despite the positive experience of applying the spectral-acoustic method for controlling the rock mass stress condition there are several tasks which are meant to be solved for increasing the integrity. They are:

Correcting the danger criteria of gas-dynamic phenomena manifestation defined by the application of a spectral-acoustic method. The idea is in taking into account the influence of other danger factors beyond the rock pressure such as inner-bedding gas pressure and coal strength in particular.

Considering the dependence of the danger criteria of gas-dynamic phenomena manifestation (relative stress coefficient) on the distance between the source and the receiver of the acoustic vibration.

IOP Conf. Series: Earth and Environmental Science **84** (2017) 012040 doi:10.1088/1755-1315/84/1/012040

Defining the sensitivity zone of the method while controlling the stress condition of the rock mass ahead of the preliminary opening when the receiver of the acoustic vibrations is installed in a borehole of the wall of the opening in a certain distance from the face.

Justifying the operating frequencies for implementing the method.

Updating the hardware-software complex for spectral-acoustic forecasting of gas-dynamic phenomena manifestation.

Working out the techniques for defining the critical value of the outburst danger index for a current state of the opening face.

The setting of the first task is vivid as the gas factor: coal gas content and non-associated gas pressure in cracks and interstices causing the development of sudden coal and gas outbursts and similar types of GDP influence greatly on the danger of gas-dynamic phenomena manifestation. The danger of GDP development also depends considerably on the coal strength and its humidity.

The setting of the second task is conditioned by the fact that the current relative stress coefficient value K_c defined as a relation of high frequency A_h and low-frequency A_l amplitude components of acoustic vibration spectrum decreases exponentially as far as the "noise" source distances from its receiver according to the expression [2]:

$$K_c = \frac{A_h}{A_l} = e^{-C\frac{\sigma_l}{\sigma_c}x}, \tag{2}$$

where σ_l and σ_c – respectively, the average limit and the average current values of stress in adjoining the working face zone of the rock mass, where the outburst danger situation is formed; x is the distance between the source and the receiver of the sound.

In this expression C parameter defines the sensitivity of the method and is equal to [2]:

$$C = \frac{\alpha_0 \beta (f_h - f_l)}{f_0}, \tag{3}$$

where f_h and f_l – the characteristic frequencies of upper and lower zones of the "noise" source, α_0– decay coefficient at a frequency f_0; β - proportionality factor determined by the rock mass property.

Thus, the critical value of the index is not only a constant quantity for the mines of a certain coal basin or for a singular mining seam but it varies as far as the face distances from the installed acoustic vibration receiver.

One of the possible solutions is studied in the paper [2]. However, it stipulates the adjustment of the transmitting coefficients in accordance with the frequencies (application of equalizer) as far as the opening face advances further and technically, it is rather difficult to do. It may be better to work out a convenient technical solution.

The third task is conditioned by the fact that while controlling the stress condition ahead of the preliminary opening face the source and the receiver of acoustic vibrations are placed behind the rock mass zone (see figure 1). However, even in this case the received acoustic signal bears the information about a stress condition of a rock mass ahead of the opening. It can be explained as follows:

Firstly, the acoustic wave while moving cannot be focused on the volume of space the cross-sectional effective radius of which is smaller than a half of the wavelength [5]. Therefore, a coal seam can be subjected to the radiation generated by the cutting unit of an operating mining equipment to at least this minimal depth.

We can estimate this value for spectral-acoustic method, which is realized by the equipment set (such as AK-1V) where the receivers of the sound are the geophones of CB-20 or CB-30 type. Their operating area is in the limits of 30-1500 Hz [6]. This range covers frequency content of wideband acoustic vibrations produced by a cutting unit of a mining combine. If the compressional wave (P-wave) velocity in the coal equals about 1600 meters per second then the operating wave length is

IOP Conf. Series: Earth and Environmental Science **84** (2017) 012040 doi:10.1088/1755-1315/84/1/012040

within the interval $\lambda \approx 1,1$-50 m. Therefore, if spherical sound waves propagated in straight lines from the wideband transmitter to the receiver then, depending on the frequency, the coal mass in the limits of the formed tube with the diameter over 0,55-25 m near the transmitter would influence on their decaying. The probing of the rock mass ahead of the preliminary opening face would go to that very depth. In reality, there appear a superposition of shear waves (S-wave) and compressional wave (P-waves) in the geophone operating area but it does not influence sufficiently both on the danger indicator of the dynamic phenomena manifestation and on the depth of the method's sensitivity zone ahead of the opening face.

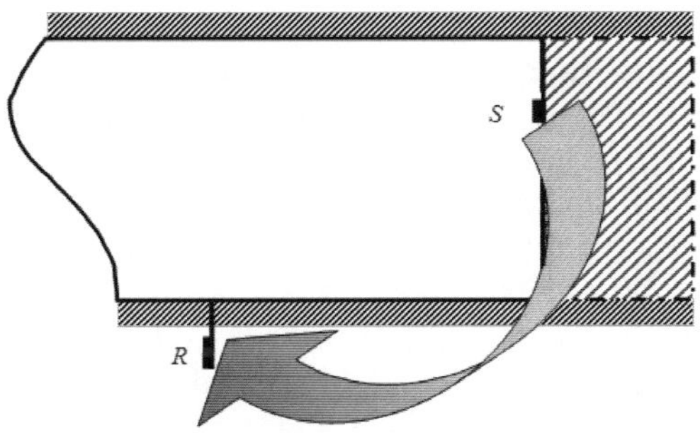

Figure 1. Scheme of the sound propagation starting from the source of the sound *S* (for example, tunneling machine) to the receiver of the signal *R* while forecasting outburst danger in the preliminary opening.

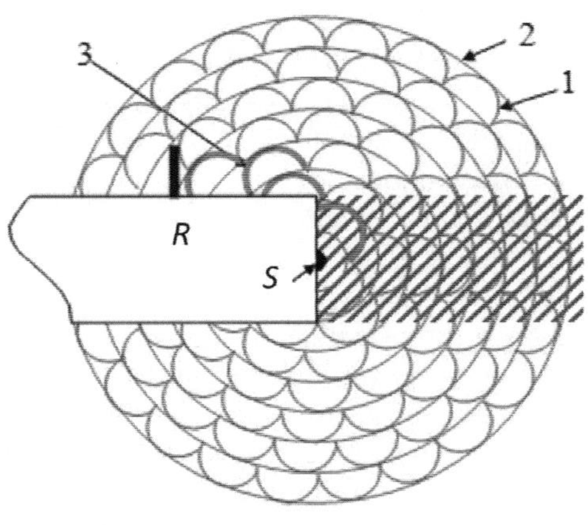

Figure 2. Scheme of a single sound harmonic propagation from the transmitter *S* to the receiver *R* according to Huygen's principle: 1- elementary spherical waves; 2 – resultant wave front; 3 – half wave propagation trajectory from the transmitter to the receiver.

However, for the scheme of the sound propagation, depicted in figure 1, the receiver is positioned in a specific acoustic shadow area. That is why the diffraction of sound in the point of changing the direction of propagation will effect on the resultant acoustic field of a receiver. A qualitative evaluation of the sound diffraction effect can be obtained through application of Huygen's principle [7].

The principle states that any wave of any shape can be considered consisting of a large number of spherical waves of similar frequency (so called elementary wave). Each of such waves is propagated from its own starting point and possesses its own initial phase and amplitude. Thereby any wave front can be considered as an envelope of all these elementary waves with their starting points in the former wave front [7]. Consequently, the scheme of a single sound harmonic propagation from the transmitter to the receiver is viewed as follows (figure 2)

Figure 2 presents a vivid idea that the farther the receiving geophone is installed into the wall of the opening and the closer to the opposite side of the wall of the opening the cutting unit of a mining combine is then the wider the zone ahead of the opening face that acoustic wave radiation covers before reaching the geophone. Thus, the width of the zone can reach 2-3 half-waves i.e. critical probing depth ahead of the opening face for this model is around 1.5 m. It is obvious that the depth of such zone for a harmonic with bigger wavelength will be even larger.

It is known that gas-dynamic phenomena as sudden coal and gas outbursts and other similar ones manifest themselves near the dislocated zones. In this case to the sound that floated from the transmitter to the receiver according to the scheme depicted in figure 2, radiation that reflected from the dislocated area of a seam is added.

Figure 3 explains the case of the reflection caused by disjunctive dislocation. The front of the reflected radiation is obtained according to geometrical acoustics principle. Regulating the line thickness it demonstrates that the farther the distance from the radiation source **S** the less the wave

IOP Conf. Series: Earth and Environmental Science **84** (2017) 012040 doi:10.1088/1755-1315/84/1/012040

amplitude. The closer the face to the dislocation the higher the amplitude of the radiation reflected from it. This radiation component goes through the face area twice before reaching the receiver (geophone) **R** installed into the borehole drilled into the wall of the opening.

Figure 4 demonstrates the scheme of formation the acoustic field around the geophone in case there is a plicative dislocation ahead of the opening.

The difference from the abovementioned case is that the reflection comes from the advanced zone of a seam in dislocated area but not from the plane interface of two structures. This peculiarity leads to broadening of the reflected signal in time.

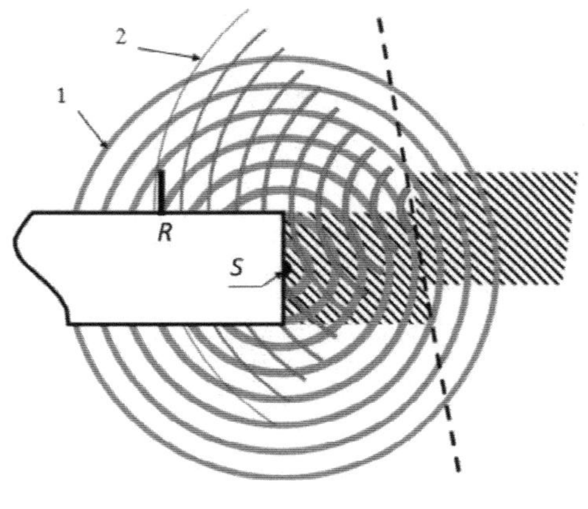

Figure 3. Scheme of a single sound harmonic propagation floated from the transmitter *S* to disjunctive dislocation and being reflected from it reached the receiver *R* according to Huygen's principle: 1 - incident wavefront; 2 – reflected wavefront.

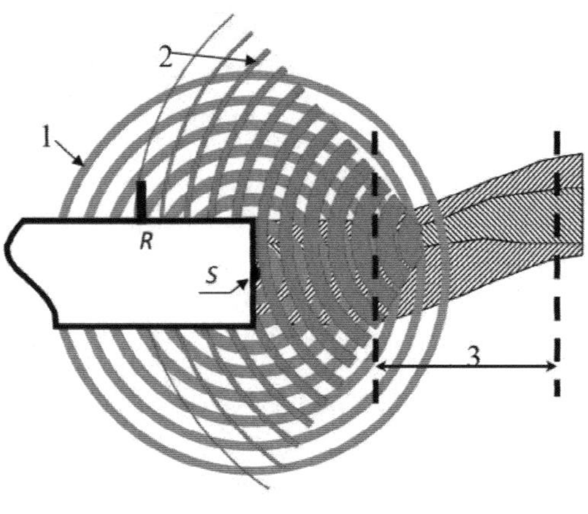

Figure 4. Scheme of formation the acoustic field around geophone in case there is a plicative dislocation ahead of the opening: 1 – incident wave; 2 – reflected wave; 3 – transition zone.

Superposition of the waves propagated according to the schemes depicted in figure 2 ("direct" wave), figure 3 or in figure 4 (the wave reflected from the dislocation) takes place in a geophone zone installation. Frequency content of the resultant signal informs about stress condition of a face area.

The importance of the forth task is in the following. On the one hand, as it is seen from the relations (2) and (3) the greater the difference (f_h-f_l) the higher the sensitivity of the method towards stress condition changes. On the other hand: the increase of the frequency f_h brings about the decrease of the sensitivity zone ahead of the preliminary opening face as the wavelength shortens in inverse proportion to the frequency (see task 3). It is necessary to keep in view that frequency content of the radiation from the cutting unit of an operating equipment depends both on the type of the equipment and on the thickness of the rock mass layers of a seam and adjacent strata. The latter is conditioned by a waveguide property, which every rock mass layer possesses. It provides minimal decaying on the frequencies with the wavelength commensurable to the layers thickness or divisible by it [8]. Frequency content interference in the measuring tract, if there any, also influences on the choice of the operating frequency band.

At present, there is no absolute answer as what operating frequency to choose. It can be explained by the fact that the known sets of stationary equipment for spectral-acoustic forecasting of dynamic phenomena manifestation AK-1M and CAKCM are produced by different companies with different principles of low and high frequency spectral range definition [4, 6].

The fifth task is set as the presently applied hardware- software complexes for forecasting dynamic and gas-dynamic phenomena by spectral-acoustic method commit a significant error in evaluating outburst danger, as first two abovementioned tasks have not been solved yet.

Another reason for errors in evaluating danger by this method is in discrepancy between basic parameters of digital signal processing (sampling frequency and, basically, sample length) and target values in some samples of the equipment meant for spectral acoustic forecasting [9].

The setting of the sixth task resulted from the acting "Instruction on forecasting dynamic phenomena". This instruction states that the critical value of the outburst danger index for spectral-acoustic method should be initially defined according to the "noise" parameters of the operating equipment in the areas of mine opening which is evaluated as " not dangerous" by other suitable methods [10]. However, the Instruction dose not say anything about the quantitative evaluation of a danger level and states that the critical value of the outburst danger index (the derivation of which is not known) should be approved by the technical director (chief engineer) of the mining company. In other words, every mine is offered to work out its own techniques for defining the critical value of the outburst danger index.

The situation with arranging the forecasting of outburst danger using this method results from the lack of scientific background justification in particular, from the failure to take into account all basic outburst danger factors and rock mass parameters. The solution is in the development of the technique which can help to set a geophysical danger parameter for a certain opening face according to the instrumental criterion which is based on measuring the parameters that define the danger of GDP manifestation while test borehole drilling [11].

4. Conclusion

At present automated method for spectral-acoustic forecasting of dynamic phenomena manifestation, for sudden gas and coal outbursts in particular, does not have any valid scientific background justification and cannot compete with the instrumental method of current forecasting outburst danger based on the initial gas emission velocity and drill fines outlet during test borehole drilling. The solving of the abovementioned tasks will heighten the validity of the forecasting made by the automated method to the value that the instrumental method provides.

Acknowledgements

The research was carried out due to the grant of Russian Science Foundation (project No.17-17-01143).

References

[1] Copyright Cert. on the Invention of the USSA # 1222853 1986 kl. E 21 F 5/00, E 21 C 39/00.
[2] Shadrin A V and Konovalenko V A 2001 *Bulletin of KuzSTU* **3** 28–31
[3] Korol V I and Skobenko A V 2013 *Acoustic Way of the Gas-dynamic Phenomena Forecast in Coal Mines: Monography* (Dnepropetrovsk: National Mining University) p 181
[4] Kopylov K N, Smirnov O V and Kulik A I 2015 *Mining Informational and Analytical Bulletin* **2** 82–8
[5] Savich A I 1971 *On the "Capture" Zones of Elastic Waves Works of the Hydroproject* No. **21** pp 29–40
[6] Mirer S V, Khmara O I and Shadrin A V 1999 *Spectral-acoustic Forecast of Coal Seams Outburst Danger* (Kemerovo: Kuzbassvuz izdat) p 92
[7] Krautkremer J and Krautkremer H 1991 *Ultrasonic Control of Materials: the Reference work* (Moscow: Metallurgy) p 752
[8] Brehovskih L M 1973 *Waves in the Layered Environments* (Moscow: Nauka) p 343
[9] Shadrin A V, Bireva Yu A, Ermolaev A M and Fomin A I 2016 *Occupational Safety in Industry* **7** 77–81
[10] Federal Norms and Rules in the Field of Industrial Security "Instruction on Forecasting the Dynamic Phenomena and Monitoring Rock Mass while Working on Coal Deposits" from August 15, 2016 No. 339 p 129
[11] Shadrin A V 2016 et al *IOP Conference Series: Earth and Environmental Science* **45** 012008

Mathematical modeling of methane migration into the mine workings during the face downtime

Yu M Govorukhin[1], A N Domrachev[2], V G Krivopalov[2] and D Yu Paleev[1]

[1]National Mine Rescue Center, 5 Gornospasatelnaya Street, Novokuznetsk, 654028, Russia
[2]Siberian State Industrial University, 42 Kirova Street, Novokuznetsk, 654007, Russia

E-mail: Govorukhin_YM@mail.ru

Abstract. For the estimation of safe distances during explosions of mixtures of coal dust, methane, and air in the process of emergency rescue operations in coal mines, it is necessary to determine the gas volumes in the mine workings. Errors in determining such volumes often lead to tragic consequences. The calculation schemes are suggested that allow the methane generation rate into the mine air to be determined on the basis of physical regularities (mine and gas pressures, gas permeability dynamics, depth of the gas drainage zone, etc.), underlying the processes of gas migration from coal and rocks into the mine workings. The following methane emission sources are considered at the site: the surface of the stopped face; walls of development opening; the gob (potential volume of the gas reservoir in the caving area). Test calculations of methane generation have been performed based on the mining, geological and technological data of one of the mines in Baydaevsky geological and economic region. In general, the results obtained are consistent with the data of long-term empirical observations. The directions of further research aimed at improving the synthesized methodology are presented.

1. Introduction

In the course of conducting mine rescue operations in mines, there is a need to determine the safe distances during explosions of coal dust methane-air mixtures. To do this, the specialized software products are used, which require the gas volume in the workings to be specified as initial data. The definition of such volumes during mine rescue operations, as well as during face standbys, depends on many factors: mining and geological parameters of the seams; schemes of opening and preparation of a mine field; mining technology; type and place of the accident, etc. Methods of their determination are very different and depend on the specific conditions. Errors in the definition lead to tragic consequences.

To estimate the gas volumes in the workings, the calculation of the total methane generation for a certain period of time is necessary. The calculation schemes given in the current normative and technical documents on design of ventilation in mines and working areas are based on the estimation of relative gas content by difference between the natural and residual gas content, as well as on the data about the actual methane release obtained during development of extraction pillar by longwalls-analogs. Absolute methane content in these cases is determined only depending on the planned daily load on the face and the speed of excavation, i.e. in the conditions of normal operation of the mining enterprise. In addition, the physical laws that underlie the processes of gas transfer from coal to mining are not taken into account. In connection with this the actual task is to develop methods for estimation

IOP Conf. Series: Earth and Environmental Science **84** (2017) 012041 doi:10.1088/1755-1315/84/1/012041

of the absolute methane content in the working area and its dynamics during emergency stops of faces. The performance of numerical modeling of geomechanical processes raises the accuracy of the estimate of absolute methane content in the working area [1 - 7].

Methane generation rate from satellite strata, coal loss and from the coal that can be located on the transport chain of the site is not considered due to certain difficulties in its determination. Further, the calculations are made for the conditions of the working area of one of the mines in Baydaevsky geological and economic region.

During emergency stops the following sources of methane emissions in the area are proposed to consider: the gob (potential volume of the gas reservoir); surface of the stopped face; walls of the prepared workings.

2. Methods of research

Assessment of the methane yield from free faces

To estimate the methane yield from all seam exposed surfaces, according to our understanding of the problem the mathematical model of B.G. Tarasova – V.A. Kolmakova is preferred [8 - 10]. It allows the change of gas content in time, depending on the rock and gas pressures to be taken into account, as well as gas permeability of the massif.

This model is based on the potential function of L.S. Leybenzon $F(H)$, which takes into account the conditions of gas filtration depending on the massif permeability $k(H)$, medium density $\rho(H)$, gas pressure and the depth of operations carried out H.

$$F(H) = \frac{1}{\mu}\int k(H)\rho(H)dH, \quad (1)$$

where μ – Dynamic viscosity of methane, kgf·s/m^2.

The magnitude and nature of gas release from the massif are determined by the following parameters: the mass velocity potential or a complex of variable magnitudes reflecting the state of the massif and gas (gas pressure, permeability, porosity, density and gas viscosity); speed of the face advance; intensity of the internal source of gas release; state of the gas bearing medium.

Mathematical model – differential equation of gas motion in an massif, containing an internal source of gas release, has the following form

$$C_1(x_i,t)\frac{\partial F}{\partial t} = \sum_{i=1}^{3}\left\{\frac{\partial}{\partial x_i}\left[a(x_i,t)\frac{\partial F}{\partial x_i}\right]\right\} + \sum_{i=1}^{3}\left\{v_{\mathrm{f}}(x_i,t)\frac{\partial F}{\partial x_i}\right\} - B(x_i,t)F + q(x_i,t), \quad (2)$$

where x_i – the distance into the depth of the massif; t – time; a – coefficient that takes into account free and sorbed gas; v_{f} – rate of face advance; q – intensity of the internal source of methane emission; C_1, B – coefficients.

Depending on the values of parameters, equation (2) can be applied to various special cases. With $C_1 = 1$ $v_{\mathrm{f}} = 0$, $B = 0$, $q \neq 0$ it allows us to calculate it from the coal massif with the stopped face.

To estimate the gas emissions from the developed seam, it is also necessary to estimate the change in depth of unloading zones $l(t)_i$, m, and gas drainage $l_{\mathrm{dr}.i}$, m, due to the influence of mining operations [8, 10]. The dynamics calculation of the unloading zone depth takes into account the changes in the rheological properties of the massif over time (figure 1), which are estimated by the coefficient $k_{t.i}$ according to the formulas:

- for the face

$$k_{t.i} = -\frac{\alpha_{\mathrm{r}}(t_0 + t_i)}{1440}, \quad (3)$$

- for development workings

$$k_{t.i} = -\alpha_{\mathrm{r}}\left(\frac{1}{2}t_{\mathrm{b}} + t_{\mathrm{sh}} + \frac{t_0 + t_i}{1440}\right), \quad (4)$$

α_r – rheological coefficient; t_b – time from beginning of tunneling up to its end, days; t_{sh} – time elapsed between the moment of finishing the development working and the face shutdown, days; t_0 – time elapsed between the moment of seam exposure (face shutdown) and the moment of emergency, min; t_i – time elapsed from the moment of emergency (value t_i changes to t_n), min.

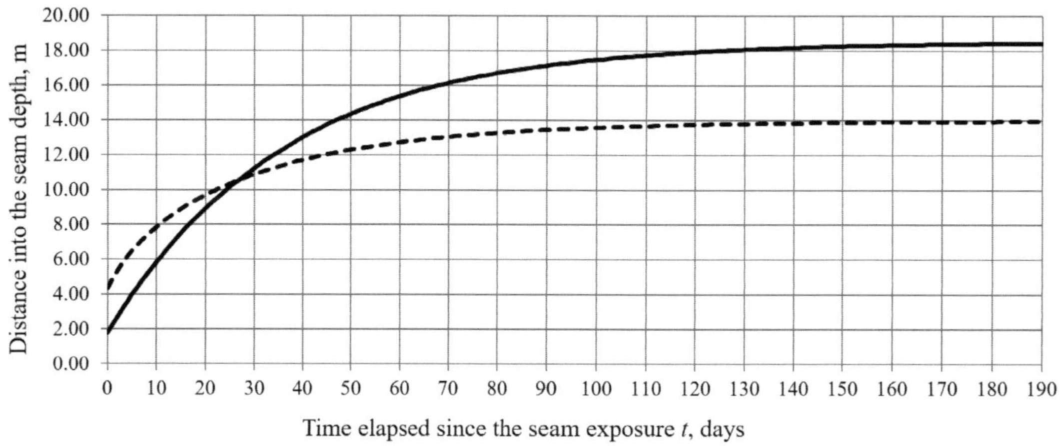

Figure1. Dynamics of the unloading zone depth $l(t)_i$ (dashed line) and gas drainage $l_{dr.i}$ (solid line) of the seam for the period from $t_0 = 0$ to t_{st} (time of stabilization of stress-strain state).

To calculate the gas emission from the seam, it is necessary to determine the initial gas pressure and the regularity of its change under the influence of mining operations. Based on the assumption that the unmined coal-bearing strata is a gas reservoir of free methane, the depth pressure P_0, kgf/cm^2, is distributed according to hydrostatic dependence [8, 10]

$$P_0 = u(H - H_0)^v + P_2, \tag{5}$$

where u, v – empirical coefficients reflecting the influence of natural factors such as gas permeability of coal and rocks, angle of strata inclination; H, H_0 – depth, respectively, of mining works and the zone of degassing, m; P_2 – gas pressure at the upper boundary of the methane zone, kgf/cm^2.

With approximate calculations the gas pressure in the unmined massif P_0, kgf/cm^2, can be estimated from the general formula [11]

$$P_0 = 0.1(H - H_0). \tag{6}$$

The gas pressure at any distance deep into the seam P_x, kgf/cm^2, at each moment of time t ($0 \leq t \leq t_{st}$) from its detection and to stabilization of the stress-strain state is calculated by the formula

$$P_x = (P_1 - P_0)e^{-\frac{x^2}{2l^2(t)}} + P_0, \tag{7}$$

where P_1 – gas pressure on the drain surface, kgs/cm^2 (due to incomparability of gas pressures in the mine working and in the massif, it is assumed that at the drain boundary the pressure equals to atmospheric pressure $P_1 = P_{atm} = 1$ kgf/cm^2); P_0 – gas pressure in the unmined massif, kgf/cm^2; x – distance deep into the massif, m.

The specific gas release from the unit area of the seam immovable surface during the formation of unloading zone $G_{t.i}$, m^3/(m^2·min), is determined by formulas [8-10]:

- for stoping

$$G_{t.i} = \frac{G_0}{\sqrt{\frac{t_0+t_i}{1440}+1}}; \tag{8}$$

- for development workings

$$G_{t.i} = \frac{G_0}{\sqrt{\frac{1}{2}t_{\text{пр}} + t_{\text{ост}} + \frac{t_0+t_i}{1440}+1}}. \tag{9}$$

where G_0 – initial gas release intensity G_0, m3/(m2·min), at the time of seam exposure, is determined depending on the difference in potential functions of gas release $\left(F_{l(t)} - F_1\right)_s$.

Gas evolution in the face, taking into account the drainage effect of the development workings $I_{\text{f}.i}$, m^3/min, at the moment of time $t_0 + t_i$, min, is determined by the formula

$$I_{\text{f}.i} = G_{t.i} m_{\text{m}} \left[l_{\text{f}} - \frac{1}{2}(l_{\text{dr}}^{\text{c}} + l_{\text{dr}}^{\text{v}})\right], \tag{10}$$

where m_{m} – mined thickness of a seam, m; l_{f} – face length, m; $l_{\text{dr}}^{\text{c}}, l_{\text{dr}}^{\text{v}}$ – depths of the zone of seam drainage, respectively, by conveyor and ventilation drifts, m.

Gas release into the development working from the seam $I_{\text{d.w.}}$, m^3/min, at the moment of time $t_0 + t_i$, is determined by the formula

$$I_{\text{d.w.}i} = 2G_{t.\text{d.w.}i} a_{\text{ex}} l_{\text{d.w.}}, \tag{11}$$

where $G_{t.\text{d.w.}i}$ – specific gas release from a unit of the area of the immovable seam surface in the development workings, m^3/(m^2·min); a_{ex} – seam exposure, m; $l_{\text{d.w.}}$ – length of the development working, m.

Figure 2 gives an example of calculation of gas release from the exposed seam surface (face). The step of calculation corresponds to days.

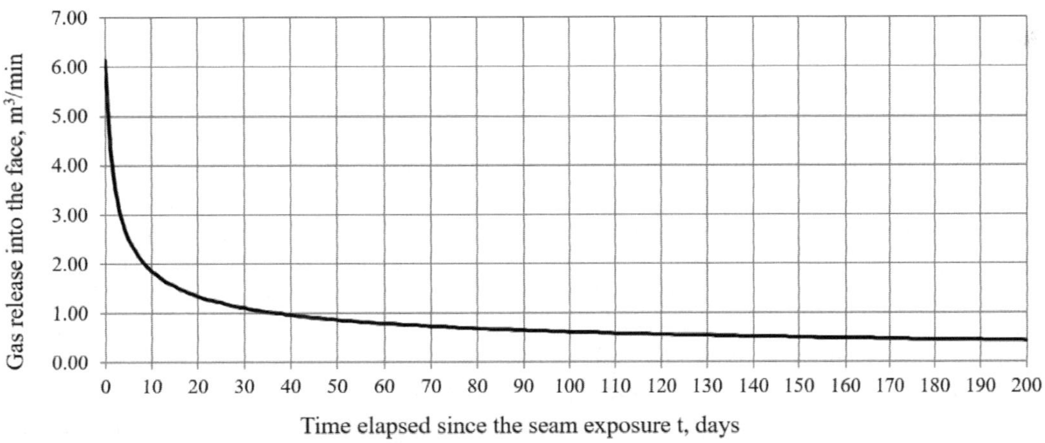

Figure 2. Dynamics of change in methane generation rate from the developed seam (face) during the period from $\boldsymbol{t_0 = 0}$ till $\boldsymbol{t_{\text{st}}}$.

Estimation of the volume of gas reservoir in the caving zone

The gob is an active component of aerogasdynamic system of working areas characterized as a medium with high aerodynamic permeability (by 2-4 times higher than in the unmined massif) [1, 2]. In the dome of caving and areas not ventilated by air leaks, as a result of gas mixture release from the coal-bearing seam its considerable volumes can be accumulated – a gas reservoir is formed. In the certain conditions the gas mixture can be squeezed into the workings. In this regard, it is necessary to evaluate the possible volumes of the gas reservoir.

To determine the formed volumes in the gob of the gas reservoir and determine the magnitude of free lowering of the main roof rocks h_{fr}, m, it is necessary to estimate the rocks fragmentation during caving. For shallow seams, the maximum fragmentation index $k_{\text{frag.max}}$ can be found from table 1.

Table 1. Maximum $k_{\text{frag.max}}$ for roof caving [2].

Seam thickness, m	Values $k_{\text{frag.max}}$
from 1.2 to 2.0	1.8
from 2.0 to 3.0	2.0
from 3.0 to 3.5 and more	2.1

As a criterion for determining the boundary between the caving zone and the zone of cracks and faults, a fragmentation index 1.1 is chosen. With a smooth reduction of k_{frag} with a decrease in the caving height of the next sublayer, it can be calculated from the linear dependence (table 2).

Table 2. Distribution of k_{frag} in the caving zone [2].

Relation of a free space to the thickness of a caved layer $h_i/m_{\text{c.l.}}$*	Value of k_{frag}		
	$k_{\text{frag.max}} = 1.80$	$k_{\text{frag.max}} = 2.00$	$k_{\text{frag.max}} = 2.10$
2.30 and more	1.80	2.00	2.10
from 1.50 to 2.29	1.65	1.80	1.90
from 0.75 to 1.49	1.40	1.50	1.50
from 0.50 to 0.74	1.20	1.30	1.30
from 0.20 to 0.49	1.15	1.15	1.20
to 0.19	1.10	1.10	1.10

*$m_{\text{c.l.}}$ – thickness of the caved sublayer of the immediate roof, m; h_i – free space after caving of the next sublayer, m.

Free space after caving of sublayers of immediate roof rocks h_i, m, can be determined by the formula

$$h_i = h_{i-1} - m_{\text{c.l.}i}(k_{\text{l.}i} - 1), \tag{12}$$

where h_{i-1} – free space remained after the caving of the previous layer, m, for the first sublayer is assumed equal to the extracted seam thickness; $m_{\text{c.l.}i}$ – thickness of the caved sublayer (chosen according to the frequency of cracks in the roof rocks), m; $k_{\text{l.}i}$ – coefficient of loosening for the caving sublayer.

Calculation of free space is performed before the layer of rock-bridge. If this layer turned out to be cleavaged, the crash of its consoles will take place without any dynamic phenomena. If the height h_i (h_{fr}) is not equal to zero, during caving of the rock-bridge layer gas can be transferred to the working site from the gas reservoir. In this case, its volume $V_{g.r.}$, m^3, can approximately be estimated by formula

$$V_{\text{g.c.}} = 0.5 h_{\text{fr}} l_{\text{w}} l_{\text{f}}, \tag{13}$$

where l_{w} – face withdrawal since the last caving of the rock-bridge, m; l_{f} – длина очистного забоя, м. length of the face, m.

3. Results and discussion

According to the methods for estimating the volume of methane release from the developed seam and the potential volume of the gas reservoir a test calculation was performed (the scheme of the working area for which the calculation is performed is shown in figure 3):

- time elapsed since the disturbance of airing $t_i = 97$ min;
- the volume of methane released in:
  face $V_{\text{f.}} = 572$ m^3;
  belt road $V_{\text{b.r.}} = 506$ m^3;
  wind roadway $V_{\text{w.r.}} = 631$ m^3;
- approximate volume of the gas reservoir (it is assumed that the length of the suspended console is close to the caving step) $V_{\text{g.c.}} = 518$ m^3.

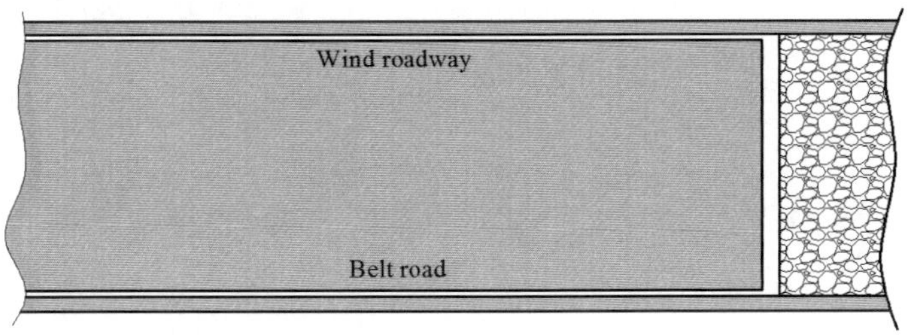

Figure 3. Scheme of the working area for which the test calculation was performed.

Currently, work is being done in order to improve this methodology, including the accounting of methane-air mixtures filtration in the gob. The method is being developed on the basis of account of interacting geomechanical and gas dynamic processes during mining operations for specific mining, geological and technological conditions (figure 4). The planned final result is a problem-oriented software package that allows to quickly the tasks in the field of gas dynamics to be solved quickly.

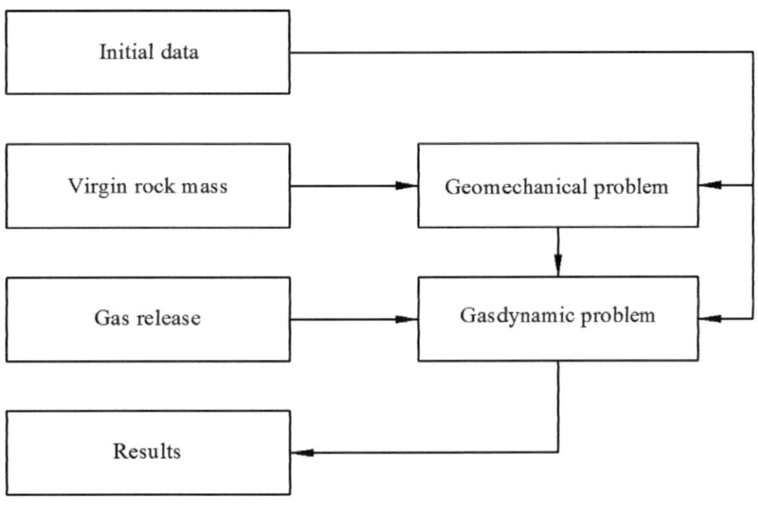

Figure 4. Conceptual scheme for solving the problem of modeling the gas release processes into the mine workings.

4. Conclusion

The results of the research presented in this paper are only a part of a very complex gas-dynamical problem. For further work, the main tasks are determination of gob and seam permeability, as well as the initial gas pressure in the massif. In general, the results of the studies (figure 2, 3) correspond to the data of long-term empirical observations. The developed technique allows the volumes of released methane for any time interval to be determined taking into account the depletion of sources and potential volumes of gas reservoir. The implementation of the results will increase the calculation accuracy of gas volumes in the mine workings and will contribute to the safe performance of mine rescue works.

Acknowledgements

The research was carried out in the research department of National Mine Rescue Center in pursuance of paragraph 6 of the Protocol of Meeting at the Deputy Chairman of the Government of the Russian Federation A.V. Dvorkovich from March 16, 2016 No. AD - P9-39pr.

References

[1] Govorukhin Yu M, Domrachev A N, Krivolapov V G, Paleev D Yu and Balagansky B Yu 2011 *Bulletin of the Sci. Center on the Safety of Works in Mining Industry* **1** 72–76

[2] Govorukhin Yu M 2012 *Extended Abstract of Cand. Sci. (Eng.) Dissertation* (Kemerovo: SC VostNII) p 21

[3] Guidelines for the Prediction and Control of Methane Emissions in Longwalls. National Institute for Occupational Safety and Health, Pittsburgh Research Laboratory 2008 http://www.cdc.gov/niosh/

[4] Whittles D N, Lowndes I S, Kingman S W, Yates C and Jobling S 2005 *Int. J. Rock Mechanics and Mining Sciences* **43** 369–87

[5] Rib S V and Domrachev A N 2015 *Proc. All-Russian Xth Conf. on Automation Systems in Education, Science and Production* pp 465–9

[6] Esterhuizen G S and Karacan C O 2007 *Proc. SME Annual Meeting 2007 and CMA 109th National Western Conf. 2007: The Power of Mining: Energy's Influence* (Denver, Colorado) http://www.cdc.gov/niosh/

[7] McPherson M J 1993 *Subsurface Ventilation and Environmental Engineering* (Netherlands: Springer) p 905

[8] Kolmakov V A 1981 *Methane Evolution and the Fight Against it in the Mines* (Moscow: Nedra) p 134

[9] Tarasov B G 1973 *Estimate of Gas Content of Mine Workings and Degassing of Mines* (Moscow: Nedra) p 208

[10] Tarasov B G and Kolmakov V A 1978 *Gas Barrier of Coal Mines* (Moscow: Nedra) p 200

[11] Petrosyan A E 1975 *Emission of Methane in Coal Mines* (Moscow: Nedra) p 188

IOP Conf. Series: Earth and Environmental Science **84** (2017) 012042 doi:10.1088/1755-1315/84/1/012042

Seismic safety in conducting large-scale blasts

I V Mashukov[1], V V Chaplygin[1], V P Domanov[2], A A Semin[1] and M A Klimkin[1]

[1]Siberian State Industrial University, 42 Kirova Street, Novokuznetsk, 654007, Russia
[2]Laboratory for Safety of Explosive Work, Scientific Center "VostNII", 3 Institutskaya Street, Kemerovo, 650002, Russia

E-mail: mashukov_nvk@mail.ru

Abstract. In mining enterprises to prepare hard rocks for excavation a drilling and blasting method is used. With the approach of mining operations to settlements the negative effect of large-scale blasts increases. To assess the level of seismic impact of large-scale blasts the scientific staff of Siberian State Industrial University carried out expertise for coal mines and iron ore enterprises. Determination of the magnitude of surface seismic vibrations caused by mass explosions was performed using seismic receivers, an analog-digital converter with recording on a laptop. The registration results of surface seismic vibrations during production of more than 280 large-scale blasts at 17 mining enterprises in 22 settlements are presented. The maximum velocity values of the Earth's surface vibrations are determined. The safety evaluation of seismic effect was carried out according to the permissible value of vibration velocity. For cases with exceedance of permissible values recommendations were developed to reduce the level of seismic impact.

1. Introduction

Kuzbass is one of the largest coal basins in the country. In 2016 in Kuzbass 209 million tonnes of coal were produced. This is more than 56 % of Russia's total coal and 80 % of coking coal. In Kuzbass 70% of coal is extracted in 80 open pit mines. For overburden operations, while preparing hard rocks for excavation, a drilling and blasting method with large-scale blasts is used. Holes are drilled and charged with explosive. The charge in large-scale blasts reaches 50, 100 and 300 tonnes. The main negative manifestations of large-scale blasts are: a shock air wave, scattering of rock pieces, seismic vibrations and impact on the protected objects, release of toxic gases, dust formation.

To improve the quality and safety of explosion operations in the mining industry, enterprises specializing in explosion operations have been established in Russia and Kuzbass: "Kuzbassvzryvprom" OJSC, "Vzryvprom Yuga Kuzbassa" OJSC, "Plant Znamya" OJSC, "Nitro-Siberia", "Azot-Chernigovets" and many others. With the increase in the number of coal mines in Kuzbass and the approach of mining operations to settlements, the negative effect of large-scale blasts is increasing. Residents of settlements that are in close proximity to mining enterprises began to complain about the impact of large-scale blasts on vibrations of houses and sound effects.

To assess the level of seismic impact of large-scale blasts on the protected buildings SibSIU together with the expert organization "Scientific Center of VostNII for Safety in the Mining Industry" OJSC (Russia, Kemerovo) performed expertise for coal mines and iron ore enterprises. Thus, seismic vibrations were registered during the production of more than 280 large-scale blasts at 17 mining enterprises in 22 settlements.

By the request of open pit mines, Open Pit Mine "Bungursky-Severny" OJSC, Open Pit Mine "Berezovsky" OJSC, Open Pit Mine "Korchakolsky" OJSC, UK "Kuzbassrazrezugol", "Taldinsky

IOP Conf. Series: Earth and Environmental Science **84** (2017) 012042 doi:10.1088/1755-1315/84/1/012042

Open Pit Mine", Open Pit Mine "Stepanovskiy" OJSC, Open Pit Mine "Yuzhny" OJSC, "Open Pit Mine Tomusinsky" the seismic vibrations of the Earth' surface from large-scale blasts conducted at the enterprises were recorded in the villages of Rassvet, Malinovka, Uspenka, Uchul, Gavrilovka, Novomoskovka, Yasnaya Polyana, Maganak, Matyushino, Pritomsky. Seismic vibrations were recorded at two observation points. They were located at a distance from 300 m to 12 km. The total mass of explosives for a blast was from 3 to 340 tonnes.

2. Methods of research

The generally accepted parameter for estimating the magnitude of seismic action is the speed of soil particles vibrations. The amplitude of vibration speed varies in wide range and depends on the mass of explosive, the distance between the place of blust and the point of registration, design of the charge, scheme of explosion, geological conditions, the direction of the breakage, the presence of free surfaces and stemming.

The problem of negative manifestation of seismic effect caused by the production of large-scale blasts is known for a long time and is relevant to the present day. There is a lot of scientific research on this topic [1 - 3], but, in general, they target the specific conditions of a particular enterprise.

The speed magnitude of seismic vibrations caused by large-scale blasts was determined using seismic receivers SM-3KV, an analog-digital converter (ADC) model E-440 with an option of recording signals to the hard disk of a laptop. The method of recording and processing was developed by the Institute of Dynamics of the Geospheres of the Russian Academy of Sciences [4].

Calculation of safe distances for buildings and structures is carried out according to the formulas given in the Safety Rules for Explosion Works [5]. The calculation is carried out taking into account the characteristics of the ground in the base of the object, its design features, the mass of explosives in the blast and the number of groups of simultaneously exploded charges.

The permissible speed of soil vibrations at the base of protected structures is regulated by the normative document RTM 36.22.91 [6]. The permissible speed of soil vibrations is determined depending on the class of responsibility of the protected structures, the type of construction of the facilities, their technical condition and soil properties at the base of the protected structures. The classes of responsibility for buildings and structures are regulated by the normative document of SNiP 2.01.07-85 [7]. The soil group is determined according to GOST 25100-95 [8].

3. Results and discussion

The performed analysis of measurements results of seismic impact shows that in most cases, when large-scale blasts were conducted, the vibrations of the Earth's surface did not exceed the permissible values (by ten or more times) [9 - 11].

However, there were abnormal manifestations of this impact, for example, in the residential sector of town Pritomsky in Mezhdurechensk and in Tashtagol. The territory with the protected buildings and facilities of the village is located at a distance less than one kilometer from the hazardous production enterprise – the mining take of Open Pit Mine "Tomusinsky" OJSC.

Figure 1 shows the speed seismogram of surface seismic vibrations in the settlement Pritomsky at the address 39 Lunacharskogo Street from the large-scale blast on September 7, 2010 in the open pit mine Tomusinsky from explosive charges 252 801 kg. The speed magnitude of seismic vibration exceeded the selected scale of measurements and by the extrapolation method is 3 cm/s, with a permissible value of 2 cm/s.

Figure 2 shows the speed seismogram of surface seismic vibrations from a large-scale blast on April 02, 2017, with a charge of explosive mass 178 000 kg for the collapse of the eastern section of block No. 6 on the level (-280) ... (-210) m at the mine in Tashtagol of Mountain Shoriya brunch of Evrazruda OJSC in the observation point School No. 1. The maximum values of seismic vibrations speed of the Earth's surface in the observation point at School No. 1 were 1.7; 1.6 and 1.9 cm/s and the horizontal component perpendicular to the direction of the mass large-scale blast exceeded the maximum permissible value of 1 cm/s by 1.9 times.

IOP Conf. Series: Earth and Environmental Science **84** (2017) 012042 doi:10.1088/1755-1315/84/1/012042

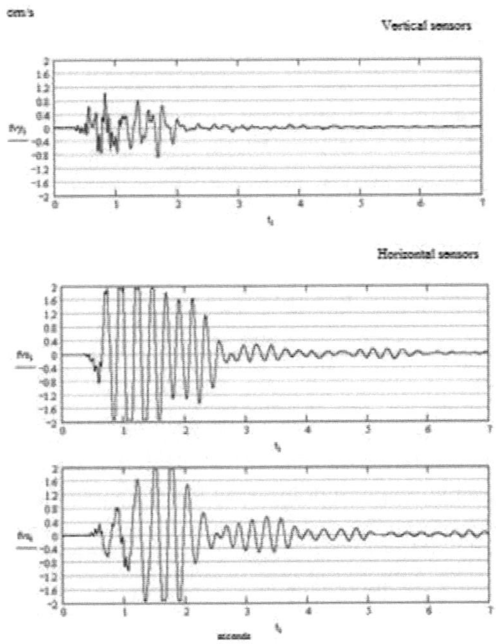

Figure 1. Speed seismogram of surface seismic vibrations in the settlement Pritomsky at the address 39 Lunacharsky Street.

To reduce the level of seismic impact it is recommended to reduce the blast charge, in the delay, to increase the delay time and the number of delays.

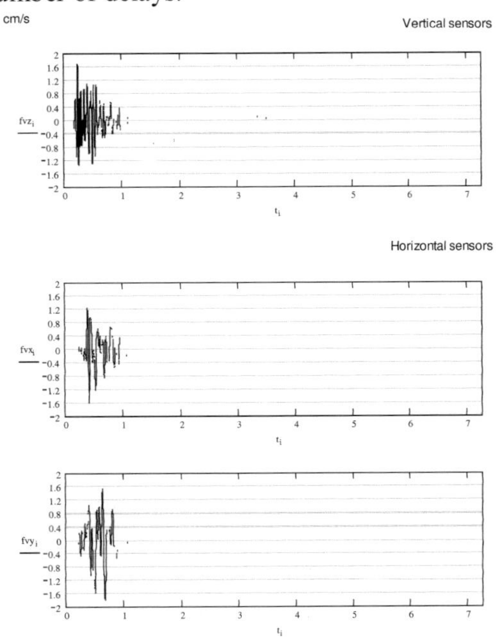

Figure 2. Speed seismogram of surface seismic vibrations caused by the large-scale blast on April 02, 2017 in the observation point School No. 1.

4. Conclusions

IOP Conf. Series: Earth and Environmental Science **84** (2017) 012042 doi:10.1088/1755-1315/84/1/012042

In the period from 2010 to 2017 450 registrations of seismic vibrations were performed during the production of 281 large-scale blasts at 16 mining enterprises in 22 settlements. The distance to the registration points was from 500 to 12 km. The mass of explosives varied from 3 to 340 tonnes. The speed maximum values of seismic vibration of the Earth's surface, basically, did not exceed the maximum permissible values. The general regularities of decrease in the vibrations speed with increase in the distances to protected objects are confirmed. In four cases, exceedance of the maximum permissible values was recorded (Pritomsky, Tashtagol). On these cases recommendations were given.

References

[1] Bogatsky V F and Pergament V Kh 1978 *Seismic Safety during Blasting Operations* (Moscow: Nedra) 1978 p 128
[2] Kutuzov B N 2009 *Safety of Blasting in Mining and Industry* (Mosow: Gornaya Kniga, MSMU) p 670
[3] Mironov P S et al 1972 *Development of the Theory and Practice of Explosives* **71/28**
[4] Eremenko A A et al 2005 *Bulletin of the RANS, West-Siberian Branch* **7** 148–58
[5] Safety Rules for Blasting Approved by the Order of Rostekhnadzor No. 605 of December 16 2013
[6] Glozman L M et al 1982 *RTM 36.22.91* (Moscow: Nedra)
[7] SNiP 2.01.07-85. *Loads and Impacts*
[8] State Standard 25100-95. *Soils. Classification*
[9] Mashukov I V 2014 *Information-analytical Bulletin on Mining* **4** 216–21
[10] Domanov V P and Mashukov I V 2013 *Bulletin of Scientific Center for Safety in the Coal Industry* **1-1** 60–4
[11] Mashukov I V et al 2013 *Bulletin of Scientific Center for Safety in the Coal Industry* **1-2** 16–21

IOP Conf. Series: Earth and Environmental Science **84** (2017) 012043 doi:10.1088/1755-1315/84/1/012043

Processing of converter sludges on the basis of thermal-oxidative coking with coals

S N Kuznetsov, M B Shkoller, E V Protopopov, S A Kazimirov and M V Temlyantsev

Siberian State Industrial University, 42 Kirova Street, Novokuznetsk, 654007, Russia

E-mail: uchebn_otdel@sibsiu.ru

Abstract. The paper deals with the solution of an important problem related to the recycling of converter sludge. High moisture and fine fractional composition of waste causes the application of their deep dehydration and lumping. To reduce environmental emissions the non-thermal method of dehydration is considered – adsorption-contact drying. As a sorbent, the pyrolysis product of coals from the Kansko-Achinsky basin – brown coal semi-coke (BSC) obtained by the technology "Thermokoks". Experimental data on the dehydration of high-moisture wastes with the help of BSC showed high efficiency of the selected material. The lumping of the dried converter dust was carried out by thermo-chemical coking with coals of grades GZh (gas fat coal) and Zh (fat coal). As a result, an iron-containing product was obtained – ferrocoke, which is characterized by almost complete reduction of iron oxides, as well as zinc transition into a vapor state, and is removed with gaseous process products. Based on the results of the experimental data a process basic diagram of the utilization of converter sludge to produce ferrocoke was, which can be effectively used in various metallurgical aggregates, for example, blast furnaces, converters and electric arc furnaces. In the basic technological scheme heat generated by ferrocoke cooling and the energy of the combustion products after the separation of zinc in the gas turbine plant will be used.

1. Introduction

In the converter production of steel, depending on the technology used and the design of the unit, from 12 to 25 kg of dust per 1 ton of steel is formed. After dust passes through the wet gas cleaning, high-moisture sludge is formed. Converter sludges are a valuable iron-containing technogenic raw material which contains up to 57 – 65 % of iron, the main part is represented by Fe_2O_3 oxide. According to the R&D Company "Energostal" the degree of cleaning of waste gases from dust exceeds 80 %, and the degree of utilization is 72 % [1, 2].

Currently, the recycling of most types of disperse metallurgical wastes is carried out by using them in the composition of agglomeration burden. However, for technological reasons the application of some of them is limited or impossible. In addition, at the sinter plants without a stockhall the use of technogenic raw materials of different composition without the possibility of its controlled dosing negatively affects the stability of the agglomerate composition and quality [3]. In terms of recycling of iron-containing dispersed wastes the process of their lumping might serve as an alternative.

The use of composite charge materials includes the production of ferrocoke and ferrocoke briquettes [4 - 6]. The concept of ferrocoke production was developed in the 30's of the last century and aimed at sintering iron ore dust unsuitable for smelting in blast furnaces with fatty or bituminous coal in coke oven batteries. Ferrocoke can be classified as an iron-carbon composition treated by heat outside the melting unit. Ferrocoke is a composition material containing mainly reduced iron and carbon. However, before ferrocoke production iron-containing dusts and sludge from gas-cleaning units must be dehydrated in advance.

A number of methods for dehydrating sludges are known consisting, as a rule, of cumbersome energy-intensive technological schemes involving preliminary removal of moisture (up to <20-25 %) by mechanical methods (thickening and filtration) and then followed by thermal drying [7, 8]. The latter allows the material to be drained to an essentially dry state, but the intensity of high temperature exposure is strictly limited by the requirements for fire and explosion hazard prevention. Thermal drying requires financial costs for the construction of separate facilities, fuel costs [9]. Currently, innovative non-thermal methods for deep drying of wet materials using adsorption processes are gaining popularity. One of the key factors in adsorption is the developed specific surface of adsorbents.

The authors developed an original technology for recycling of converter sludges, the conceptual basis of which is their adsorption dehydration in combination with thermo-chemical lumping [10, 11]. The symbiosis of these methods of converter sludges processing provides the technology with competitive advantages in terms of energy saving (reduction of energy consumption for moisture removal) during drying of high-moisture sludges and in terms of resource-saving (there is no need to use the binder) during lumping.

2. Methods of research

To study the adsorption dehydration of sludge the experiments were carried out with a high-moisture (W^r = 50.0 %) converter sludge of JSC "EVRAZ ZSMK". The chemical composition of dry sludge is given in table 1.

Table 1. Chemical composition of converter sludge.

Mass fraction of the element, % wt								
Fe_{tot}	Fe_2O_3	FeO	MgO	CaO	SiO_2	K_2O	V_2O_5	Cr_2O_3
46.81	64.05	1.82	4.59	16.68	5.75	0.19	0.069	0.10

Mass fraction of the element, % wt										
C	S	ZnO	CuO	PbO	MnO	Al_2O_3	Na_2O	P_2O_5	Fe_{met}	TiO_2
0.63	0.24	1.11	0.061	0.11	1.08	1.93	0.88	0.32	abs.	0.21

The content of zinc and lead in the slutty is different and depends not only on the composition of the raw material but also on the dust dispersion. Evaluation of the granulometric composition of thermally dried sludge was performed on a laser size-distribution device MALVERN-2000 in the Materials Science Center of Siberian State Industrial University (Novokuznetsk, Russia).

Solid pyrolysis residue of brown coal – fine-grained BSC was used as an adsorbent which is currently produced according to the technology of "Termokoks-KS" at the pilot plant of open-pit mine Berezovsky-1. According to [12], BSC has a highly developed and well-accessible porous structure and, accordingly, high adsorption capacity and energy properties. The micropores volume of the porous BSC structure is tenfold higher than the volume of micropores in the structure of coal coke. In this regard the adsorption capacity of BSC is close in this indicator to traditional activated carbons. In terms of adsorption capacity BSC is slightly inferior in activity to the sorbent brand ABG-D (TU600209591-443-95) and significantly exceeds the sorbent brand DAK (TU0321002-51577712).

At the same time, it should be noted that the density of BSC particles even if the entire porous space is filled with adsorbed moisture (1.42 g/cm3) will be more than 2.5 times lower than the density of converter sludge particles (3.8 g/cm3). The general BSC characteristics are given in table 2.

As it follows from the data in table 2 BSC is a low-ash and low-sulfur high-calorie product with a highly developed and well-accessible porous structure.

Table 2. Characteristics of BSC from open pit coal mine Berezovsky-1.

Indicators	Unit of measurement	Value indicator
Total moisture	%	3. 0
Ash, dry condition	%	9.7
The yield of volatile substances	%	9.9
Elemental composition, combustible mass:	%	
Carbon		92.8
Oxygen		4.45
Hydrogen		1.52
Sulfur (dry state)		0.24
Phosphorus (dry state)		0.026
Highest heat of combustion	kcal/kg	7830
Lower heat of combustion, working state	kcal/kg	6700-6800
Adsorption activity of iodine	%	not less than 42
Total pores volume	cm^3/g	0.47
Specific surface of pores	m^2/g	500
Density		
absolute	kg/m^3	1880
apparant		974
Bulk density	kg/m3	550

The results of dehydration were evaluated according to the free-flowing index of material (GOST 25139-93). The basis for BSC ratio determination: sludge, which is necessary for the production of a free-flowing mixture, is data on sludge moisture and adsorption capacity of BSC.

After gaining free-flowing properties, the mixture of sludge and BSC is supposed to be subjected to pneumatic separation, after which the BSC is sent for technological and energy use, the sludge – for thermo-chemical lumping.

During the experiments on the study of thermo-chemical lumping of converter sludges two series of samples were used. Samples of series No. 1 concentrate CPP Kuznetskaya GZh + Zh – 50 % wt + 50 % by weight converter sludge and samples of series No. 2 concentrate of coal grade Zh of the Mezhegeyskoye deposit – 50 % wt + 50 % wt converter sludge (to obtain a material containing carbon, coal concentrates are added in excess). The characteristics of coal concentrates are given in table 3. The choice of such ration of mixture components is based on the notion that in this case the sludge is an additive to coking coal with a high volatile yield and for the production of a strong lump material it is necessary to have a certain level of mixture sintering.

Table 3. Characteristics of coal concentrates.

Sample	Technical analysis, %				Plastometry, mm		Petrography		
	W^r	A^d	V^{daf}	S^d	X	Y	Vt,%	S_R	R
GZh+Zh	10.5	7.8	38.0	0.56	17	24	85.0	0.56	0.864
Zh	8.6	8.1	38.2	0.67	–2	34	93.0	0.045	0.853

The study of sludge lumping by the thermo-chemical method was carried out in two stages. In the first stage the mixtures were heated in the unit of plastometric analysis to temperature 730 °C. In the second stage the samples subjected to plastometric analysis were heated in a Tamman furnace for 30 minutes to the temperature of completion of coking process (1050-1100 °C).

Determination of the mechanical compressive strength of briquettes from coke batch and mixtures containing coals from the Mezhegeysky and Kuznetsky deposits with converter sludge was carried out in accordance with GOST 21289-75 "Coal briquettes. Methods for determination of mechanical strength". The hydraulic press 2PG-10 was used for the test.

3. Results and discussion

Evaluation of the granulometric composition of both samples revealed the similarity of the composition of converter sludge and BSC (figure 1).

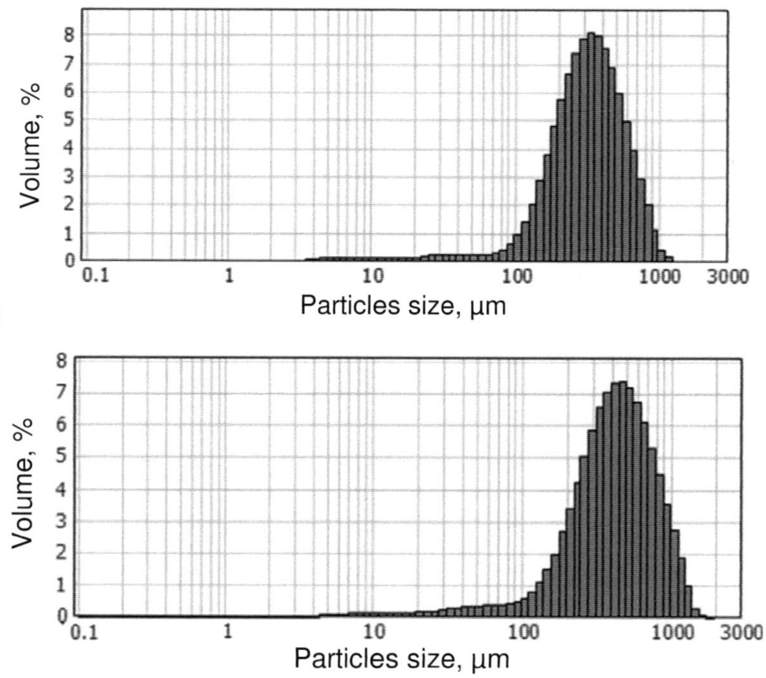

Figure 1. Granulometric composition of sludge (a) and BSC (b).

From these data it follows that the grain-size distribution of the solid part of the materials is represented by small particles in the range 0.5-1000 μm, and the maximum of the grain-size distribution curve is about 500 μm. Experimental observations confirmed the correctness of the calculated adsorption capacity of semicoke (table 4).

Table 4. Quality and bulk properties of sludge and sludge mixture with BSC.

Product	Moisture contents, $W^r{}_t$%	Free-flowing, g/s
Initial sludge (mud-like condition)	50.0	0.0
BSC	3.0	25.28
Thermally dried sludge	1.35	0.0
Sludge + BSC*	16.55	33.34

*Mixing time – 2 min.

When mixing BSC with the wet waste it absorbs moisture giving the mixture a high free-flowing, which is important for transportation of the product in process streams. At the same time the moisture adsorbed in the pores passes into a bound state and becomes an active participant in oxidation-reduction processes.

As a result of laboratory experiments on thermo-chemical lumping it was established that the thickness of the plastic layer for the samples in series No. 1 was 10 mm, for the samples in series No. – 2-17 mm. Heating of the experimental mixtures in the apparatus of plastometric analysis to a temperature of 730 °C was accompanied by the removal of volatile substances from the coal portion of the samples, while the content of sludge in the samples increased to 56-57 % by weight. An increase in the FeO content in the chemical composition of the sludge from 1.82 to 14.3 % wt and the appearance in the amount of 2.03 % wt of Fe_{met} were established The zinc oxide content in the samples was 0.48 % wt The results of plastometric analysis confirmed the results of thermodynamic modeling [13]. At temperatures of heating the test mixtures up to 730 °C the iron reduction, even in the presence of excessive carbon, practically does not occur.

Calcination in the Tamman furnace was accompanied by solid-phase reduction of iron. In particular, the degree of reduction to Fe_{met} in the samples of series No. 1 was 84.9 % wt, in the samples of series No. 2 – 94.4%.

The product of ferrocoke (figure 2) obtained as a result of a series of experiments contains 35-39 % of Fe_{met} and 45-49 % of C, content of ZnO does not exceed 0.017 %, the compressive strength is 2.8 MPa.

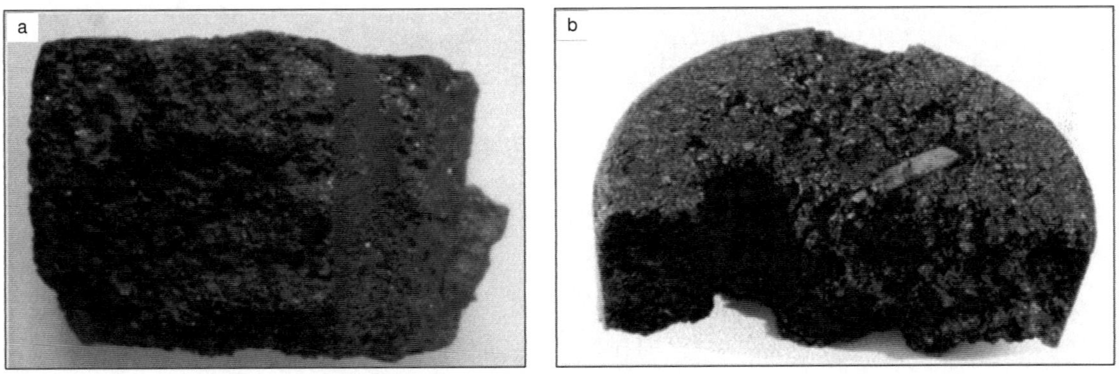

Figure 2. Samples of ferrocoke (a, b).

Based on the results of experimental studies, a technological scheme for recycling of high-moisture wastes was developed that includes non-thermal adsorption dehydration of converter sludges in combination with the subsequent thermo-chemical lumping (figure 3).

The converter sludge from the sludge reservoir 1 enters the thickener 2 and is then transferred to the mixer-adsorber 5 to contact with BSC serving as a moisture adsorbent. Then the mixture of products is transferred to the separation into pneumatic classifier 7, from where the lighter BSC through the dust separating system (cyclone 6, bag filter 8) enters the hopper 9, from where it is taken for energy-engineering needs, and the dust-free air is discharged into the atmosphere. The heavier sludge from the pneumatic classifier 7 passes through the dispenser into the mixer 12, into which the coking coal from the hopper 11 is also fed through the dispenser. The mixture formed in the specified ratio is subjected to thermo-oxidizing coking in the rotary hearth furnace 13. The ferrocoke obtained at a final temperature of 1100-1150 °C is cooled in the dry quenching unit 15 with a waste heat boiler 17 and sorted into classes 0-10 mm, 25-10 mm and +25 mm. Heat for coking is formed by burning above the bed of charge in the furnace with a rotating hearth of evolved gaseous products. The combustion products from the furnace are forwarded for later use in a gas turbine plant 14.

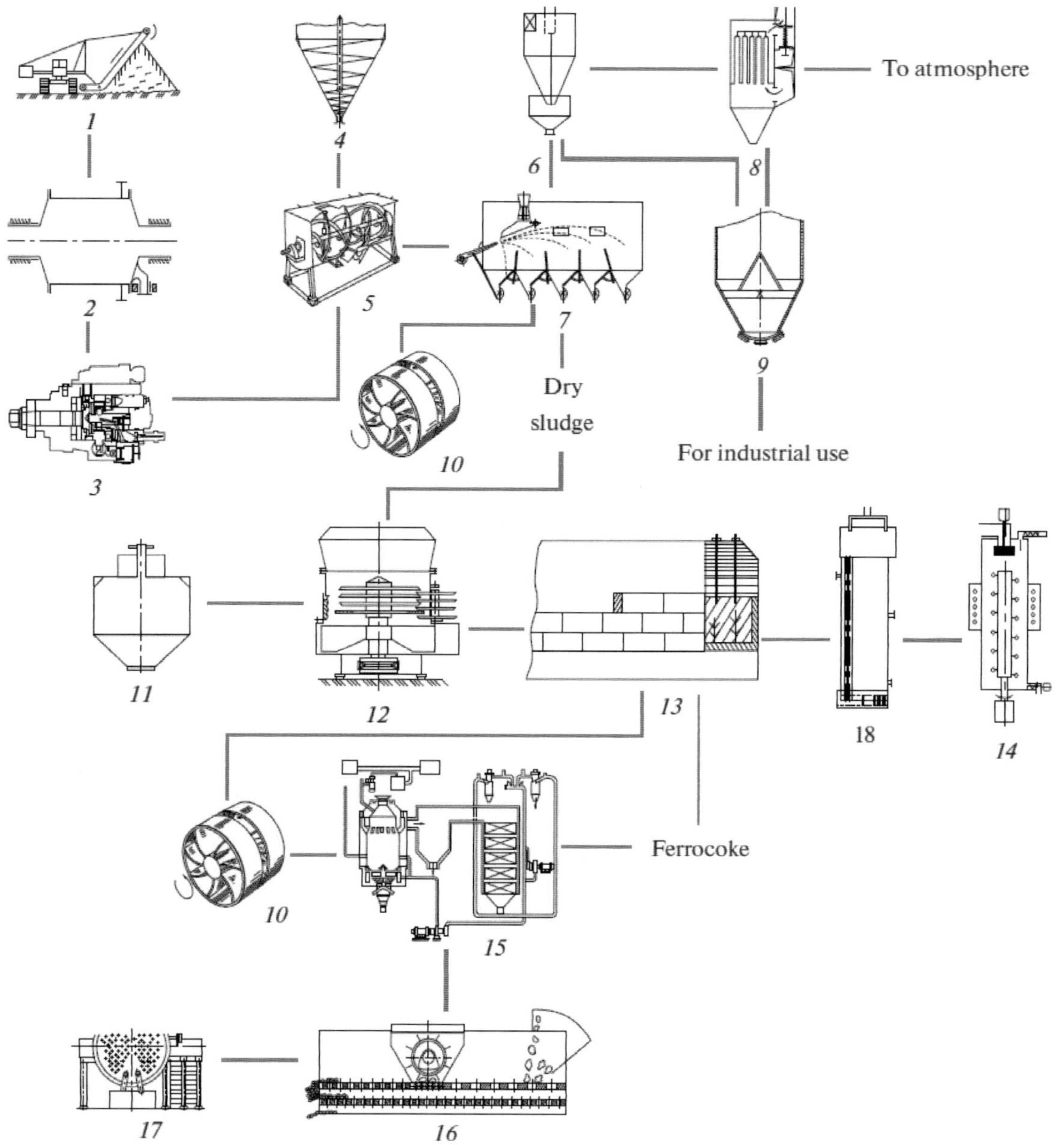

1 – sludge reservoir; 2 – thickener; 3 – sludge pump; 4 – BSC bunker; 5 – mixer-adsorber; 6 – cyclone; 7 – pneumatic classifier; 8 – filter bag; 9 – hopper of wet BSC; 10 – blower; 11 – hopper of coal GZh + Zh; 12 – mixer; 13 – coke oven with a rotating hearth; 14 – top-pressure recovery turbine; 15 – dry coke-quenching installation; 16 –ferrocoke separation; 17 – waste-heat boiler

Figure 3. Principal process scheme for conditioning of wet converter sludge by adsorption dehydration and thermo-chemical lumping.

4. Conclusions

The results obtained in this paper indicate that on the basis of principles of conditioning of wet converter sludges by adsorption dewatering and thermo-chemical lumping it is possible to develop an efficient technology for their utilization with the production of ferrocoke. Ferrocoke is a composite material containing mainly reduced (metallic) iron and carbon. Such product has a wide field of

IOP Conf. Series: Earth and Environmental Science **84** (2017) 012043 doi:10.1088/1755-1315/84/1/012043

application as an iron and carbon-containing charge component. In addition to its use in blast furnace production, its application for converter melting technology, including with the liquid-phase reduction elements, is of practical interest.

References

[1] Dobrovolsky I Ya et al 2000 *Processing and Utilization of Industrial Wastes of Chelyabinsk Region* (Chelyabinsk: ChelSU) p 256
[2] Tolochko A I et al 1990 *Utilization of Dust and Sludge in the Ferrous Metallurgy* (Moscow: Metallurgy) p 206
[3] Noskov V A et al 2002 *Metallurgical and Mining Production* **1** 119–22
[4] Ulyanov V P et al 2004 *Integrated Technologies and Energy Saving* **3** 48–53
[5] Ulyanov V P et al 1985 *Discoveries. Inventions* **15** 115
[6] Taits E M 1965 *Coke and Iron Coke on the Basis of Briquetting* (Moscow: Metallurgy) p 173
[7] Baryshnikov V G et al 1986 *Secondary Material Resources of Ferrous Metallurgy* (Moscow: Economika) vol 2 p 344
[8] Dobrovolsky et al 2010 *Bulletin of Chelyabinsk State University* **8** 40–5
[9] Kirillov K M et al 2015 *Coal* **6** 54–6
[10] Shkoller M B, Protopopov E V, Kuznetsov S N, Temlyantsev M V and Ivanov V P 2016 *Vestnik of the Mining and Metallurgical Section of the Russian Academy of Natural Sciences* **37** 46–53
[11] Kuznetsov S N, Shkoller M B, Protopopov E V, Temlyantsev M V and Feyler S V 2017 *Ferrous metallurgy* **4** 268–75
[12] Shkoller M B et al 2014 *Coke and Chemistry* **57(12)** 471–6
[13] Kuznetsov S N et al 2017 *Bulletin of SibSIU* **1** 25–8

Contemporary trends in improvement of organizational-economic mechanism of environmental management

T V Kiseleva[1], V G Mikhailov[2] and G S Mikhailov[2]

[1]Siberian State Industrial University, 42 Kirova Street, Novokuznetsk, 654007, Russia
[2]T. F. Gorbachev Kuzbass State Technical University, 28 Vesennyaya Street, Kemerovo, 650000, Russia

E-mail: kis@siu.sibsiu.ru

Abstract. The article deals with the effective functioning of ecological and economic systems of various levels on the basis of an adequate organizational and economic management mechanism. The compliance matrix of the presented innovative elements in the structure of organizational and economic mechanism of environmental management is developed. The practical component of the conducted study can be recommended to municipal, regional and federal authorities, as well as industrial enterprises, to support effective, environmentally reasonable management decisions that are consistent with the global concept of sustainable development.

1. Introduction

Development of the real sector of economy determines the high level of technogenic impact on the environment, conditioning the need to address the problems of resource and nature saving with minimization of financial costs [1]. Increase in the effectiveness of environmental protection activities depends on the process of managing territorially distributed ecological and economic system with its unique features. In this case formation of an organizational and economic mechanism of management of environmental activities, which corresponds to high modern requirements caused by the international integration of enterprises and environmental law enforcement [2], is of a decisive importance here. This situation determines the relevance and practical significance of the present study.

2. Materials and methods

The object of the study is a system for managing a geographically distributed ecological and economic system. The subject of the study is the organizational and economic mechanism for managing environmental protection activities. The study is based on an analysis of domestic and foreign literary sources on this issue. Particular attention is paid to the study of innovative elements of organizational and economic mechanism of environmental management, which have practical significance. Verification of the adequacy of theoretical developments is based on the factual data from environmental reports of enterprises. Elements of system analysis and the results obtained by specialists in the field of management of ecological and economic systems are also used in the work.

3. Results and discussion

In the conditions of a region with environmental problems due to the presence of developed industry, the process of effective management of ecological and economic system, the simplified scheme of which is presented in figure 1 [3 - 6], is relevant.

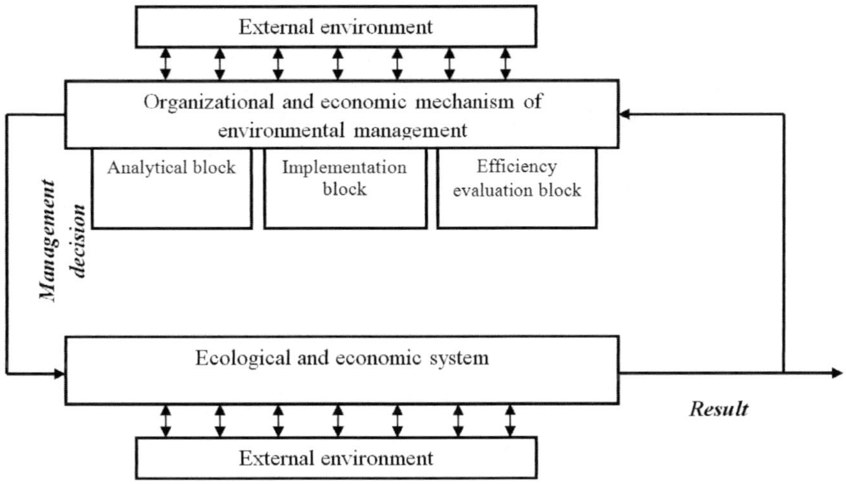

Figure 1. Simplified scheme of the management mechanism of ecological and economic system.

Figure 1 shows that the key element in the presented scheme of mechanism for managing the geographically distributed ecological and economic system is the organizational and economic mechanism of environmental management, in which three main elements are distinguished:

- analytical block;
- implementation block;
- efficiency evaluation block.

In order to identify the developed innovative elements of organizational and economic mechanism of environmental management, three main blocks have been identified, for which a compliance matrix presented in table 1 has been developed.

Table 1. Compliance matrix of the developed innovative elements to the structure of organizational and economic mechanism of environmental management.

Innovative elements	Analytical block	Block of implementation	Efficiency evaluation block
Reduced integrated indicator of the aquatic ecosystem state	+		
Penalization system of enterprises for providing unreliable information on the level of negative impact on the environment		+	
System of universal ecological and economic indicators of the enterprise operation	+		+
System of ecological and economic indicators of production and consumption wastes	+		+
Diversification of the enterprise production program due to environmental restrictions		+	

Table 1 shows that the largest number of the developed elements belongs to the analytical block, which additionally reflects its significance. To determine the practical feasibility in usage of the proposed innovative elements, it is necessary to study them in details [7, 8].

Estimation of the quality of aquatic ecosystem can be carried out using an integral indicator, the main disadvantage of which is processing time due to the need to use multiple input data. The carried out sensitivity analysis allows us to conclude that a number of components of the complex indicator is of little informative value because of the use of strongly correlated components. Therefore, some of them can be neglected in order to simplify the integral indicator and its more effective practical application. As a result, the integral index was reduced [9].

The implementation block (figure 1, table 1) is represented by the development of a penalization system of enterprises for providing unreliable information on the level of negative impact on the environment [9]. Below there are different levels of penalties for providing unreliable data with differentiation according to hazard classes and depending on the range of deviations between the results of pollutants sampling performed by the officially authorized state environmental management body and the enterprise. In formulas (1) – (4), the percentage indicates the additional level of penalties relative to its base value for pollutants of different hazard classes.

$$P_{AD\_I} = \begin{cases} 10\ \%\ P_{BASE}\ , if\ \Delta D^{EX} > (1-5)\ \% \\ 15\ \%\ P_{BASE}\ , if\ \Delta D^{EX} > (5-10)\ \% \\ 30\ \%\ P_{BASE}\ , if\ \Delta D^{EX} above\ 10\ \% \end{cases} \quad (1)$$

$$P_{AD\_II} = \begin{cases} 8\ \%\ P_{BASE}\ , if\ \Delta D^{EX} > (2-10)\ \% \\ 15\ \%\ P_{BASE}\ , if\ \Delta D^{EX} > (10-20)\ \% \\ 20\ \%\ P_{BASE}\ , if\ \Delta D^{EX} above\ 20\ \% \end{cases} \quad (2)$$

$$P_{AD\_III} = \begin{cases} 6\ \%\ P_{BASE}\ , if\ \Delta D^{EX} > (3-10)\ \% \\ 10\ \%\ P_{BASE}\ , if\ \Delta D^{EX} > (10-20)\ \% \\ 15\ \%\ P_{BASE}\ , if\ \Delta D^{EX} above\ 20\ \% \end{cases} \quad (3)$$

$$P_{AD\_IV} = \begin{cases} 4\ \%\ P_{BASE}\ , if\ \Delta D^{EX} > (4-10)\ \% \\ 5\ \%\ P_{BASE}\ , if\ \Delta D^{EX} > (10-20)\ \%\ , \\ 10\ \%\ P_{BASE}\ , if\ \Delta D^{EX} above\ 20\ \% \end{cases} \quad (4)$$

where P_{AD} – additional value of penalties for provision of false information; P_{BASE} – base value of penalties for pollution of an element of environment;

$$\Delta D^{EX} = \left|\frac{C_{SEMA} - C_{ENT}}{C_{SEMA}}\right| 100\ \%\ , \quad (5)$$

where ΔD^{EX} – exceedance of pollutant concentration values according to the data of the state environmental management authority, %; C_{SEMA} –actual concentration of polluting substances according to sampling data of state environmental management authority; C_{ENT} – actual concentration of polluting substances according to sampling data of an industrial enterprise. The conducted calculations showed that the proposed incentive system allows the amount of penalties to be increased by an average of 18 %, which should motivate economic entities to provide the most reliable information in connection with the increase in costs included into the cost of production or attributed to the net profit of the enterprise.

The most important aspect of the analytical block is the development of a system of universal environmental and economic performance of an enterprise, which can also be used to formulate a comprehensive assessment of its environmental safety [10 - 13]. If we consider the current activity of the enterprise, then the developed schemes for production environmentalization, including the ecological and economic assessment of anthropogenic impact on the environment, can be used to make an environmentally oriented economic decision [14]. Specialists working in this field have developed a system of criteria for environmental and economic sustainability of an enterprise, including the ecological efficiency of the enterprise as a complex characteristic of the effectiveness of nature management – resource, energy, and waste intensity, efficiency of using the consumed "external" and "internal" resources of the enterprise [15]. A number of authors identify environmental performance indicators that characterize negative impact on the environment [16].

In some cases it is advisable to adapt the environmental and economic indicators to a particular industry or sub-sector that has its own characteristics. For example, for coal mining and coal processing, the environmental performance indicators are divided into three groups: ecological, social and economic [17, 18]. Due to the fact that economic entities within the coal cluster have a significant burden on the soil cover, it is possible to calculate the economic damage from soil pollution by production and consumption wastes, including per unit of the production capacity of the enterprise [19].

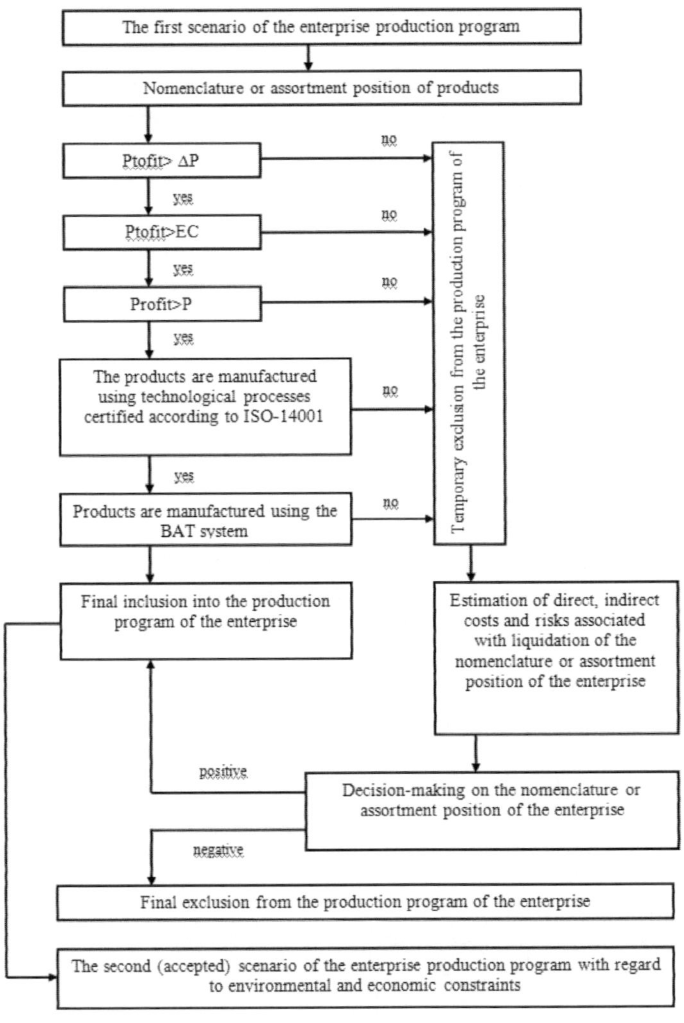

Figure 2. Formation scheme of the enterprise production program with regard to environmental and economic constraints.

The authors of this study have developed a system of environmental and economic indicators, which also takes into account the features of the enterprise under investigation [6, 20]. The final innovative element of this block is the system of ecological and economic indicators of production and consumption wastes [21, 22].

The implementation block of the organizational and economic mechanism of environmental management is also shown by the diversification of the enterprise production program due to environmental restrictions, it is schematically depicted in figure 2, where it is seen that after the formation of a first scenario of environmentally safe production program, the main technical and economic indicator (profit) is compared with various environmental and economic components (ΔP – increase of payment for negative impact on the environment; EC – economic costs of environmental pollution; P – penalties). The decision on the final inclusion into the production program is made on the basis of the best available techniques (BAT) criterion, which is conditioned by modern environmental requirements [23].

The considered innovative elements of organizational and economic mechanism of environmental management can be used in the construction of an economic and mathematical model that ensures the selection of an optimal complex of management decisions, the implementation of which will reduce the negative impact of enterprises on the environment, reduce the prime costs of primary and secondary products through recycling of production and consumption wastes and minimization of environmental payments [24].

4. Conclusions

The study of the development of innovative architecture of organizational and economic mechanism for environmental management allowed the following conclusions to be made:

- the necessity of improving the organizational and economic mechanism for managing environmental activities is substantiated, taking into account external requirements and environmental law enforcement;
- a simplified scheme for management of ecological and economic system with differentiation of the blocks of organizational and economic mechanism for environmental management is developed;
- a compliance matrix of the developed innovative elements in the has been the structure of organizational and economic mechanism of environmental management is formed;
- • the analysis of various approaches to the assessment and interpretation of environmental and environmental-economic indicators of an enterprise is performed, including the study of negative impact on all elements of the environment;
- the verification of the developed innovative elements of the organizational and economic mechanism of environmental management for the adequacy was performed in order to implement it in practical activities.

References

[1] Burkov V N, Novikov D A and Shchepkin A V 2009 *Problems of Security and Emergency Situations* **4** 30–40
[2] Zenkov I V 2016 *Gornyi Zhurnal* **10** 96–9
[3] Burkov V N, Novikov D A and Shchepkin A V 2008 *Mechanisms for managing ecological and economic systems* (Moscow: Fizmatlit) p 244
[4] Burkov V N, Novikov D A, Schepkin A V 2009 *Problems of Management* **1** 2–7
[5] Tretyakova E A 2014 *Studies on Russian Economic Development* **25(4)** 423–30
[6] Kiseleva T V, Mikhailov V G and Karasev V A 2016 *IOP Conf. S.: Earth and Environmental Science* **45** 012013
[7] Bubnova M B, Ozaryan Y A 2016 *J. Mining Science* **52 (2)** 401–09
[8] Avdeev V P, Kiseleva T V and Burkov V N 2001 *Automation and Remote control* **62(10)** 1645–50
[9] Kiseleva T V, Kulakov S M, Mikhailov V G and Mikhailov G S 2005 *Control systems and information technology* **2(19)** 84–7
[10] Dolzhenko E N, Monich A I, Kudryakov A G and Sazykin V G 2016 *Int. Research J.* **5-1(47)** 75–8
[11] Savon D Yu and Tibilov D P 2014 *Gornyi Zhurnal* **12** 31–5
[12] Zolotukhin V M, Gogolin V A, Yazevich M Yu, Baumgarten M I and Dyagileva A V 2017 *IOP Conf. S.: Earth and Environmental Science* **50** 012027
[13] Galanina T V, Baumgarten M I, Mikhailov V G, Koroleva T G and Mikhailov G S 2017 *IOP Conf. S.: Earth and Environmental Science* **50** 012030
[14] Epifantseva E I 2003 *Transactions of the TSTU* **9(3)** 538–43
[15] Burkov V N and Burkova I V 2014 *Automation and Remote Control* **75(3)** 470–80
[16] Burkov V N and Burkova I V 2015 *Game Theory and Application* **17** 17–36
[17] Ageev I A, Burkov V N, Zinchenko V I and Kiseleva T V 2005 **66(6)** 995–1002

[18] Agienko M I, Bondareva E P, Chistyakova G V, Zhironkina O V and Kalinina O I 2017 *IOP Conf. S.: Earth and Environmental Science* **50** 012022.
[19] Tyulenev M, Lesin Y, Tyuleneva E and Murko E 2017 *E3S Web of Conferences* **15** 02003
[20] Mikhailov V G, Koryakov A G, Mikhailov G S 2015 *J. Mining Science* **51(5)** 930–6
[21] Mikhailov V G, Golofastova N N, Galanina T V, Koroleva T G and Mikhailova Ya S 2017 *IOP Conf. S.: Earth and Environmental Science* **50** 012038
[22] Efimov V I, Sidorov R V and Korchagina T V 2015 *Ugol* **12** 73–6
[23] Kiseleva T V and Mikhailov V G 2016 *Proc. IVth All-Rus. Conf. On Modeling and Knowledge-intensive Information Technologies in Technical and Socio-economic Systems* (Novokuznetsk: SibSIU) part 2 pp 27–31
[24] Shorokhova A V, Dmitrieva O V and Fryanov V N 2014 *Information and Analytical Bulletin on Mining* **12** 294–6

Use of overburden rocks from open-pit coal mines and waste coals of Western Siberia for ceramic brick production with a defect-free structure

A Yu Stolboushkin, A I Ivanov, G I Storozhenko, V A Syromyasov and D V Akst

Siberian State Industrial University, 42 Kirova Street, Novokuznetsk, 654007, Russia

E-mail: ivanovaliv1989@gmail.com

Abstract. The rational technology for the production of ceramic bricks with a defect-free structure from coal mining and processing wastes was developed. The results of comparison of physical and mechanical properties and the structure of ceramic bricks manufactured from overburden rocks and waste coal with traditional for semi-dry pressing mass preparation and according to the developed method are given. It was established that a homogeneous, defect-free brick texture obtained from overburden rocks of open-pit mines and waste coal improves the quality of ceramic wall materials produced by the method of compression molding by more than 1.5 times compared to the brick with a traditional mass preparation.

1. Introduction

Western Siberia along with the richest reserves of oil and energy resources has also a great mineral resource base. Intensive industrial development of the region in the XX century was accompanied by an uncontrolled development of natural resources which led to a constant growth and accumulation of industrial waste. The average utilization rate of wastes as secondary raw materials is estimated approximately at one third, which is 2.5 times lower than in the countries with the developed environmental policies [1]. As a result, Russia's industry suffers significant losses in raw materials and energy resources, the intensive accumulation of unused wastes in the environment continues – every year approximately 2 – 2.5 billion tonnes per year [2].

The problem of rational consumption of resources is especially acute in Kemerovo region, in the territory of which half of the country's wastes from mineral mining and processing are formed and stored. The main "suppliers" of wastes are the mining, energy and metallurgical industries. The annual volumes of coal mining and processing wastes stored on the territory of Russia equals to approximately 15 million tonnes [3]. As a rule, in case of open-cast mining one tonne of coal accounts for 4 tonnes of overburden the storage of which leads to significant environment pollution [4].

In the long term the solution of the problem is possible only through introduction of waste-free enrichment technologies and complex processing of accumulated wastes using their mineral part for manufacturing building materials. Overburden rocks, mining wastes, mining waste of various minerals are an inexhaustible source of raw materials for the production of building materials and products. The use of industrial wastes will cover up to 40% of construction industry's needs in raw materials and will reduce by 30% the costs for manufacturing of construction materials in comparison with their production from natural raw materials [5]. The overburden rocks at the open pit coal mines in West Siberian region are mainly represented by lean dusty loams, the wastes from processing plants – by argillites and aleurolites of variable composition. These types of raw materials do not provide a high-quality (in terms of compressive strength and frost resistance) ceramic bricks when traditional production technology is applied. In this regard, the purpose of the present studies was to check the working hypothesis on the necessity to form a defect-free ceramic brick structure by using new approaches and methods for mass preparation of wastes from coal production and processing.

2. Methods and materials

The authors carried out research into the production of bricks from coal byproducts [6], and also developed a scheme for supplying brickworks with technological fuel due to secondary enrichment of waste coal [7].

The release of bricks from these types of raw materials in semi-dry pressing plants during the Soviet era showed a low quality of the products manufactured for strength and frost resistance [8]. This was facilitated by insufficient shredding of raw materials (up to 3 mm inclusive). The heterogeneity of the composition of press powders obtained from coarse-grained masses leads to the production of loose raw material of low strength. The method of formation of ceramic bricks with optimal structures by fine dry grinding, granulation, compression molding, drying and baking developed by the authors allows the structure defects to be avoided [9]. The rational granulometric composition of the press masses is the decisive prerequisite for good pressing and the compact packing of granules associated with it to a large extent determines the structure, physical and mechanical properties of ceramic bricks.

Implementation of the stated principles of rational mass preparation for the production of ceramics using the method of compression molding was performed on the example of overburden loamy rock and waste coal from Abashevskaya processing plant (Kemerovo region, Russia).

Novokuznetsk loam refers to the most common clay rocks on the territory of Siberia and Krasnoyarsk Krai. Raw material is low-dispersed, non-caking, highly sensitive to drying, by its mineral composition refers to polymineral rocks of hydromica-kaolinite-montmorillonite type.

The waste coal is close to clays and consists of a mineral part represented by hydromuscovite, quartz, plagioclase, montmorillonite, chlorite, siderite, calcite and a carbon residue containing free carbon in an amount of 6 to 8% after secondary enrichment. The raw material is coarse-dispersed with predominance of sand fraction, non-caking, insensitive to drying. The content of coloring oxides (Fe_2O_3+TiO_2) is 5.5-7 %. The chemical composition of waste coal is given in table 1. Drying and grinding of the raw material were carried out in a vortex mill-drier up to class of -300 μm. After grinding, the plasticity of the loam was 12.5 (moderately plastic), the waste coal – 6.0 (low-plastic).

Table 1. Chemical composition of waste coal.

Raw material	Mass fraction of the components on the dried substance, %								
	SiO_2	Al_2O_3	TiO_2	Fe_2O_3	CaO	MgO	K_2O	Na_2O	LOI*
Waste coal from Abashevskaya processing plant	57-64	14-17	0.5-1	5-6	3-5	1.5-2	2-2.5	1.5-2	6-8
Clay loam	61-64	14-15	0.8	4-5	4-4.5	2-2.5	3.5-4		5-6

*LOI – loss on ignition

3. Experimental research

In order to study the structure and properties of ceramic bricks from waste coal and coal enrichment using the developed method of mass preparation, the pilot baths of bricks were manufactured at Berdsky Brick Factory LLC. The composition of the batch for factory tests is given in table 2.

The technology of obtaining granulated press-powder for batch No. 1 consisted in the following: at the first stage the dry grinding of loam to a grade of -300 μm was carried out, and then the loam was granulated in the intensive mixer to obtain granules fraction 1-3 mm with the simultaneous press powder moistening up to the molding moisture 10.5-11.3 %. To obtain granulated press powder No. 2 waste coal with the fraction +13 mm was used.

Table 2. Batch composition for factory tests.

No. batch	Batch components	Content, wt. %
1	Waste coal	100
2	Waste coal	80
	Clay loam	20

The granulated press powder of batch No. 2 was obtained as follows: a separate dry grinding of loam and wastes to a class of -300 μm was performed, and then wastes were granulated in the intensive mixer to obtain granules fraction 1-3 mm with the simultaneous press powder moistening up to the molding moisture 12.5-13.0 % and subsequent powdering with dry grinded loam (20 wt.%).

Dry grinding of loam and wastes was carried out on the unit USP-S-04.55, granulation in the intensive mixer was carried out on the production base of Baskey Keramik LLC (Chelyabinsk). The finished press powders were packed in the bulk bags with polyethylene liners to preserve moisture and sent to Berdskiy Brick Factory LLC for compression molding of ceramic bricks.

Molding of bricks was carried out on presses SM-1085B at a pressure of 17-18 MPa. The moisture content of the granulated press powders was 9.9 % and 10.5 %, respectively. The pressed bricks in quantity of 800 pieces had a normal appearance without cracks of re-pressing. The drying and firing time in the tunnel furnace was 42 hours at a maximum temperature of 950-1000 °C. Defective bricks after firing was 1.5 % from the total amount, the fired bricks had a terracotta color, homogeneous uniformly-granular structure can be noted on the fracture. On the ceramic bricks firing cracks, swellings and curvatures were not observed. Bricks for physical and mechanical tests were selected from the fired products and the results are presented in table 3. The study of the structure and properties of ceramic bricks produced with the use of the developed technology was carried out in comparison with the samples of ceramic bricks produced from this raw material during the Soviet period (currently the plants are closed [9]).

Table 3. Physical and mechanical properties of ceramic bricks.

No. batch	Composition of batch, wt. %	Average density, kg/m^3	Compressive strength, MPa	Water absorption,%	Frost-resistance, cycle
1	Clay loam 100	1980	19.3	11.6	>50
1*	Clay loam 100	2010	11.2	14.5	35
2	Waste coal 80 Clay loam 100	1680	15.4	15.4	50
2*	Waste coal 100	1590	10.5	18.7	25

*Brick produced at the factories of the USSR

4. Examination of the brick structure

Figure 1 shows the macrostructure of ceramic bricks from loam prepared by the traditional method at Novokuznetsk brick factory in the 1970's (figure 1, a-c) and by the developed method with fine grinding and granulation (figure 1, d-f), fired at 1000 °C. The examination of the structure of ceramic compression-molded cork showed that the bricks prepared in the traditional way have a euporphyric texture (fugure 1, a) [10, 11] with more intensely colored, against the background of the main mass, metamorphosed and decomposed fragments of the original minerals (figure 1, a). On their perimeter there are macropores unfilled by the main amorphized mass and indicating the structural defects at the macrolevel occurring during firing (figure 1, a-c).

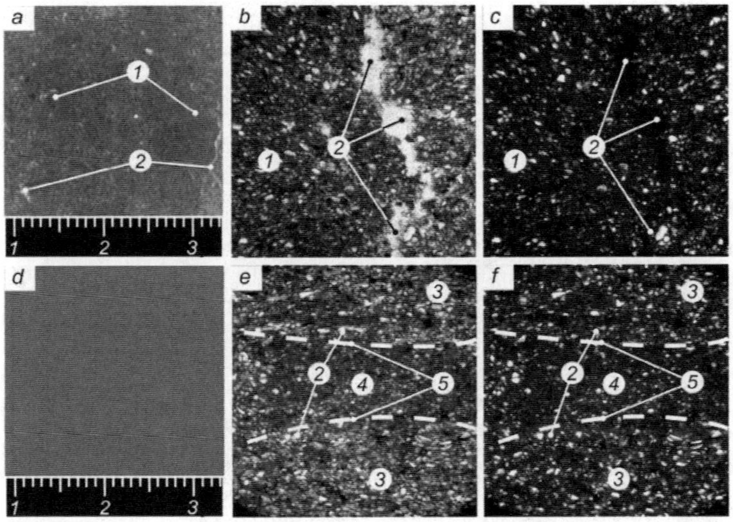

Figure 1. The macrostructure of ceramic brick made from loam prepared by the traditional method (a-c) and finely grinded granulated loam (d-f). Polished section, reflected light (a, d); the thin section, transmitted light, ×10, nicoli II (b, e) and + (c, f): 1 – metamorphosed initial minerals; 2 – the pores; 3 – the granule body; 4 – the boundary layer of the granule; 5 – conditional boundary between the granules.

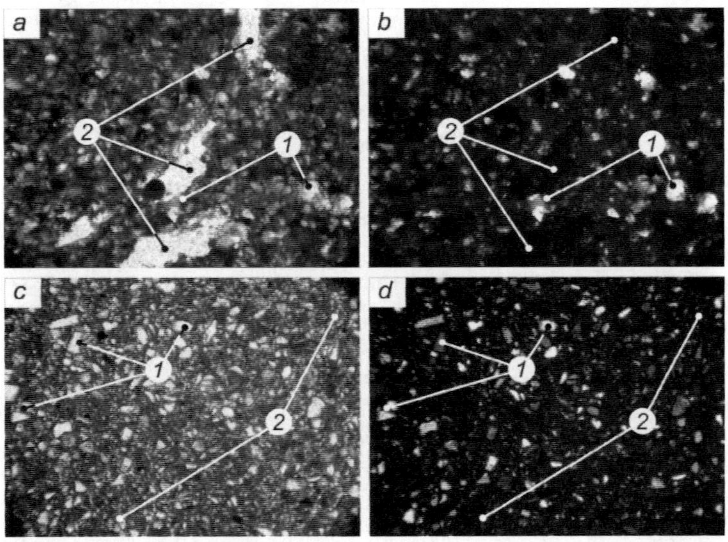

Figure 2. Microstructure of ceramic bricks made from loam prepared by the traditional method (a, b) finely grinded granulated loam (c, d). Polished section, transmitted light, magnification ×20, nicoli II (a) and + (b); magnification ×20, nicoli II (c) and + (d): 1 – fragments of relict minerals; 2 – pores.

In the brick from finely grinded granulated material the structure is afanite (cryptocrystalline) (figure 1, d). Fine-grained mineral phases are immersed into a full-crystalline bulk (figures 1, e, f, and figures 4.3, c, d). The bulk is colored with iron oxides, fine-grained quartz-feldspar, in which the idiomorphic quartz particles are uniformly distributed. Visually the appearance of the boundary layer between the granules on the transverse section of the brick from granulated batch is not expressed (figure 1, d).

A more intense color of the boundary layer is due to the concentration of colloidal clay fraction evolving at the boundary of the granules during pressing, in which the oxygen ions bind to the ions of aluminum, magnesium, iron and other metals located in the octahedral layer [12]. Similarly, the boundary layer is saturated with water-soluble minerals [13] which form low-melting eutectics during firing of ceramic materials. At the boundary of granules a eutectic melt is abundantly formed, in which the ions of alkali and alkaline-earth metals actively interact with iron, silicon and aluminum oxides [14, 16]. As a result during cooling ferriferrous dark brown (up to black) glass-crystalline formations appear. The amount and size of the pores are much less than in the samples prepared by the traditional method (figure 1-2) due to the formation of liquid-phase interlayers with a low viscosity filling the pores. The total area of pores along the section of the polished section decreases by 50-70% (figure 2), the average pore size in the brick from loam prepared by the traditional method is 100-200 μm, in the brick of granulated loam is less than 5 μm. In crossed nicols the uniform distribution of mineral formations and relict minerals is observed (figure 2, c, d).

The decrease in the porosity in the brick from finely grinded granulated loam is explained, firstly, by the dense structure of the granulated material obtained on the mixer of intensive action at a rotational speed of the blades at 800-1000 rpm. Secondly, by rational packing of grains without air pressed inside during molding due to the established optimal granulometric composition and pressing parameters (pressing pressure and moisture of granulated powders). Thirdly, by the greater reactivity of finely dispersed mineral grains during sintering. As a result during firing a liquid phase is formed connecting relic and newly formed minerals.

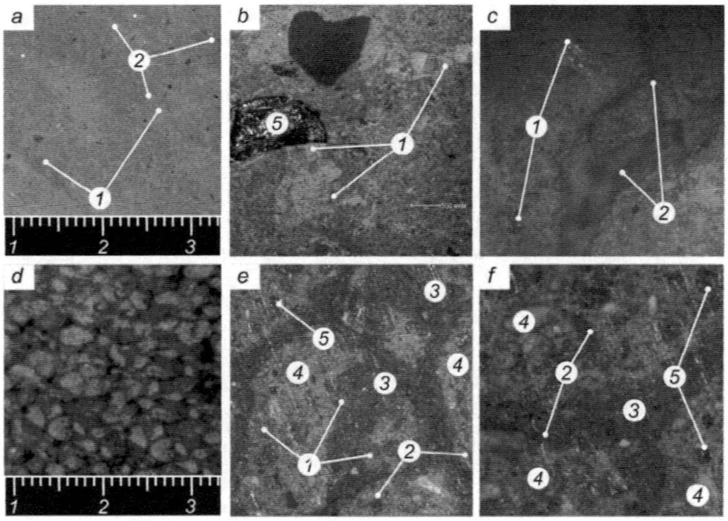

Figure 3. The macrostructure of ceramic brick made from waste coal prepared by the traditional method (a-c) and finely grinded granulated waste coal powdered by loam (d-f). Polished section, reflected light (a, d); the thin section, transmitted light, ×10 (b, e) and ×20 (c, f): 1 – metamorphosed initial minerals; 2 – pores; 3 – boundary layer of the granule from loam; 4 – the body of a granule from waste coal waste; 5 – unburned carbon.

Figure 3 shows the macrostructure of ceramic bricks from waste coal prepared by the traditional method produced at Abashevskiy Brick Factory in the 1990's (figure 3, a-c) and by the developed method with fine grinding and granulated waste coal and subsequent powdering with loam (figure 3, d-f) fired at 1000 °C.

The ceramic sample produced with the traditional mass preparation has macrocracks (figures 3, b, c) and a more porous texture at macro and micro levels (figures 3, 4). The macrostructure of ceramic bricks made from finely grinded granulated powdered with loam … что… has a pronounced spatial framework (figure 3, d-f). The cells of the frame, formed from granulated waste coal are mostly 1-2 mm in size and filled with a fine-grained material of pinkish-cream color (figure 3, d). The boundary layer between the granules is distinguished by a brick-red color and has a thickness of 80-200 μm. When studying polished sections under a microscope in the reflected light, a uniform distribution of large inclusions (on average 50-200 μm) of brown and gray color with an angular shape is observed. The examination results of thin sections in the transmitted plane-polarized light for parallel and crossed nicols are shown in figure 4.

On all photomicrographs of bricks made from finely grinded granulated waste coal, powered by loam, the characteristic separation of the boundary layer by the red color due to fine-dispersed hematite formations can be noted. At the microlevel the boundary layer between the granules also does not have clearly defined boundaries and passes into granules with the formation of a transition layer (figure 4, d-f).

The concentration of *carbon* in the brick from waste coal produced according to the developed method is significantly (more than three times) higher than that of a brick from waste coal prepared by the traditional method, which is caused by the secondary enrichment and fine grinding of industrial raw materials.

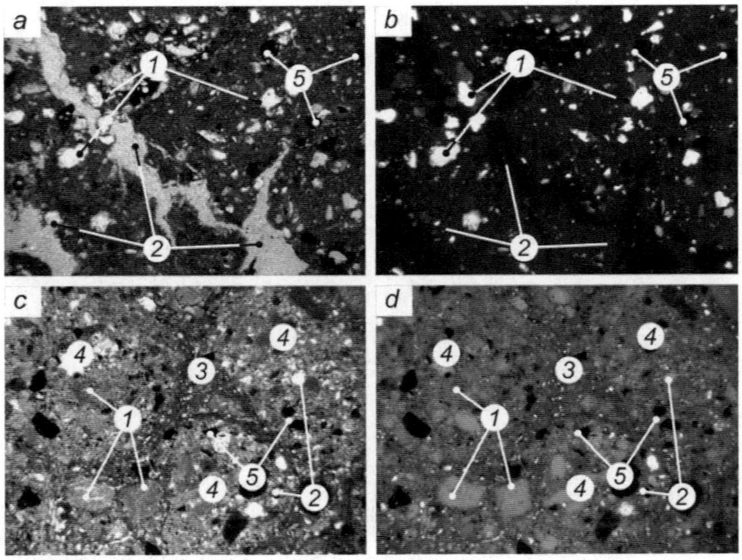

Figure 4. The microstructure of ceramic brick made from waste coal prepared by the traditional method (a, b) and finely grinded granulated waste coal powdered by loam (c, d). Polished section, transmitted light, magnification ×20, nicols II (a) and + (b); magnification ×20, nicols II (c) and + (d): 1 – metamorphosed initial minerals; 2 – pores; 3 – boundary layer of a granule from loam; 4 – the body of a granule from waste coal; 5 – unburned carbon.

5. Conclusion

Based on the study of the mechanical properties and structure of ceramic bricks made from coal mining and processing wastes prepared by the traditional method and using the developed method of mass preparation, it can be concluded that the surface energy of the grains and the surface area of their contact are the determining factors for the sintering of ceramic materials. The homogeneous, defect-free brick texture provides an improvement in the quality of compression-molded ceramic wall materials (compressive strength increases by 1.7 times from 11.2 to 19.3 MPa for bricks from overburden loamy rocks extracted at open-pit mines and by 1.5 times from 10.5 up to 15.4 MPa for bricks made of waste coal from Abashevskaya processing plant).

During firing in the boundary layer an intensive formation of pyroplastic phase takes place due to the predominance of the fusible component of the colloidal clay fraction migrating to the surface of granules during pressing, which leads to the formation of continuous surface contact sintering zones that increase the degree of interaction between the granules, which increases the compressive strength of ceramic bricks.

Acknowledgements
The results of the study were obtained within the framework of the state assignment from the Ministry of Education and Science of the Russian Federation, project code No. 7.7285.2017 / 8.9 "Fundamental research in the field of building ceramic composite materials with a matrix structure based on technogenic and natural raw materials".

References

[1] Feuerborn H J 2011 *Conference "World of Coal Ash (WOCA)"* (Denver) vol 1 pp 125–148
[2] Chernyshov E M 2010 *Proc. of Int. Sci. Tech. Conf on Achievements and Problems of Materials Science and Modernization of the Construction Industry* (Kazan: KGASU) vol 1 8–9
[3] Shchadov V M 2007 *Coal* **8** 28–31
[4] Lukina N V, Chibrik T S and Glazyrina M A 2008 *Biological Reclamation and Monitoring of Disturbed Lands* (Ekaterinburg: USUy) p 256
[5] Stolboushkin A Yu 2011 *Building Materials* **4** 43–46
[6] Storozhenko G I et al 2013 *Building Materials* **4** 57-61.
[7] Stolboushkin A Yu, Ivanov A I et al 2016 *IOP Conf. Series: Earth and Environmental Science* **45** 012017
[8] Ivanov A I, Stolboushkin A Yu and Storozhenko G I 2015 *Building Materials* **4** 65-69

IOP Conf. Series: Earth and Environmental Science **84** (2017) 012045 doi:10.1088/1755-1315/84/1/012045

[9] Stolboushkin A Yu, Ivanov A I et al 2017 *Building Materials* **4** 20–24
[10] Polovinkina Yu I 1966 *Structures and Textures of Igneous and Metamorphic Rocks* (Moscow: Nedra) vol 1 p 423, vol 2 p 271
[11] Isaenko M P 1983 *Dterminant of Textures and Structures of Ores* (Moscow: Nedra) p 261
[12] Betekhtin A G et al 1964 *Structural and Textural Features of Endogenous Ores* (Moscow: Nedra) p 598
[13] Tarasevich Yu I et al 1975 *Adsorption on Clay Minerals* (Kiev: Naukova Dumka) p 351
[14] Shchukin E D 1988 *Surface Water Films in Disperse Structures* (Moscow: MSU) p 279
[15] Geguzin Ya E 1967 *Physics of Sintering* (Moscow: Science) p 360
[16] Abdrakhimova E S et al 2009 *J. of Inorganic Chemistry* **54(1)** 42–46